AF441830

USE OF AGRICULTURALLY IMPORTANT GENES IN BIOTECHNOLOGY

NATO Science Series

A Series presenting the results of activities sponsored by the NATO Science Committee.
The Series is published by IOS Press and Kluwer Academic Publishers in conjunction with the NATO
Scientific Affairs Division.

General Sub-Series

A Life Sciences	IOS Press
B Physics	Kluwer Academic Publishers
C Mathematical and Physical Sciences	Kluwer Academic Publishers
D Behavioural and Social Sciences	Kluwer Academic Publishers
E Applied Sciences	Kluwer Academic Publishers
F Computer and Systems Sciences	IOS Press

Partnership Sub-Series

1 Disarmament Technologies	Kluwer Academic Publishers
2 Environmental Security	Kluwer Academic Publishers
3 High Technology	Kluwer Academic Publishers
4 Science and Technology Policy	IOS Press
5 Computer Networking	IOS Press

The Partnership Sub-Series incorporates activities undertaken in collaboration with NATO's Partners in
the Euro-Atlantic Partnership Council - countries of the CIS and Central and Eastern Europe - in Priority
Areas of concern to those countries.

NATO-PCO-DATA BASE

The NATO Science Series continues the series of books published formerly in the NATO ASI Series.
An electronic index to the NATO ASI Series provides full bibliographical references (with keywords
and/or abstracts) to more than 50 000 contributions from international scientists published in all
sections of the NATO ASI Series.
Access to the NATO-PCO-DATA BASE is possible in two ways:
- via online FILE 128 (NATO-PCO-DATA BASE) hosted by ESRIN,
Via Galileo Galilei, I-00044 Frascati, Italy;
- via CD-ROM "NATO-PCO-DATA BASE" with user-friendly retrieval software in English, French and
German (© WTV GmbH and DATAWARE Technologies Inc., 1989).
The CD-ROM of the NATO ASI Series can be ordered from PCO, Overijse, Belgium.

Series A: Life Sciences - Vol. 319   ISSN: 1387-6686

Use of Agriculturally Important Genes in Biotechnology

Edited by

Geza Hrazdina

NYSAES Food Biotechnology Laboratory, Cornell University, Geneva, NY, USA

IOS
Press

Ohmsha

Amsterdam • Berlin • Oxford • Tokyo • Washington, DC

Published in cooperation with NATO Scientific Affairs Division

Proceedings of the NATO Advanced Research Workshop on
Use of Agriculturally Important Genes in Biotechnology
Szeged, Hungary
17–21 October 1999

ISBN 1 58603 019 1 (IOS Press)
ISBN 4 274 90377 X C3045 (Ohmsha)
Library of Congress Catalog Card Number: 00-103806

Publisher
IOS Press
Nieuwe Hemweg 6B
1013 BG Amsterdam
Netherlands
fax: +31 20 620 3419
e-mail: order@iospress.nl

Distributor in the UK and Ireland
IOS Press/Lavis Marketing
73 Lime Walk
Headington
Oxford OX3 7AD
England
fax: +44 1865 75 0079

Distributor in the USA and Canada
IOS Press, Inc.
5795-G Burke Centre Parkway
Burke, VA 22015
USA
fax: +1 703 323 3668
e-mail: iosbooks@iospress.com

Distributor in Germany
IOS Press
Spandauer Strasse 2
D-10178 Berlin
Germany
fax: +49 30 242 3113

Distributor in Japan
Ohmsha, Ltd.
3-1 Kanda Nishiki-cho
Chiyoda-ku, Tokyo 101
Japan
fax: +81 3 3233 2426

Contents

Introduction

Agriculture during the 45 years of communist regimes in Hungary, Poland and Slovakia was centrally directed without regard to quality factors and market need, and was also heavily subsidized. Democratization of these countries and adoption of market driven economies, including agriculture, created conditions that require new approaches to find market areas that will fill the needs of each country and provide produce that is not redundant for the Common Market. As a first step towards integration into the European Union, Hungary and Poland have joined NATO. To qualify for member status, these countries have to integrate their economies, including agricultural production.

To help establish these conditions for the improvement of agricultural production, produce quality and market economy, Cornell University's International Agriculture Program initiated a dialogue between researchers and economists from Hungary, Poland and Slovakia. The first step in this dialogue was the organization of a Planning Workshop during May 28-30, 1998 in Sielinko, Poland, with the theme: "Agriculture and Rural Development: restructuring, modernization and competitiveness". This workshop had five sessions: 1. Rural Development; 2. Environmental Issues Related to Land and Water Management; 3. Quality and Safety Assurance of Agricultural and Food Products; 4. Market Economics; 5. Biotechnology. In the Biotechnology session ten leading scientists participated from Hungary, Poland, Slovakia and the US. A follow-up discussion of these five program areas took place at the campus of Cornell University, Ithaca, New York in the spring of 1999, at which an agreement was reached on the expanded discussion of these five program areas in the form of workshops.

In the area of Biotechnology, five topics of mutual interest were identified for further discussion to establish a common platform for future cooperation. These were: Genetic engineering of crops; Genome analysis; Plant transformations; Breeding for resistance and Legal aspects of biotechnology. It was decided that a workshop shuld be conducted to address these areas. This NATO-funded workshop, "Use of agriculturally important genes in agricultural biotechnology," was held in Szeged, Hungary on October 17-21, 1999. The first four sessions were introduced by a key speaker. The fifth session, dealing with the legal aspects of biotechnology, had four key speakers, one each from Hungary and the USA, and two from Poland. In the Slovak Republic, where biotechnology is in its infancy, legal aspects of biotechnology had not yet been dealt with, and therefore they did not present a position paper on this subject.

The workshop's scientific program started on Monday morning with Professor D. Dudits's key lecture on "Crop Improvement by Transgenic Technology". In his lecture Prof. Dudits discussed the limiting effects of biotic and abiotic stress on crop production, and gave an overview of the role of Reactive Oxygen Species in the damage caused in crop plants under biotic and abiotic stress. He also highlighted new developments at his Institute using a ferritin and a novel NADPH-dependent aldose/aldehyde reductase system to limit necrotic damage in plants that is caused by viral or fungal infections. The keynote lecture was followed by contributed papers on transgenic potato plants carrying truncated cDNA copy of the severe viral strain PSTVd; on Modification of sulfur metabolism in transgenic plants overexpressing bacterial genes; on Down regulation of ethylene production in transgenic apples; and on Elements of biotechnology applied to potato breeding at IHAR-Mlochow.

The second session, Genome Analysis, was introduced by Professor A. Kubek from the Slovak Agricultural University, Nitra, who spoke on "Mapping of the wheat genome for genetic diversity". His discussion of the current standing of the wheat mapping project was followed by contributed papers on linkage studies in *Pisum* and *Lupinus* genomes; on use of molecular markers in wheat in breeding for resistance; on identification of molecular markers linked to gene resistance and marker assisted introduction in barley and wheat; on identification and mapping of scab resistance genes in apple; on genetic mapping of alfalfa; on application of gene probes and maps of *Arabidopsis thaliana* in chromosomal mapping and gene isolation in *Brassica*; on micromanipulation of gametes in higher plants; and on identification of storage protein genes as markers for breadmaking quality of wheat.

The third session on Plant Transformations was introduced by Dr. G. Horvath, discussing the role of gene-transformations in crop improvements. His overview of the topic highlighted research from his own lab and was followed by papers on the transformation of carnation, maize, results of ten generations of transgenic triticale, of potato, on transformations of foreign genes and sexual transfer of transgenes in wheat, and on the development of a highly regenerable germplasm and genetic transformations in alfalfa.

The fourth session of the workshop dealt with Breeding for Resistance. The session was introduced by Professor Z. Kiraly, who is a recognized authority in this area of research. In his paper, Prof. Kiraly discussed the presently available strategies for improving resistance. Contributed papers discussed the evaluation of transgenic cucumbers expressing the thaumatin gene; enhancement of resistance in apples; performance of virus resistant transgenic potatoes; plant transformations to improve resistance to potato virus y; and production of herbicide tolerant rice by in vitro selection.

In the fifth session of the workshop, the legal aspects of biotechnology were discussed by four invited speakers representing Hungary, Poland and the US. Professor A. Aniol from IHAR-Radzikow/ Poland discussed the "Current procedures for applying risk assessment in genetically modified crops. Prof. Erwin Balazs, the representative on GMOs for Hungary, discussed "Questions arising from the implementation of the Hungarian gene technology law" that was recently passed by the parliament. Professor T. Twardowski from the Polish Academy of Sciences discussed the "Public perception and legislation of Biotechnology in Poland. R. Cahoon from Cornell University, Ithaca, NY/USA discussed the "Legal aspects of biotechnology". Each session was followed by a discussion period.

Wednesday afternoon was dedicated to a cultural event with an excursion to the historic site dedicated to the occupation of the Carpathean Basin by the Hungarian tribes during 950-980 AD.

Thursday morning was dedicated to a general discussion on collaborative projects. The discussion was preceded by a presentation by J. Haldeman on the topic of "Opportunities for a successful collaboration". This was an essential part of the workshop because the idea behind the workshop was to provide a platform for scientists from Hungary, Poland and the Slovak Republic to address problems that they perceive as common to all three countries.

This book contains the reviews presented by the invited speakers, and the original research articles contributed by the attendants. The organizers of the workshop hope that the contributions in this book provide an authentic picture of the status of biotechnology in the two new NATO countries and the Slovak Republic that is a partner country, of the agricultural problems that are being addressed there, and of the status of GMO regulation and labeling in these countries.

The organizers and participants express their thanks here to NATO's Science Program for funding this Advanced Research Workshop and the printing of its proceedings. Thanks are also due to Cornell University's International Agriculture Program and the OMFB for their partial support.

Geneva, March 8, 2000.

ACKNOWLEDGEMENTS

The organizers would like to thank to NATO's Science Program for funding this Advanced Research Workshop and the printing of its proceedings. Thanks are also due to Cornell University's International Agriculture Program, the OMFB, Nancy Long , Elizabeth Keller and all of the participants for making this workshop so successful.

*Use of Agriculturally Important
Genes in Biotechnology*
G. Hrazdina (Ed.)
IOS Press, 2000

Gene Technology in the Hungarian Plant Biology Research and Crop Improvement

Dénes Dudits

*Institute of Plant Biology, Biological Research Center,
Hungarian Academy of Sciences
H-6701 Szeged, P.O.Box 521. Hungary*

Abstract: Biotic and abiotic stress belong to the major limiting factors in crop production. Plants possess a diverse defence systems to overcome the damages caused by variety of environmental factors. The gene isolation and transformation technologies provide new ways to improve plant adaptation. Rapid accumulation of reactive oxygen species (ROS) and their toxic reaction products with lipids and proteins significantly contributes to the damage of crop plants under biotic and abiotic stresses. We have identified several stress activated alfalfa genes, including the gene of the alfalfa ferritin and a novel NADPH-dependent aldose/aldehyde reductase enzyme. Transgenic tobacco plants that synthesize alfalfa ferritin in vegetative tissues-either in its processed form in chloroplast or in the cytoplasmic nonprocessed form-retained photosynthetic function upon free radical toxicity generated by paraquat treatment and exhibited tolerance to necrotic damage caused by viral and fungal infections. We propose that by sequestering intracellular iron involved in generation of the very reactive hydroxyl radicals through a Fenton reaction, ferritin protects plant cells from oxidative damage. Our preliminary results with the other stress-inducible alfalfa gene (a NADPH-dependent aldo-keto reductase) indicate, that the encoded enzyme may play role in the stress response of the plant cells. These studies reveal new pathways in plants that can contribute to the increased stress resistance with a potential use in crop improvement.

1. Introduction

Significant discoveries resulting in the improvement of methodology of scientific research can essentially restructure research strategies, approaches and bring original new concepts to understand basic scientific problems. In the mid seventies, the development of recombinant DNA technology initiated a new dimension in biology, especially highlighting the structural bases of genetic control of cellular events and inheritance. The use of restriction nucleases, ligases, and the construction of cloning vectors, gene libraries have opened the way for artificial designing of genes and their functional elements. The DNA molecules generated *in vitro* can be introduced into cells where the expression of the transgene results in synthesis of new proteins, enzymes and development of new traits. The introduced new genes can be inherited during cell divisions and transferred to sexual progenies. It is evident that the methods of recombinant DNA technology have also been used in plant molecular biology research. The first conference in this field was organized in 1979 in Edinburgh [1]. Simultaneously with cloning of the first plant genes, the applications of this methodology in plant breeding has been initiated by several laboratories. The breeders alter the gene pool of crop species by crossing of desired genotype and selection of the required traits in segregating populations. Normally, these activities are based on the evaluation of phenotypic characters reflecting indirectly the contribution of the genes. The recombinant DNA technology allows the physical isolation of defined genes and their use in highly specific modification of agronomic traits. The previous successes in crop improvements were significantly supported by the use of scientific results from different disciplines such as genetics, botany, plant physiology and plant pathology. Following this tradition, the breeders want to utilize the advantages offered by the recombinant DNA techniques. Nowadays, we can experience the growing use of cultivars improved by this methodology. The thousands of field trials and cultivation of crops advanced by gene transformation in the area of 40 millions hectares clearly show the significance of this technology. Despite of the fact that the first products were developed in the USA and primarily used in the American continent and presently the European countries did not favour the extended use of GM crops, we can expect a significant increase in the use of cultivars derived from molecular plant breeding. This development will be especially stimulated by introduction of new products with high quality, lower price and guarantees for the lack of risk for consumer's health and environment.

Seeing the advantages of gene transformation of plant breeding and special needs of Hungarian seed industry, it became necessary to integrate these methods into the breeding programs in the country. Here, we briefly summarize the major achievements in establishing of the transformation methods and cloning of plant genes for the use in academic research and crop improvements.

2. Advanced tissue culture systems for production of transgenic plants

In the everyday practice of plant breeding, the new gene combinations are generated in sexual progenies by random distribution of parental genes in zygotes. The reproductive organs, female and male gametes mediate the formation of individual plants with altered gene pool. In contrast, the isolated genes are preferentially introduced into somatic cells, therefore, after integration of transgene into the recipient genome, the regeneration of fertile plants from the transformed cells is a prerequisite for further use of the new genotypes. The methodology for re-differentiation of plants from cultured cells has been extensively studied in several crop species in tissue culture laboratories organized by the Hungarian breeding institutes already in the seventies. We can hardly find breeding unit that lacks this facility. Consequently, the Hungarian breeders can rely on this methodical background. Efficient tissue culture systems for wheat and maize have been developed in the Cereal Research Institute, Szeged and in the Agricultural Institute of the Hungarian Academy of Sciences [2, 3, 4]. Selection of the responsive genotypes and the performance of a successful breeding program to improve morphogenic potential turned out to be a key factor in production of wheat and maize cultures with high embryogenic and organogenic capabilities. The re-differentiation of rice plants from callus tissues has been extensively studied in the Department of Genetics and Plant Breeding, Agriculture University and Agricultural Biotechnology Center, Gödöllö [5, 6]. The techniques of plant regeneration from *in vitro* cultures of various grasses have also been established [7]. In the case of potato, plants were successfully grown from cultured leaf mesophyll protoplasts [8]. Different genotypes of alfalfa provide an outstanding experimental material for studies of somatic embryogenesis. Embryos can be obtained from root, petiol or leaf explants and cultured protoplasts. The trigger of the embryogenic program in somatic cells is depends on the reactivation of cell division and induction of an overall stress responses [9, 10]. Out of tree species, horse-chestnut, plum and apricot plants were generated in tissue cultures [11, 12]. Sporadic experiments were carried out with eggplant and sugarbeet species [13, 14].

3. Successes in transformation of crop species in Hungary

The first transgenic plants derived from *Agrobacterium*-mediated gene transfer were reported in the years of 1983/1984 [15, 16, 17]. Two-three years after these reports, the Hungarian scientists from the Biological Research Center of the Hungarian Academy of Sciences (Szeged) published the production of transformed alfalfa plants [18]. This was the first successful transformation experiment with this leguminous crop. Molecular and biochemical data showed the presence of a functional foreign gene in the regenerated alfalfa plants. Transgenic tobacco plants were produced to express integrated gene IV. of cauliflower mosaic virus [19]. Later, potato protoplasts were used for direct DNA uptake [20]. Subsequently, the PVX coat protein gene was expressed in transgenic potato plants under the control of extensin-gene promoter [21]. According to the virological tests carried out by Ervin Balázs, these transformants exhibited virus resistance. An important achievement was the development of a highly efficient maize transformation system by using the protoplasts of the unique HE/89 embryogenic genotype [22, 23]. These experiments demonstrated that direct, PEG-mediated uptake of plasmid DNA molecules into suspension culture protoplasts can result in fertile transgenic maize plants. Parallel with this academic research, Günther Donn at Hoechst A.G. (Frankfurt) used the same system for production of phosphinotricine resistant genotypes. Based on these material, hybrids were developed that are already in crop production in the USA. The *Agrobacterium*-mediated gene transfer allowed the detailed characterization of different promoter functions in transgenic rapeseed plants [24]. Production of wheat transformants expressing of phosphinotricine resistance gene required a more extensive experimentation. The use of helium-drived particle bombardment of embryo-derived callus tissues resulted in fertile transformed wheat plants (J. Pauk personal communication). Virus resistant tobacco plants were also used in virological research [25, 26].

The availability of variety of transgenic crops encouraged the molecular biologists and the plant breeders to organize jointly a nation wide risk assessment program for the period of three years. With the financial help OMFB (National Committee for Technological Development, Budapest) experiments were carried out to monitor the movement of transgenes in larger plant populations under field conditions. Transgenic tobacco, potato, rapeseed, alfalfa and maize served as experimental material. These transformants were grown in isolated areas and destroyed at the end of the experiments. Analysis of distribution of marker gene showed limited spread of foreign gene by outcrossing [27].

4. Gene isolation projects in basic science and in production of improved genotypes for plant breeding

In Hungary, the recombinant DNA methods were first used to clone mitochondrial plasmid like DNA from maize. [28]. These molecules were considered as potential constituents of a transformation vector for monocot species. In the mid-eighties, an extensive gene isolation program was initiated in alfalfa. The primary goal of these studies was to generate molecular markers for the analysis of initiation of somatic embryogenesis. It was especially attempting to characterize the transcriptional reprogramming of somatic cells during the transition to embryogenic state. A unique culture system was developed for differential screening of cDNA libraries. As a cell division marker, first histone H3 genes were cloned from genomic library constructed by Gy.B. Kiss [29, 30]. Later, this work resulted in the identification of variant histone promoter with S-phase specific activity [31]. Based on the observations that embryo development from cultured cells requires the activation of the division cycle, concentrated efforts were devoted to clone a set of cell cycle control genes (cdc2-related kinases, cyclins, phosphatases) and to characterize the kinase complexes involved in cell cycle progression [32, 33, 34, 35]. Activation of stress responses during induction of embryogenesis was reflected by up-regulation of varieties of stress gene such as heat shock, ferritin, aldose reductase, annexin genes [36, 37]. The isolation of alfalfa ferritin cDNA and its overexpression in tobacco plants allowed to demonstrate that the reduction of free iron level can cause oxidative stress resistance and reduce the symptoms after necrotic infections by virus, bacteria and fungi [38]. In the Institute of Genetics, Biological Research Center, Szeged Gy. B. Kiss's group constructed the genetic map of *Medicago* and generated large number of genetic markers for this crop [39].

An internationally recognized research project headed by Ferenc Nagy (Institute of Plant Biology, Biological Research Center, Szeged) has been carried out from the beginning of the nineties to identify the major regulatory elements in light-dependent gene expression [40]. Recently, this group showed the regulated translocation of phytochrome into the nucleus [41].

In the Agricultural Biotechnology Center, potato and its wild relatives (*S. brevidens, S. chacoense*) have been used for molecular genetic studies on tuber development and stress responses. Cloning of tuber specific cDNA/GM7 and a water-stress-inducible cDNA (DS2) allowed extensive gene expression studies with this crop [42]. In this center the recombinant DNA techniques are extensively used in virus research. Expression of an altered form of plum pox virus helicase gene resulted in virus resistance [43]. Genome analysis of tombusvirus, tobacco necrosis virus and cymbidium ringspot virus defective interfering RNA generated novel informations. [44].

References

[1] Genome Organization and Expression in Plants. In: C.J. Leaver (ed.). Plenum Press New York-London, 1980.

[2] K.Z. Ahmed and F. Sági, Culture and fertile plant regeneration from regenerable embryogenic suspension cell-derived protoplasts of wheat (*Triticum aestivum* L.), *Plant Cell Reports* 12 (1993) 175-179.

[3] J. Pauk *et al.*, Fertile wheat (*Triticum aestivum L.*) regenerants from protoplasts of embryogenic suspension culture, *Plant Cell, Tissue and Organ Cult.* 38 (1994) 1-10.

[4] S. Mórocz *et al.*, An improved system for the obtention of fertile regenerants via maize protoplasts isolated from a highly embryogenic suspension culture, *Theor. Appl. Genet.* 80 (1990) 721-726.

[5] L.S. Nam and L.E. Heszky, Rice tissue culture and application to breeding: 1. Induction of high totipotent haploid and diploid callus from the different genotypes of rice, *Cereal Research Communication* 14 (1986) 197-203.

[6] B. Jenes and J. Pauk, Plant regeneration from protoplast derived calli in rice (*Oryza sativa* L.) using Dicamba, *Plant Science* 63 (1989) 187-198.

[7] L.E. Heszky *et al.*, Increase of green plant regeneration efficiency by callus selection in *Puccinella limosa* (Schur.) Holmbg, *Plant Cell Reports* 8 (1989) 174-177.

[8] A. Fehér *et al.*, Differentiation of potato (*Solanum tuberosum* L.) plants from cultured leaf protoplasts, *Acta Biologica Hungarica* 40 (1989) 369-380.

[9] D. Dudits *et al.*, Molecular and cellular approaches to the analysis of plant embryo +development from somatic cells *in vitro*, *J. Cell Science* 99 (1991) 473-482.

[10] D. Dudits *et al.*, Molecular Biology of Somatic Embryogenesis. In: In Vitro Embryogenesis in Plants, T.A. Thorpe (ed.), Kluwer Press, Dordrecht, 1995, pp. 267-308.

[11] Zs. Jekkel *et al.*, Cryopreservation of horse-chestnut (*Aesculus hippocastanum* L.) somatic embryos using three different freezing methods, *Plant Cell Tissue and Organ Culture* 52 (1998) 193-197.

[12] M. Csányi *et al.*, Tissue culture of stone fruit plants basis for their genetic engineering, *J. Plant Biotechnology* 1 (1999) 91-95.

[13] M. Fári *et al.*, *Agrobacterium* mediated genetic transformation and plant regeneration via organogenesis and somatic embryogenesis from cotyledon leaves in eggplant (*Solanum melongena* L. cv. 'Kecskeméti lila'), *Plant Cell Reports* 15 (1995) 82-86.

[14] O. Toldi *et al.*, Antiauxin enhanced microshoot initiation and plant regeneration from epicotyl-originated thin-layer explants of sugarbeet (*Beta vulagris* L.), *Plant Cell Reports* 15 (1996) 851-854.

[15] R.T. Fraley *et al.*, Expression of bacterial genes in plant cells, *Prod. Natl. Acad. Sci. USA* 80 (1983) 4803-4807.

[16] R.B. Horsch *et al.*, Inheritance of functional foreign genes in plants, *Science* 223 (1984) 496-498.

[17] M. De Block *et al.*, Expression of foreign genes in regenerated plants and in their progeny, *The EMBO Journal* 3 (1984) 1681-1689.

[18] M. Deák *et al.*, Transformation of *Medicago* by *Agrobacterium* mediated gene transfer, *Plant Cell Reports* 5 (1986) 97-100.

[19] E. Balázs, Disease symptoms in transgenic tobacco induced by integrated gene IV. of cauliflower mosaic virus. *Virus Genes* 3 (1990) 205-211.

[20] A. Fehér *et al.*, PEG-mediated transformation of leaf protoplasts of Solanum tuberosum L. cultivars, *Plant Cell, Tissue and Organ Culture* 27 (1991) 105-114.

[21] A. Fehér *et al.*, Expression of PVX coat protein gene under the control of extensin-gene promoter confers virus resistance on transgenic potato plants, *Plant Cell Reports* 11 (1992) 48-52.

[22] V.M. Golovkin *et al.*, Production of transgenic maize plants by direct DNA uptake into embryogenic protoplasts, *Plant Science* 90 (1993) 41-52.

[23] S. Omirulleh *et al.*, Activity of a chimeric promoter with the doubled CaMV 35S enhancer element in protoplast-derived cells and transgenic plants in maize, *Plant Molecular Biology* 21 (1993) 415-428.

[24] I. Stefanov *et al.*, Differential activity of the mannopine synthase and the CaMV 35S promoters during development of transgenic rapeseed plants, *Plant Science* 95 (1994) 175-186.

[25] Á. Kollár *et al.*, Efficient pathogen-derived resistance induced by integrated potato virus Y coat protein gene in tobacco, *Biochimie* 75 (1993) 623-629.

[26] L. Palkovics *et al.*, Pathogen-derived resistance induced by integrating the plum pox virus coat protein gene into plants of *Nicotiana benthamiana*, *Acta Horticulturae* 386 (1995) 311-317.

[27] J. Pauk *et al.*, A study of different (CaMV 35S and *mas*) promoter activities and risk assessment of field use in transgenic rapeseed plants, *Euphytica* 85 (1995) 411-416.

[28] Cs. Koncz *et al.*, Cloning of mtDNA fragments homologous to mithochondrial S2 plasmid-like DNA in maize, *Mol. Gen. Genet.* 183 (1981) 449-458.

[29] S.C. Wu *et al.*, Isolation of an alfalfa histone H3 gene: structure and expression, *Plant Molecular Biology* 11 (1988) 641-649.

[30] S.C. Wu *et al.*, Polyadenylated H3 histone transcripts and H3 histone variants in alfalfa, *Nucleic Acids Research* 17 (1989) 3057-3063.

[31] T. Kapros *et al.*, Differential expression of histone H3 gene variants during cell cycle and somatic embryogenesis in alfalfa, *Plant Physiology* 98 (1992) 621-625.

[32] H. Hirt *et al.*, Complementation of a yeast cell cycle mutant by an alfalfa cDNA encoding a protein kinase homologous to p34^{cdc2}, *Proc. Natl. Acad. Sci. USA* 88 (1991) 1636-1640.

[33] H. Hirt *et al.*, Alfalfa cyclins: Differential expression during the cell cycle and in plant organs, *The Plant Cell* 4 (1992) 1531-1538.

[34] Z. Magyar *et al.*, Active cdc2 genes and cell cycle phase-specific cdc2-related kinase complexes in hormone-stimulated alfalfa cells, *Plant Journal* 4 (1993) 151-161.

[35] D. Dudits *et al.*, Cyclin-dependent and Calcium-dependent Kinase Families: Response of Cell Division Cycle to Hormone and Stress Signals. In: D. Francis, D. Dudits, and D. Inzé (eds.), Plant Cell Division: Molecular and Developmental Control. Portland Press, London, 1998, pp. 21-46.

[36] J. Györgyey *et al.*, Alfalfa heat shock genes are differentially expressed during somatic embryogenesis. *Plant Molecular Biology* 16 (1991) 999-1007.

[37] M. Deák *et al.*, Involvement of stress related and cell cycle genes in transition from somatic to embryogenic cells of alfalfa, *Acta Phytopathologica et Entomologica Hungarica* 30 (1995) 119-122.

[38] M. Deák *et al.*, Plants expressing the iron-binding protein, ferritin ectopically are tolerant to oxidative damage and pathogens, *Nature Biotechnology* 17 (1999) 192-196.

[39] G.B. Kiss *et al.*, Construction of basic genetic map for alfalfa using RFLP, RAPD, isozyme, and morphological markers, *Mol. Gen. Genet* 238 (1993) 129-137.

[40] E. Fejes *et al.*, A 268 bp upstream sequence mediates the circadian clock-regulated transcription of the wheat Cab-1 gene in transgenic plants, *Plant Molecular Biology* 15 (1990) 921-932.

[41] S. Kircher *et al.*, Light quality-dependent nuclear import of the plant photoreceptors phytochrome A and B, *Plant Cell* 11 (1999) 1445-1456.

[42] Zs. Bánfalvi *et al.*, Starch synthesis- and tuber storage protein genes are differently expressed in *Solanum brevidens*, *FEBS Letters* 383 (1996) 159-164.

[43] A. Wittner *et al.*, *Nicotiana benthamiana* plants transformed with the plum pox virus helicase gene are resistant to virus infection, *Virus Research* 53 (1998) 97-103.

[44] Z. Havelda *et al.*, Secondary structure dependent evolution of *Cymbidium* ring spot virus defective interfering RNA, *J. General Virology* 78 (1997) 1227-1234.

Improvement of Tolerance against Dehydration Stress in Cereals

B. Jenes, J. Pauk *, O. Toldi, Z. Bánfalvi, R. Dóczi, A. Oreifig and G. Dallmann
Agricultural Biotechnology Center, Gödöllö, Hungary
**GK KHT, Szeged, Hungary*

Abstract: The breeding of major monocot crop plants - like wheat (*Triticum aestivum* L.) or rice (Oryza sativa L.) - for drought tolerance has been going on for a long time, but so far there is no real break through.

After the recent results of plant physiology and plant molecular genetic research there is more chance to find successful approaches to raise the drought tolerance *via* gene technology. Our first approach is as follows: The *mtlD* gene coding for the mannitol-1-phosphate dehydrogenase of *E. coli* has been built in into a chimeric plasmid. Transformed into tobacco plants it was expressed and initiated an increase of the synthesis of mannitol and its accumulation. On the basis of these findings it seemed to be wise to continue the improvement of wheat and rice for drought tolerance using these genes - especially the mt1D gene- for genetic transformation of the selected genotypes. To make the stress tolerance more specific we continuously searched for dehydration stress specific regulatory elements. The stress gene DS2 is originally isolated from the wild potato species, *Solanum chacoense*. It represents a very rare type of water-stress-inducible genes whose signalling pathway is not related to abscisic acid. Abiotic stresses other than drought, as high salinity, cold treatment or anaerobiosis, cannot induce the expression of the DS2 gene. We started to introduce mtlD gene and to check out the DS2 promoter in a test construct in rice.

1. Introduction

Higher plants are permanently subjected to abiotic events. The most important environmental stimuli for these plants are as follows: extreme temperatures, unbalanced supply of minerals and the lack of water.

During the effect of these environmental stimuli higher plants suffer physiological stress. According to the self-defence response we classify the plants into two main groups, the stress sensitive species and the stress tolerant species.

One of the most important stresses is the lack of water and water deficit is the most intensively studied stress factor in plants. The most frequent factors causing water deficit are the dry environment, high temperature and high salt content of the soil.

During the different environmental stresses the plants loose some of their fitness. Reduction of the general fitness of plants occurs most often by the following physiological changes: difficulties in the uptake of minerals and metabolite transport; reduction of photosynthetic activity; a lot of energy consuming processes are activated on the cell level as well as on the whole plant level, which are suitable to restore the balance of water household. The biochemical events during water deficit are summarised on Figure 1.

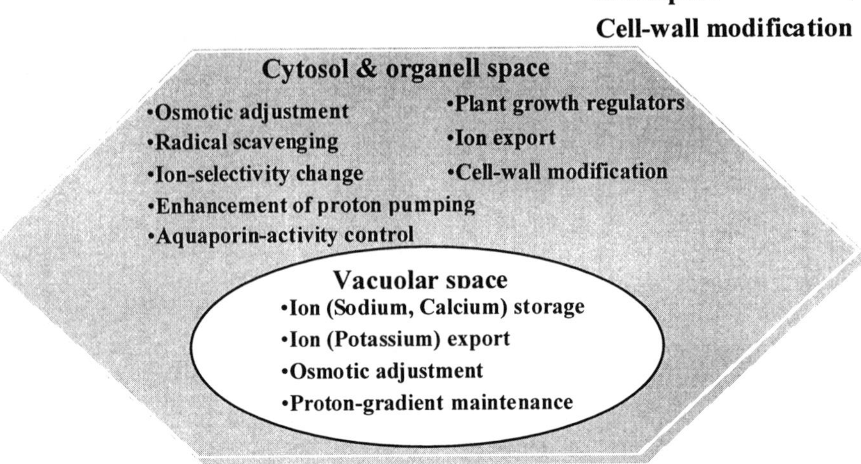

Figure 1. Biochemical functions associated with tolerance to plant water-deficit (After Bohnert & Jensen, 1996)

Even the water deficit sensitive plants can tolerate about 10-15% loss of their water content by osmoregulation capacity and stress releasing factors. What are the elements of this complex self defence system?

The essential physiological processes of the cell occur an aqueous phase. For this reason when water deficit happens on the whole plant level, the cells try to keep back the needed water by synthesis (special proteins, prolin, glycin-betain, mannitol, sorbitol, etc.) and accumulation of special compounds (mineral ions, mostly potassium) which can immobilise intracellular water.

In the mean time there are severe morphological and physiological changes on the whole plant level, which promote strong reduction of evaporation.

The basic physiological processes for self-protection are: the plant gets rid of certain portion of its leaves - the most dramatic and most often seen reaction; stomata of the remaining leaves are kept closed because of the intensive sunshine and high temperature; the production of waxy compounds is increased on the surface of the leaves.

There are considerable differences in the sensitivity to this stress between different organs of the plant. The vegetative organs of annual plants contain about 70% water and the water content of seeds is usually only about 10-15% and they are viable for years.

The so-called tolerant plant species fight more effective against dehydration stress. A very typical protection is the special morphology: shoots and leaves are covered with hard and thick epidermis, the "water-storage" organ of succulent species, and another protection is when the strong, narrow leaves turn their sharp edges towards the sun.

There are major differences in the photosynthetic activity and the method of photosynthesis, which considerably effectuate the water-household of the plants. The most important cultivated species such as wheat, barley, sugarbeet, bean, potato, etc., have "C3" type photosynthesis - the end-product of the Calvin cycle is a sugar with 3 carbon atoms.

There are species (corn, millet, sorghum, etc.) that have the "C4" type photosynthesis - the end product of the "dicarbonic-acid" cycle is a sugar with 4 carbon atoms. These require about 50% less water for photosynthesis than the "C3" plants.

The so-called glycophyte plants have no special mechanisms to curb too high concentration of some mineral ions. For this reason they control their osmotics of the cytoplasm by complex processes [4]. Such controlling mechanisms are the reduction of ion-transfer toward shoots, the transport of ions through the root system, redistribution of ions among different organs of plants as well as among different cell organelles.

The main osmotic components are potassium-salts of organic acids and sugars. Sodium and chloride is always present but it can be found in considerable amounts only in halophytes. Uptake of inorganics is decreasing when their concentration in the cell increases (negative feed-back mechanism). In many glycophyte species the important role of sugars is osmoregulation because of the low concentration of mineral ions in the soil.

2. Some genes for potential use to increase drought tolerance

When we started this project we had to find suitable and available genes to induce drought tolerance in our transgenic plants. Some of these genes are listed here with a short introduction of their original source and way of action.

2.1 Cholin oxidase gene (CodA) from Arthrobacter globiliformis

Cholin oxidase converts cholin into glycine-betain (N-trimetilglycin), its activity results in the increase of betain level. Betain is accumulated in several species of higher plants as a response to salt or cold stress. Probably it is synthesized in the chloroplast (e.g. barley, spinach) and most probably it reduces the inhibition of "photosystem II by dehydration or cold and helps the rapid repair mechanisms (Hayashi et al. 1997).

2.2 Trehalose-6-phosphate synthase 1st subunit gene (TPS1) from yeast

Trehalose is a non-reducing disaccharide. It showed stabilisation of the dehydrated lipid membranes in *In vitro* experiments. Driven by the Rubisco promoter of *Arabidopsis thaliana* it produced increased level of trehalose, which resulted in increased drought tolerance in tobacco (Holström et al., 1996).

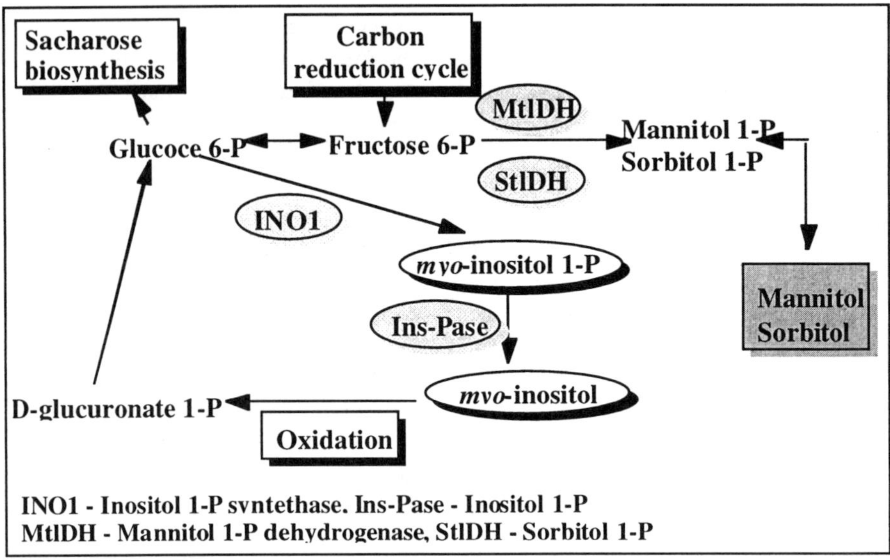

Figure 2. Gluconeogenesis in plants (after Bohnert et al. 1995).

2.3 Mannitol-1-phosphate dehydrogenase gene (MtlD/MtlDH) from Escherichia coli

It is active in *E. coli* in the transformation of mannitol into fructose-6-phosphate. The effect is opposite in transgenic tobacco plants: it increases the biosynthesis of mannitol (Tarczinsky et al., 1992).

These are not sensitive for the plants feedback mechanisms. In case of a raise in fructose-2,6-bisphosphate (F-2,6-P_2) level the carbon biosynthesis moves to gluconeogenesis (increased starch production in tobacco). If the F-2,6-P_2 level, decreases the carbon biosynthesis moves to glycolysis in transgenic tobacco plants.

At the beginning of the project we chose the *Mannitol-1-phosphate dehydrogenase* (MtlD/MtlDH) from *Escherichia coli* to start with. The role of mannitol-1-phosphate dehydrogenase in the mannitol pathway is shown in Figure 2.

3. An ABA-independent drought-inducible promoter - DS2 - from potato

Controlling the gene which provides drought tolerance is suggested by the regulatory elements specific for dehydration stresses. For this reason it seems to be a good idea to involve such a gene for the rice and wheat improvement project.

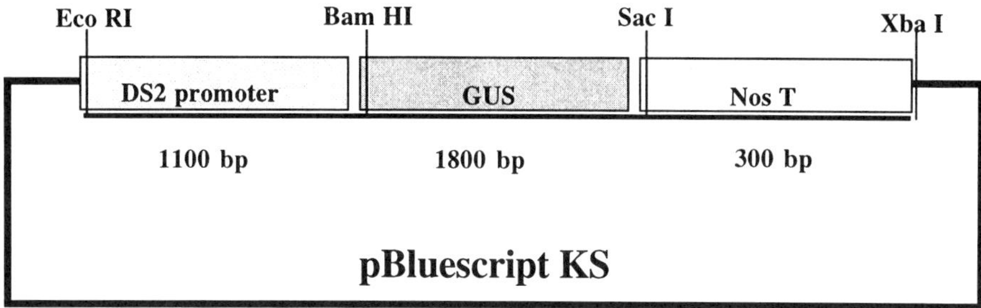

Figure 3. Plasmid construct based on pBluescript KS vector carrying the DS2 promoter, the *GusA* gene and the *nos* terminator sequences.

The stress gene DS2 was originally isolated from the wild potato species, *Solanum chacoense* (Silhavy et al., 1995). It represents a very rare type of water-stress-inducible gene whose signalling pathway is not related to abscisic acid. It was shown that external treatment with other hormones that influence plant growth and development does not result in activation of DS2 transcription.

Abiotic stresses other than drought, as high salinity, cold treatment or anaerobiosis, can not induce the expression of the DS2 gene.

A 1 kb region of the DS2 promoter was isolated from a commercial potato cultivar and analysed at DNA sequence level. No known *cis* elements characteristics for other drought-inducible promoters could be detected.

The 1 kb promoter region of the DS2 gene was fused to the GusA reporter gene. Conservation of the signal transduction pathway leading to expression of the DS2 gene was tested in rice protoplasts by measuring the activity of the DS2 promoter-GusA fusion in transient assay. The plasmid construct of DS2 promoter and GusA gene are described in Figure 3.

In preliminary experiments where dehydration was induced by PEG in transformed rice protoplasts the specific regulation by DS2 promoter seems to be correct (see Figure 4.).

Figure 4. Histochemical staining of wild type, non-PEG treated and transformed (DS210:10 AM:GUS), PEG treated (at several concentration) rice protoplast suspension.

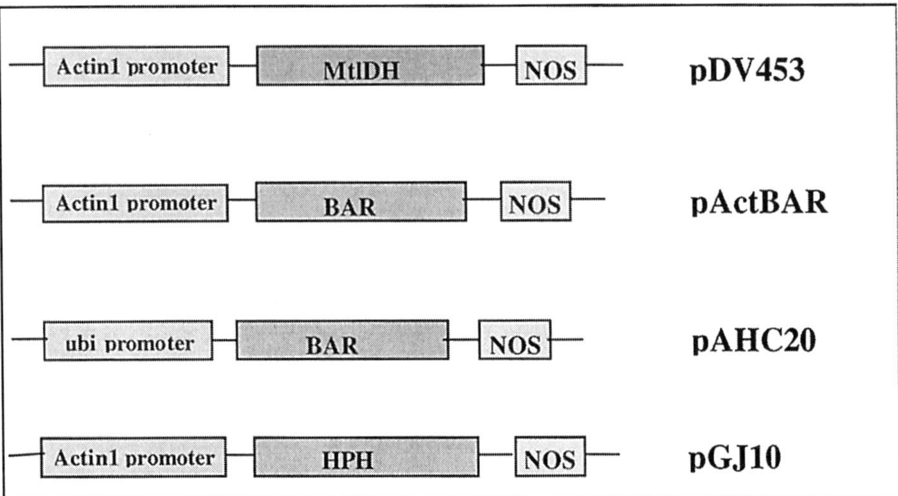

Figure 5. Gene constructs used in drought tolerant cereal project.

4. Wheat transformation experiments

Two thousand two hundred immature embryos and scutella were excised from wheat cv. GK Tavasz and CY 45 and laid on MSD4 medium. The embryo-scutellum complexes were bombarded twice with different plasmid constructs carrying the *mtlD* gene as target gene and the bar gene which provided selection following the gene transfer (see Figure 5.).

The selected calli are placed on MS medium with out a hormone for regeneration. Some of the surviving calli were screened for the presence of *mtlD* sequences in the chromosomal DNA. Figure 6 shows a PCR test made on 5 selected wheat calli. As the regeneration process is progressing, the obtained wheat plantlets will undergo PCR test and Southern blotting to verify the presence of the introduced gene. Later on, the production of mannitol will be tested.

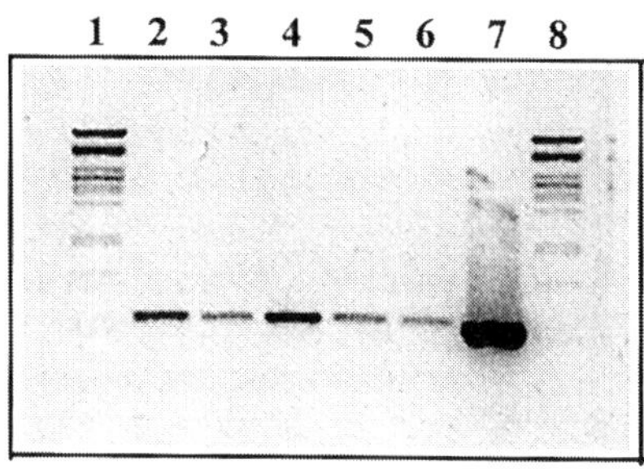

Figure 6. PCR for the mtlD gene. Lanes 1,8= size marker, Lanes 2, 3, 4, 5, 6=DNA of transgenic wheat calli, Lane 7 =pDV454 plasmid.

References

[1]   H.J. Bohnert *et al.*, *Mesembryanthemum crystallinum*, a higher plant model for the study of environmentally induced changes in gene expression. *Plant Mol Biol Rep* **61** (1998) 10-28.

[2]   H.J. Bohnert *et al.*, Adaptations to environmental stresses, *The Plant Cell* **7** (1995) 1099-1111.

[3]   H.A. Hayashi *et al.*, Transzformation of *Arabidopsis thaliana* with the *codA* gene for cholin oxidase; accumulation of glycin betaine and enhanced tolerance level to salt and cold stress, *The Plant Journal* **12** (1997) 133-142.

[4]   J.A. Hellebust, Osmoregulation. *Ann. Rev. Plant Physiol.*, **27** (1976) 485-505.

[5]   K. Holström *et al.*, Drought tolerance in tobacco, *Nature* **379** (1996) 683-684.

[6]   D. Silhavy *et al.*, Isolation and characterisation of water-stress-inducable cDNA clone from *Solanum chacoense*, *Plant Mol. Biol.* **27** (1995) 587-595.

[7]   M.C. Tarczynski *et al.*, Expression of bacterial mt1D gene in transgenic tobacco leads to production and accumulation of mannitol. *Proc. Natl. Acad. Sci.* USA **89** (1992) 2600-2604.

*Use of Agriculturally Important
Genes in Biotechnology
G. Hrazdina (Ed.)
IOS Press, 2000*

Transgenic Potato Plants Carrying Truncated cDNA Copy of a Severe Strain of PSTVd

A. Góra-Sochacka, I. Rosa, W. Podstolski, *T. Candresse, W. Zagórski
*Institute of Biochemistry and Biophysics, Polish Academy of Sciences, Pawinskiego
5A, 02-106 Warsaw, Poland. *Virologie Végétale, UMR GD2P, IBVM, INRA, BP81,
33883 Villenave d'Ornon cedex, France*

Abstract. In order to confer resistance against potato spindle tuber viroid
(PSTVD), we introduced into the plant genome a truncated cDNA copy (355bp)
of the severe S23 strain. The 355bp cDNA was cloned into the binary plasmid
pKYLX71-35S^2 under control of the double CaMV 35S promoter. Constructs
with inserts in two orientations (+ and -) were obtained and used in
Agrobacterium mediated transformation of *Solanum tuberosum* cv. Irga leaf
discs. A total of 113 plant lines were regenerated and grown *in vitro*. None of the
fifty tested plant lines was found to be resistant to PSTVd. Further analysis of the
transformants lead to an unexpected observation. Sap extracted from several
transgenic potato plants (expressing (+) RNA of the truncated transgene) was a
source of *bona fidae* PSTVd infection. The analysis of the viroid progeny from
these transgenic plant lines showed the appearance of full-length, repaired
infectious pathogen molecules. The mechanism underlying such an unusual
repair remains to be elucidated.

1. Introduction

The potato spindle tuber viroid (PSTVd) is a non-encapsidated, circular single-
stranded RNA molecule of approximately 360 nt. Because of high base pairing, the
pathogen molecules assume a rod-like secondary structure. PSTVd does not code for
proteins and is completely dependent on the host cells enzyme system for its replication
and for the expression of its pathogenecity [1-3]. Comparative sequence analysis of
different PSTVd family members has indicated the presence of five structural domains
designated as: T_L, P, C, V and T_R (T_L -left terminal, P -pathogenicity, C -central, V -
variable, T_R -right terminal), each of them being responsible for different functions [4].
The symptoms of potato spindle tuber viroid disease may vary considerably
depending on PSTVd strain, potato cultivar and environmental conditions. The symptoms
often are undetectable in primary infected plants and increase in consecutive generations
of infected host. The typical foliar symptoms include stunting, uprightness, and small
leaflets. The tubers are smaller, fewer in number, elongated, with numerous shallow eyes,
and have abnormal skin colour and texture. PSTVd infections cause significant decrease in
potato production and result in serious economic losses [5].
So far none of the commercially grown potato cultivars has been found to be resistant
to PSTVd. Construction of pathogen resistant plants by means of molecular biology
techniques seems to be the simplest way to overcome the problem posed by this disease.
Many successful approaches have been undertaken in constructing transgenic plants
resistant against viral pathogens. Most of them were based on use of viral genes or their
antisense. These techniques cannot be applied to PSTVd since viroid RNA does not code
for any proteins and all pathogen functions are totally dependent on host cell machinery.
The only possible strategies involve targeting the RNA molecule itself. The most

promising approaches included use of antisense RNA fragments [6, 7], yeast ribonuclease *pac*1 specific for double-stranded RNA [8] and recombinant hammerhead ribozyme [9].

2. Materials and Methods

2.1. Cloning of PSTVd into a binary vector

The ends of a 355 bp SmaI-BamHI restriction fragment of the S23 PSTVd cDNA [19] were filled by the Klenow fragment of DNA polymerase I and ligated into a HincII cleaved pUC1813 vector. This step was performed to add symmetric HindIII sites on both sides of the cDNA. The resulting HindIII fragment was re-cloned into the HindIII-site of the binary pKYLX71-35S^2 vector under control of the double 35S CaMV promoter. The resulting constructs with PSTVd cDNA in plus (+) or minus (–) orientation were used to transform *Agrobacterium tumefaciens* LBA 4404.

2.2. Potato transformation and regeneration

Transformation of *Solanum tuberosum* cv. Irga was carried out according to the method of Martini *et al.* [20] with some modifications. An *Agrobacterium tumefaciens* culture was spread on the MS medium in Petri dishes. Freshly harvested leaves and internodes were cut in several places and put into Petri dishes with the MS medium and bacteria. After 2 days incubation, the explants were transferred to the MCI medium (MS salts and vitamins, 2mg/L glycine, 1,6% glucose, 5mg/L NAA, 0.1mg/L BAP, 08% agar) with antibiotics: 50mg/L kanamycin and 500mg/L cefotaxime. Ten days later the plant tissues were transferred to the GR_2 medium (MS salts and vitamins, 2mg/L glycine, 1,6% glucose, 0.02mg/L GA_3, 0.02 mg/L NAA, 2mg/L zeatine riboside, with both antibiotics. The procedure was repeated every two weeks. Regenerating shoots were moved to MS medium [21] containing both antibiotics for rooting. Regenerated plants were micropropagated and grown *in vitro*. The plants were tested for the presence of introduced cDNA by PCR assays. The reactions with primers specific to PSTVd sequence were performed under following conditions: 30sec denaturation at 92^{0}C, 30sec annealing at 64^{0}C and 1min polymerisation at 72^{0}C. Fifty selected plant lines with introduced PSTVd cDNA were planted in the soil and subjected to infectivity assay.

2.3. Infectivity assays

Four to five plants from each selected transgenic line were tested to evaluate their resistance against PSTVd infection at the Plant Breeding and Acclimatisation Institute, Unit Mlochów. Plants were maintained in a greenhouse at 25-32^{0}C and 16h day length. Plants were mechanically inoculated with sap extracted from tomato plants showing severe symptoms of PSTVd infection. Six weeks after inoculation the symptoms were observed. Each potato plant inoculated with PSTVd was compared to mock inoculated control plant of the same line. Similar tests were also performed on plants that had been transformed with the empty pKYLX71-35S^2 vector or not transformed at all. Leaf samples from inoculated plants were collected and used for biological indexing tests carried on tomato cv. Rutgers [13]. The tomato test plants were mechanically inoculated with the sap

                    A. Góra-Sochacka et al. / Transgenic Potato Plants

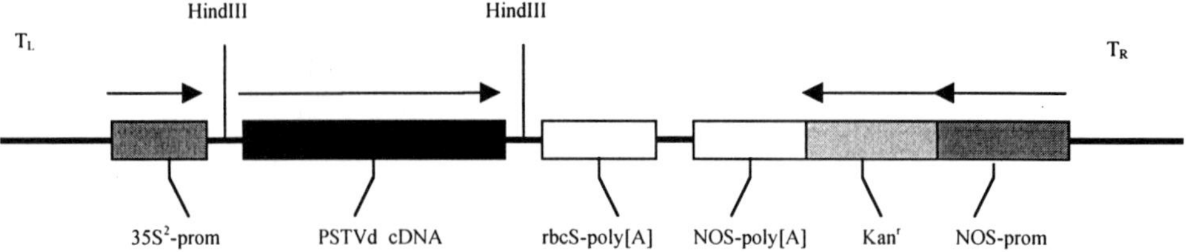

Fig. 1 Schematic representation of insertion of truncated S23 PSTVd cDNA into the binary plant transformation vector pKYLX-71S².
T_L ,T_R- left and right T-DNA borders , 35S²-prom - double 35S cauliflower mosaic virus promoter, PSTVd cDNA –truncated cDNA of S23 PSTVd, rbcS-poly[A] – ribulose-1,5 –bisphosphate carboxylase small subunit gene terminator , NOS-poly[A] - nopaline synthase gene terminator, Kan^r - kanamycin resistance gene, NOS-prom - nopaline synthase gene promoter.

extracted from the potato leaf samples (diluted in 0,1M K_2HPO_4). After another 6 weeks the symptoms on tomato plants were observed.

3. Results and Discussion

We have aimed to construct pathogen resistant transgenic plants by employing antisense RNA technology [10]. In order to confer resistance against PSTVd w e introduced into the plant genome a truncated cDNA copy (355 bp) of the S23 severe strain. Due to cloning procedures (see Materials and Methods), the cDNA contains a two nucleotides ($C_{93}C_{94}$) deletion in the central conserved region (CCR), known to be crucial for the pathogen replication [1, 2, 3]. According to the present knowledge of the PSTVd biology this deletion should assure that the RNA expressed from the transgene should not be viable [11, 12]. The modified cDNA was finally cloned into the binary pKYLX71-35S² vector under the control of the double cauliflower mosaic virus 35S promoter (Fig. 1). Constructs with the cDNA inserted in both orientations (+ and -) were obtained, designated as No 36 and No 37 respectively, and used in *Agrobacterium* mediated transformation of *Solanum tuberosum* cv. Irga leaf discs. A total of 113 (53 plants No 36 and 60 plants No 37) transgenic lines were regenerated and grown *in vitro*. PCR analysis of the plant DNA was performed to confirm the presence of introduced constructs.

Fifty selected independent transformants (22 for construct No 36, 24 for construct No 37 and 4 transformed with the empty vector) were micropropagated, planted in the soil and kept under controlled greenhouse conditions. These transgenic plants (4-5 plants for each line) were tested to evaluate their degree of resistance to PSTVd infection (see Materials and Methods). Potato plants were inoculated with sap extracted from tomato plants severely infected with PSTVd. Six weeks after inoculation leaf samples were collected and subjected to biological tests carried out by back-inoculating on tomato cv Rutgers [13]. After another 6 weeks symptoms on tomatoes were observed.

Of the 50 transformants initially tested (data not shown) four (36/11, 36/32, 36/34 and 37/2) seemed to be resistant since no symptoms developed on the test tomatoes inoculated with sap derived from these challenged plants. However further tests showed that none of them was really resistant to PSTVd (see Table 1).

Table 1. Results of biological test performed with sap from the challenged transgenic potato plants on tomato 'Rutgers'.

Transgenic potato line	Biological test					
	1	2	3	4	5	Control
36/11	+	+	+			+
36/32	+	x	+	+	+	-
36/34	+	+	+	-	x	+
37/2	+	x	x	+	+	-

(+) - symptoms of PSTVd infection, (-) - no symptoms of PSTVd infection, (x) - ambiguous symptoms, numbers 1-5 refer to separate tomato test plants. "control"- tomato 'Rutgers' test plants inoculated with sap from non-inoculated transgenic potato plant of the relevant line.

The analysis of these biological assays led, however, to an unexpected observation. A few control, non-inoculated transformed plants showed distinctive morphological changes - growth stunting and leaf malformation. This phenotype was strikingly similar to the disease symptoms caused by PSTVd. This suggested that transformants *per se* could produce and accumulate infectious viroid molecules. Similarly, in biological tests carried out on Rutgers tomato plants, sap extracted from 10 non-inoculated transgenic potato plants (expressing (+) RNA of transgene) caused the development of PSTVd infection (see Table 1, lines 36/11 and 36/34). This result confirms that some non-inoculated transgenic potatoes were able to accumulate infectious viroid RNA despite the two base pair deletion introduced into the fully conserved central region of the molecule. The presence of viroid molecules in transgenic potato plants showing disease symptoms has been confirmed by gel analysis of RNA extracts, supporting again the observation of the symptoms. It should be stressed that the accumulation of infectious PSTVd molecules in transformants was not a random event since it was only observed in transformants carrying the transgene in the (+) orientation (10 out of 22 plants) and never in the plants carrying the transgene in the (-) orientation (0 out of 24 plants).

The observation that only the (+) sense construct gave rise to infective progeny is not surprising in view of similar previous observations involving either agroinfection assays or plant transformation [14, 15]. Similarly, the fact that these techniques may allow the recovery of viable PSTVd molecules from mutated, non-infectious constructs has previously been documented [16, 17]. In the most striking example to date, a construct containing an 8 nucleotide deletion (nt 70-78) gave rise to a symmetrically deleted (nt 282-290) PSTVd molecule of only 341 nt long [17]. In all those cases, as in the results reported here, the most likely explanation is that random mutations may appear in the course of transcription of the transgene (or, possibly of its aborted replication) and that in *vivo selection* acting on this population of mutants may ultimately favour the apparition of a viable molecule which then leads to a full-blown infection of the transgenic plant [16, 17, 18]. However, our observation appears to be the first case in which viable molecules are recovered from constructs containing a deletion in the highly conserved central region of PSTVd.

Acknowledgements

This work was partly supported by the Polish-French Plant Biotechnology Center (CNRS-KBN).

References

[1]   T.O. Diener, The viroids, Plenum, New York, 1987.
[2]   J.S. Semancik, Viroids and viroid-like pathogens, CRC Press, Boca, 1987.

[3]   R. H. Symons, The intriguing viroids and virusoids: what is their information content and how did they evolve?, *Mol. Plant - Microbe Interactions* **4** (1991) 111-121.

[4]   P. Keese and R. H. Symons, Domains in viroids: evidence of intermolecular RNA rearrangements and their contribution to viroid evolution, *Proc. Natl. Acad. Sci. USA* **82** (1985) 4582-4586.

[5]   M. A. Phannenstiel and S. A. Slack, Response of potato cultivars to infection by potato spindle tuber viroid, *Phytopathology* **70** (1980) 922-926.

[6]   J. Matousek *et al.*, Inhibition of viroid infection by antisense RNA expresion in transgenic plants, *Biol. Chem. Hoppe-Seyler* **375** (1994) 765-777.

[7]   D. Atkins *et al.*, The expression of antisense and ribozyme genes targeting citrus exocortis viroid in transgenic plants, *J. Gen. Virol.* **76** (1995) 1781-1790.

[8]   T. Sano *et al.*, Transgenic potato expressing a double-stranded RNA-specific ribonuclease is resistant to potato spindle tuber viroid, *Nature Biotechnology* **15** (1997) 1290-1294.

[9]   X. Yang *et al.*, Ribozyme-mediated high resistance against potato spindle tuber viroid in transgenic potatoes, *Pro. Natl. Acad. Sci. USA* **94** (1997) 4861-4865.

[10]   M. Prins and R. Goldbach, RNA-mediated virus resistance in transgenic plants, *Arch. Virol.* **141** (1996) 2259-2276.

[11]   T. Candresse *et al.*, The role of the viroid central conserved region in cDNA infectivity, *Virology* **175** (1990) 232-237.

[12]   M. Wassenegger *et al.*, RNA-directed de novo metylation of genomic sequences in plants, *Cell* **76** (1994) 567-576.

[13]   K. H. Fernow, Tomato as a test plant for detecting mild strains of potato spindle tuber virus, *Phytopathology* **57** (1967) 1347-1352.

[14]   R.C. Gardner *et al.*, Potato spindle tuber viroid infections mediated by the Ti plasmid of *Agrobacterium tumefaciens*, *Plant Mol. Biol.* **6** (1986) 221-228.

[15]   J. Matousek, *et al.*, Instabile expression of potato spindle tuber viroid cDNA in transformed potato (*Solanum tuberosum* L.), *Arch. Phytopathol. Pflanzenschutz* **27** (1991) 167-173.

[16]   R.A. Owens *et al.*, Site-specific mutagenesis of potato spindle tuber viroid cDNA : alterations within premelting region 2 that abolish infectivity, *Plant Mol. Biol.* **6** (1986) 179-192.

[17]   M. Wasseneger *et al.*, An infectious viroid RNA replicon evolved from an *in vitro*-generated non-infectious viroid deletion mutant via a complementary deletion *in vivo*, *EMBO J.* **13** (1994) 6172-6177.

[18]   R.W. Hammond, *Agrobacterium*-mediated inoculation of PSTVd cDNAs onto tomato reveals the biological effect of apparently lethal mutations, *Virology* **201** (1994) 36-45.

[19]   A. Góra *et al.*, Analysis of the population structure of three phenotypically different PSTVd isolates, *Arch. Virol* **138** (1994) 233-245.

[20]   N. Martini *et al.*, Promoter sequences of potato pathogenesis-related gene mediate transcriptional activation selectively upon fungal infection, *Mol. Gen. Genet.* **236** (1993) 179-186.

[21]   T. Murashige and F. Skoog, A revised medium for rapid growth and bioassays with tobacco tissue cultures, *Physiol. Plant* **15** (1962) 473-497.

Modification of Sulfur Metabolism in Plants by Overexpression of Bacterial cysE and cysK Genes

A. Blaszczyk, F. Liszewska, R. Brodzik, A. Sirko

*Institute of Biochemistry and Biophysics, Polish Academy of Sciences,
ul. Pawinskiego 5A, 02-106 Warszawa, Poland*

Abstract. Plant expression cassettes containing either the *Escherichia coli* cysE gene (encoding SAT) or cysK gene (encoding OAS-TL) were constructed. After the Agrobacterium-mediated transformation of tobacco we identified stable transformed plants containing several-fold higher SAT or OAS-TL activity in comparison to the control plants. Selected plants were further characterized. Determination of non-protein thiol content indicated 2- to 3-fold higher cysteine and glutathione levels in some of these transgenic plants and their progeny. The maximal elevation of the cysteine level was about fourfold while that of GSH was about twofold higher than in the controls. The most striking physiological consequence of the modification of sulfur metabolite levels in the transgenic plants, however, was their increased resistance to oxidative stress generated by exogenous hydrogen peroxide.

1. Introduction

Sulfur is one of the six macronutrients required for all living organisms. It is essential constituent of cysteine, methionine, glutathione (GSH), several co-enzymes, thloredoxins, phytochelatins, sulfolipids and all the proteins which contain sulfur amino acids. When one considers that sulfur in plants is only 3-5% as abundant as nitrogen, it is perhaps understandable that sulfur assimilation has been less well studied. Recently, it has been the subject of several reviews [1, 2, 3]. Sulfur metabolism in plants includes uptake of sulfate from the environment, assimilation into cysteine and channeling into proteins and secondary metabolic compounds.

Formation of cysteine from sulfide and 0-acetyl-L-serine (OAS) is catalyzed by Ó-acetylserine (thiol) lyase (OAS-TL). OAS is synthesized by serine acetyltransferase (SAT) from acetyl-coenzyme A and serine. Bacterial and plant cytosolic isoform of SAT are subjects to feedback inhibition by cysteine (Fig.1). It is suggested that SAT and OAS are the major regulatory factors in the biosynthesis of cysteine. The availability of OAS, of which the cellular concentration is strictly controlled by the action of SAT regulates the biosynthetic flow of cysteine. Moreover, in bacteria OAS acts also as an inducer of the genes involved in sulfate assimilation.

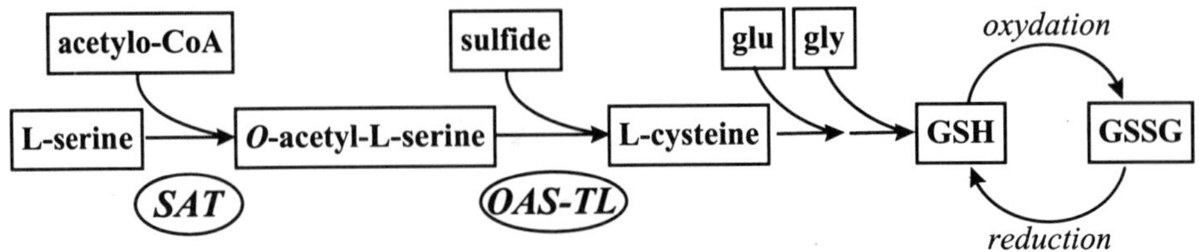

Fig. 1. Pathway of cysteine and glutathione biosynthesis. Enzymes: SAT-serine acetyltransferase; OAS-TL - 0-acetylserine-(thiol)-lyase.

Both OAS-TL and SAT seem to be required in all cellular compartments that carry out protein synthesis. They have been found in the cytosol, chloroplasts and mitochondria from various plants [1], possibly due to an inability of these organelles to transport cysteine across their membranes [4]. In pea leaves, the majority of the total SAT activity (76%) is present in the mitochondria with only 10 and 14% being present in chloroplasts and the cytosol, respectively [5]. This is different when compared to the relative distribution of OAS-TL, which is predominantly located in chloroplasts (42%) and the cytosol (44%) and only 10% of the activity is present in mitochondria [4]. SAT and OAS-TL exist in an enzyme complex known as cysteine synthase. It is suggested that OAS-TL functions as regulatory subunit that regulates SAT activity in response to OAS and sulfide levels [I].

GSH serves not only as the main storage form of reduced sulfur but it also plays an important role in the control of the thiol-disulfide status of the cell, in the detoxification of xenobiotics and finally, in plant responses to abiotic stress and pathogens. It has been shown that the ascorbate-glutathione cycle has major significance in the chloroplasts and outside the chloroplasts as an enzymatic scavenging mechanism for the removal of active oxygen species [6]. Moreover, GSH levels or redox status of glutathione was suggested to have a regulatory impact on the expression of genes whose products are directly involved in scavenging of H_2O_2 in the cytosol [7]. GSH is synthesized in two steps which are catalyzed by gamma-glutamylcysteine synthetase and glutathione synthetase (Fig. 1.).

The steps of cysteine biosynthesis in plants and bacteria are so similar that it was possible to complement *Escherichia coli* SAT- and OAS-TL-deficient cysteine auxotrophs by the corresponding cDNAs from plants [8, 9].

A major aim of our current work was to manipulate the cellular thiol levels by increasing the availability of cysteine precursor, OAS, in the two cellular compartments, chloroplasts and cytosol. The discrepancy between the relative distribution of SAT and OAS-TL in the mitochondria, chloroplasts and cytosol [4, 5] suggested that the process of cysteine synthesis in these compartments might be limited by SAT activity. We wanted to enhance it by overexpression of the bacterial *cysE* gene, encoding SAT. Since the feedback inhibition of SAT by cysteine could contribute to a reduction in SAT activity, a mutant *cysE* gene [10] encoding SAT that is insensitive to feedback inhibition by cysteine was also employed. Moreover, recently we transformed our transgenic tobacco plants over expressing SAT with another bacterial gene from cysteine biosynthesis pathway - cysK gene, encoding OAS-TL. This enzyme was also targeted either to cytosol or chloroplasts.

2. Results

2.1. Overexpression of the recombinant SAT in tobacco plants

Four different constructs were used to obtain transgenic tobacco plants expressing *E. coli cysE* alleles encoding SAT. The products of recombinant genes from pRCEM and pRCE contain the chloroplast-targeting leader sequences and, therefore, the bacterial SAT from these groups of transformants is located in the chloroplasts. The groups of plants transformed with pCEM and pCE contain bacterial Sat located in the cytoplasm. The wild type copy of cysE is present in plasmids pRCE and pCE. To minimize the effect of feedback inhibition of SAT activity by cysteine a mutant allele, insensitive to such inhibition [10] was used to derive plasmids pRCEM and pCEM.

About 30 transformants from each group were screened for SAT activity. Plants (3-5 transformants per group) with highest levels of SAT determined by two independent assays were subsequently transferred to a green house and used for further experiments. The level of SAT activity in the chosen transformants measured in the absence of an inhibitor (cysteine) is shown in Fig. 2a. The assay confirmed that the higher activity of SAT in most of the tested plants was maintained. On the average, the activity of SAT in the transformants was 2-3-fold higher in comparison to the control plants, however, one plant from the RCE group (RCE-5) had 15-fold higher SAT activity. Except for the very high level of activity in this plant, no obvious correlation of the SAT activity levels with a particular group of transfortmants was noted (Fig. 2a).

2.2. Analysis is of cysteine and glutathione levels in transgenic plants

We expected that the increased SAT activity observed in the transgenic plants would cause the differences in levels of cysteine and GSH which is the most common non-protein thiol in tobacco and the main storage compound of cysteine. Hence, we determined the non-protein thiol content in all the selected transformants by separation by reverse-phase HPLC after derivatisation with monobromob 1 mane. The cysteine levels (Fig. 2b) were higher than in the control plants in all except two (RCE-8 and CEM-12) transformants. In most analyzed plants a variation of between a two to seven- fold increase of cysteine level in comparison to the controls was observed. Moreover, in each group of plants overexpressing SAT we were able to identify at least one transformant with approximately 2-fold increased GSH level in comparison to the controls (Fig. 2c). However, in all transgenic plants, as in the controls, about 10% of the total glutathione pool was present in the oxidized form.

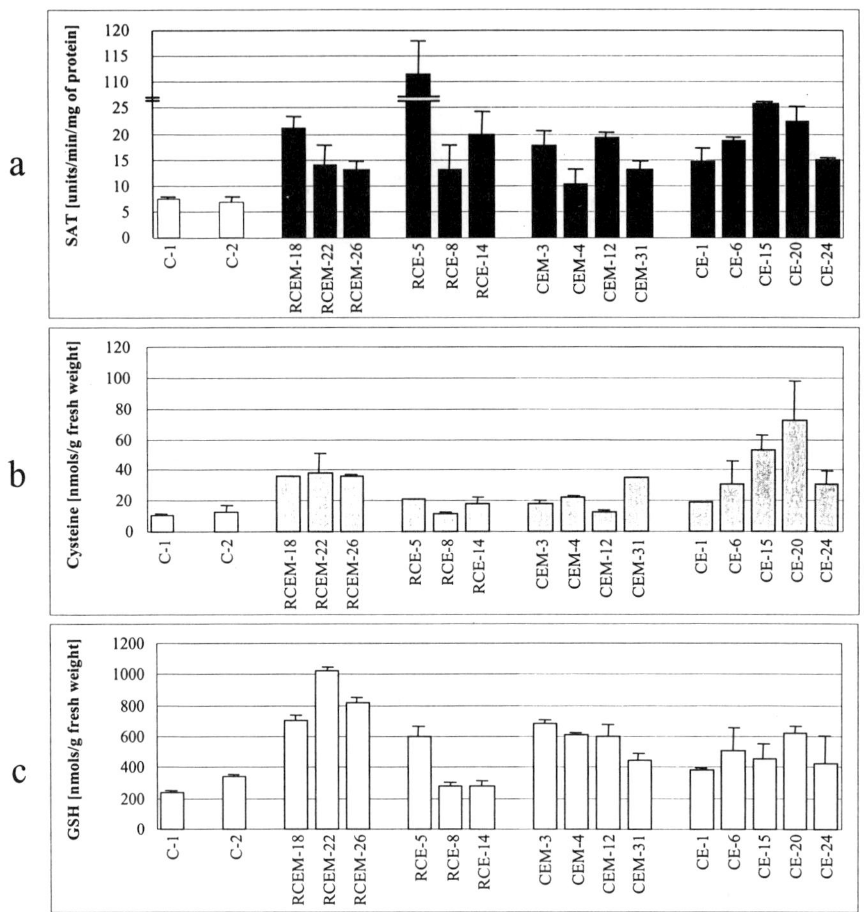

Fig. 2. SAT activity (a), cysteine (b) and glutathione (c) levels in the four groups of transgenic plants. C-1, non transformed LAB21; C-2, plant transformed with pBISN1 vector; RCEM18, RCEM22 and RCEM 26, plants transformed with pRCEM containing the mutated allele of *cysE* whose product is targeted to the chloroplasts; RCE5, RCE8 and RCE14, plants transformed with pRCE containing the wild type *cysE* whose product is targeted to the chloroplasts; CEM3, CEM4, CEM12 and CEM31, plants transformed with pCEM containing the mutated *cysE* whose product is targeted to the cytosol; CE1, CE6, CE15, CE20 and CE24, plants transformed with pCE containing the wild type cysE whose product is targeted to the cytosol. Values are the mean + standard deviation of triplicate (SAT activity) or duplicate (cysteine and GSH levels) samples taken within a one week period. The scale brake has been introduced in the case of SAT activity in RCE-5 in the panel (a).

2.3. Analysis of oxidative stress resistance in transgenic plants

Elevated GSH level in plants was reported to correlate well with their increased resistance to abiotic stresses, causing oxidative stress [11, 12]. We decided to check whether modification of the thiol content in SAT-overproducing plants results in increased resistance to oxidative stress. The oxidative stress was generated by exposing plant leaf tissue to 1 M H_2O_2 and high light. The percentage of chlorophyll remaining after such treatment was then determined (Fig. 3.). In each group of transformants, individuals with significantly higher resistance to H_2O_2 (higher % of residual chlorophyll) in comparison to

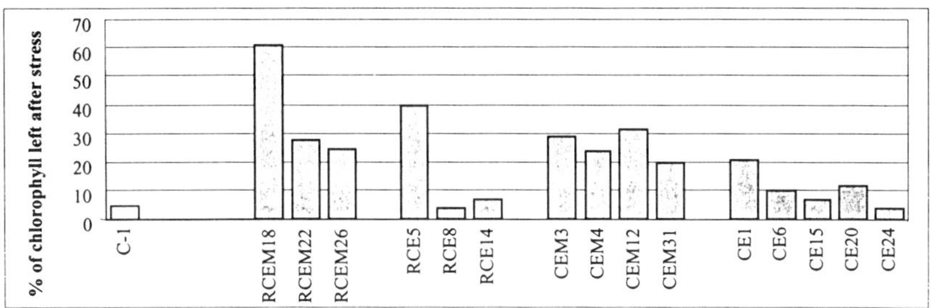

Fig. 3. Resistance of transgenic plants to oxidative stress. Resistance is shown as a percentage of the pigments left after the stress treatment. Transgenic plants numbers and descriptions as in Fig. 2, LAB 2 1, non transformed plant.

the controls were found. These results were essentially in correlation with the level of GSH what indicates the presence of a positive correlation between these two factors.

2.4. Analysis of the progeny of primary transformants

All primary transformants except CE-6 were fertile and the stability of the reported effects could be examined in their progeny. For these purposes the evaluation of the resistance to oxidative stress in the selected individuals of T_1 generation was carried out. The test was performed with leaves of the three months old plants of two lines producing bacterial SAT: RCE-5 and CE-20 and the control lines transformed with pBISN1. The % of chlorophyll remaining after stress generated by H_2O_2 and light was significantly higher in the progeny of RCE-5 and CE-20 than in the control plants (results not shown.

2.5. Overexpression of the recombinant OAS-TL in tobacco plants expressing SAT

Two constructs were used to obtain double transformants expression *E. coli cysE* alleles encoding SAT and *cysK* gene encoding OAS-TL. The selected transgenic plants containing bacterial SAT (either wild-type or mutated allele) located in the cytoplasm were transformed with the plasmid enabling production of cytosol-located bacterial OAS TL while the selected transgenic plants producing bacterial SAT in chloroplasts were transformed with the construct enabling targeting bacterial OAS-TL to chloroplasts. The aim of these studies was to select plants with simultaneous overproduction of both bacterial enzymes, OAS-TL and SAT, in either chloroplasts or cytosol.

So far, about 10 transformants from each group were screened for OAS-Tl activity. The assay confirmed that the level of OAS-TL activity in most of the tested plants was increased. A variation of between a two- to seven- fold increase of the activity level in comparison to the controls was observed. Further analysis of these transgenic plants is in a progress.

3. Summary and perspectives

Overexpression of bacterial *cysE* gene encoding SAT in transgenic plants leads to the increased levels of cysteine and glutathione under non-stress conditions. Furthermore, leaf tissues of these transgenic plants exhibited increased resistance to oxidative stress generated by H_2O_2. These results are in an accord with previously reported data that (i) feeding with OAS increased cysteine level in isolated chloroplasts [13], (ii) feeding cysteine to leaf discs increased the GSH level in leaf tissue [14] and (iii) the levels of reduced GSH *in Arabidopsis thaliana* suspension cells strongly correlated with production from oxidative injury [15].

During their life cycle plants may have to deal with periods of low temperature, frost, heat, high radiation, drought or pollutants. All these unfavorable conditions are the cause of elevated production of active oxygen species. In this view an increment of the levels of antioxidants such as GSH, ascorbate or tocopherol would be of a great benefit. We demonstrated that the increased level of GSH in the transgenic plants was correlated with their elevated resistance to the oxidative stress generated by hydrogen perioxide.

Metabolism of sulfur compounds in plants is very complex. There are many possible interactions between organelles and potential points of control by intermediates and feedback mechanisms. Our results provide further evidence for the regulatory role of SAT in controlling both, cysteine and GSH accessibility in plants. We would like to emphasize that our transgenic plants are a good starting point for the identification of the necessary targets to be modified in order to develop plants resistant to a variety of biotic and abiotic stresses. Additionally, plants with increased cysteine and GSH levels could be used as a valuable source of sulfur amino acids in food and feed stuffs. Transgenic plants with enhanced sulfur demand may also be used for detoxifying soil from excess sulfur or air from sulfur dioxide.

References

[1]　T. Leustek and K. Saito, Sulfate Transport and Assimilation in Plants, *Plant Physiology* **120** (1999) 637-643.

[2]　C. Brunold and H. Rennenberg, Regulation of sulfur metabolism in plants: first molecular approaches. *Progresses in Botany* **58** (1997) 164-186.

[3]　G. Noctor *et al.*, Manipulation of Glutathione and Amino Acid Biosynthesis in The Chloroplasts, *Plant Physiology* **118** (1998) 471-482.

[4]　J.E. Lunn *et al.*, Localization of ATP Sulfurylase and O-Acetylserine -(thiol)-Lyase in Spinach Leaves, *Plant Physiology* **94** (1990) 1345-1352.

[5]　M.L. Ruffet *et al.*, Purification and Kinetic Properties of Serine Acetyltransferase Free of 0-Acetylserine-(thiol)-Lyase from Spinach Chloroplasts. *European Journal of Biochemistry* **227** (1995) 500-509.

[6]　A. Jim6nez *et al.*, Evidence for the Presence of the Ascorbate-Glutathione Cycle in Mitochondria and Peroxisomes of Pea Leaves, *Plant Physiology* **114** (1997) 275-284.

[7]　S. Karpinski *et al.*, Photosynthetic Electron Transport Regulates the Expression of Cytosolic Ascorbate Peroxidase Genes in Arabidopsis During Excess Light Stress, *Plant Cell* **9** (1997) 627640.

[8]　N. Bogdanova *et al.*, Cysteine Biosynthesis in Plants: Isolation and Functional Identification of a cDNA Encoding a Serine Acetyltransferase from *Arabidopsis thaliana, FEBS Letters* **358** (1995) 43-47.

[9]　J.R. Howarth *et al.*, Cysteine Biosynthesis in Higher Plants: a New Member of the Arabidopsis thaliana Serine Acetyltransferase Small Gene-Family Obtained by Functional Complementation of an *Escherichia coli* Cysteine Auxotroph, *Biochimica et Biophysica Acta* **1350** (1997) 123-127.

[10] D. Denk and A. 136ck, L-cysteine Biosynthesis in *Escherichia coli:* Nucleotide Sequence and Expression of the Serine Acetyltransferase *(cysE)* Gene from the Wild-Type and a Cysteine Excreting Mutant, *Journal of General Microbiology* **133** (1987) 515-525.

[11] G. Kocsy *et al.,* Glutathione Synthesis in Maize Genotypes with Different Sensitivities to Chilling, *Planta* **198** (1996) 365-370.

[12] D. O'Kane *et al.,* Chilling, Oxidative Stress and Antioxidant Responses in *Arabidopsis thaliana* Callus, *Planta* **198** (1996) 371-377.

[13] K. Saito *et al.,* Modulation of Cysteine Biosynthesis in Chloroplasts of Transgenic Tobacco Overexpressing Cysteine Synthase [0-Acetylserine-(thiol)-Lyase], *Plant Physiology* **106** (1994) 887-895.

[14] G. Noctor *et al.,* Synthesis of Glutathione in Leaves of Transgenic Poplar Overexpressing γ-Glutamylcysteine Synthetase, *Plant Physiology* **112** (1996) 1071-1078.

[15] M.J. May and C.J. Leaver, Oxidative Stimulation of Glutathione Synthesis in *Arabidopsis thaliana* Suspension Cultures, *Plant Physiology* **103** (1993) 621-627.

Use of Agriculturally Important
Genes in Biotechnology
G. Hrazdina (Ed.)
IOS Press, 2000

Down Regulation of Ethylene Production in Apples

G. Hrazdina[1], E. Kiss[2], C. Rosenfield[1], J. L. Norelli[3], H. S. Aldwinckle[3]
*[1]Department of Food Science and Technology, Cornell University,
Geneva, NY 14456, USA;*
[2]Department of Plant Genetics, Agricultural University, Gödöllö H6701, Hungary;
[3]Department of Plant Pathology, Cornell University, Geneva, NY 14456, USA

Abstract. Apples are an important agricultural commodity in the US. Most apples are produced for the fresh market, and have to be stored under controlled atmospheric conditions to avoid softening. The softening of fruits is the result of structural changes in the cell walls. These structural changes are caused by the hydrolytic activity of such enzymes as polygalacturonase, cellulase and the diverse hemicellulases, that are under the control of the ripening hormone ethylene. Ethylene is synthesized in plants from S-adenosyl methionine by a short pathway that consists of two enzymes: 1-aminocyclopropane -1-carboxylic acid synthase(ACS) and 1-amino-cyclopropane-1-carboxylic acid oxidase (ACO). To interfere with ethylene synthesis in plants we have cloned two ACS genes from ripening apples (McIntosh). The gene showing the closest similarity to ripening - related ACS gene in other fruits was used to make antisense constructs. Royal Gala and McIntosh plants were transformed with these antisense constructs using a *Agrobacterium* mediated transformation system. Transgenic plants were propagated on antibiotic -containing agar, transferred to the green house for conditioning and later to the field. Transgenic Royal Gala fruits were evaluated for morphological characteristics, ethylene production, and ripening parameters. Data indicate differing ethylene production in the transgenic lines.

1. Introduction

Apples are an important crop in the United States, annual production is approximately 251.5 million bushels (1 bushel = ca. 42 lb =ca. 20 kg), that is approximately 5 million tons. Fom this New York State produced last year 25.5 million bushels, or nearly 0.5 million tons. New York States apple production consists mainly of the apple cultivars listed in Table 1. The cultivar produced in largest amount is McIntosh. McIntosh apples have a deep red color, pleasant aroma, flavor and firm texture. However, the fruits soften rather soon after harvest and have a relatively short storage time. Present practice to extend shelf life is the use of controlled atmospheric storage where gas composition and temperature is kept under strict control. This storage method is effective for up to four months for McIntosh, but after this time fruit texture rapidly deteriorates. Also, controlled atmospheric facilities are rather expensive.

Strategies are available that permit an alternative approach to prevention of softening of fruits. One of these strategies is the use of antisense constructs of genes or gene fragments to interfere with the fruits softening mechanism [5].

Table 1. Total production of apples by variety, 1989-1998

Variety	1989	1990	1991	1992	1993	1994	1995	1996	1997	1998
Cortland	70	65	80	80	60	80	80	75	80	75
Crispin (Mutsu)	20	15	20	15	10	20	25	30	40	45
Delicious	95	110	120	140	85	115	115	115	120	95
Empire	50	55	55	65	65	95	95	110	130	110
Golden Delicious	45	55	50	55	40	40	40	40	40	55
Gala 2/										**5**
Gingergold 2/										5
Idared	75	85	80	90	75	85	80	70	80	95
Jerseymac 1/							10	10	10	5
Jonagold 1/							15	20	25	20
Jonamac 1/							15	15	20	20
Macoun 1/							10	15	15	10
McIntosh	**195**	**190**	**240**	**280**	**175**	**250**	**235**	**220**	**215**	**230**
Northern Spy	17	15	20	20	25	20	15	10	15	10
Paula Red 1/							15	10	15	15
R.I. Greening	135	95	125	135	110	115	105	90	75	75
Rome	110	155	130	145	110	150	155	120	155	135
Spartan 1/							15	10	20	20
Twenty Ounce	50	45	40	40	40	40	35	25	25	25
Other	98	105	90	105	75	95	50	45	40	30
All Varieties	960	990	1,050	1,170	870	1,100	1,110	1,030	1,120	1,070
Thousand Bushels (42 pounds per bushel)										
All Varieties	22,857	23,571	25,000	27,857	20,714	26,190	26,429	24,524	26,667	25,476

1/ Estimates began in 1995.
2/ Estimates began in 1998.

From New York Agricultural Statistics

2. The Cause of Fruit Softening

The crisp texture of apples and other pome fruits results from the composition and intactness of their cell walls. Developing and maturing fruits have firm, tightly linked cell walls where the main constituents such as cellulose, pectin, the diverse hemicelluloses and some lignin are tightly bound to each other. Upon ripening, the organization of the cell wall undergoes changes, and these changes are the main cause of fruit softening. The cell wall changes in turn are the result of the activity of such enzymes as polygalacturonase, pectin methyl esterase, cellulase and the diverse hemicellulases [7]. The activity of these enzymes in turn is governed by ethylene, the "fruit ripening hormone" [8]. In the ripening process of fruits ethylene is produced. The presence of ethylene triggers the activation of genes for the cell wall modifying enzymes. The enzymes, in turn, degrade the cell walls and the softening of the fruits ensues. Therefore, down regulatio of ethylene in fruits will most likely result in diminished softening of the fruits. Transgenic experiments with tomatoes [10] and melons [1], where ethylene synthesis was down regulated by transformation with antisense constructs of the ACS gene that controls the activity of the rate limiting enzyme of the ethylene pathway have shown that this strategy works. Although apples were thought of as having a different ripening mechanism that doesn't involve an endopolygalacturonase in the cell wall degradation processes [2], the presence of this enzyme has been shown recently [14].

The ethylene biosynthetic pathway is a relatively short pathway that is composed of two enzymes [15]. The first enzyme is 1-aminocyclopropane carboxylic acid synthase, commonly abbreviated in the literature as ACS. This enzyme uses S-adenosyl methionin as substrate and converts it to 1-aminocyclopropane carboxylic acid (ACC). The second enzyme, 1-amino cyclopropane carboxylic acid oxidase, oxidizes ACC to ethylene. In the pathway ACS is the rate limiting enzyme. Therefore, if the activity of this enzyme is reduced, the pathway will produce less ethylene. The antisense method has been succesfully used in tomatoes [10], melons [1], and has now been applied to apples.

3. Antisense Gene Construction and Transformation

To obtain an apple ACS gene we have chosen the following method. We have isolated mRNA from the cortical tissue of ripening McIntosh apples and using primer sequences that have been conserved in ACS genes isolated from tomatoes [9] and apples [4] we synthesized a 1093 bp DNA fragment by RT-PCR. This DNA fragment had an 82% sequence homology to an ACS gene reported from peppers [6], a 67% homology to the ripening and wound induced ACS gene from tomato (LeACS4) and a 56% homology to the ACS gene reported from apple. This DNA fragment showed the characteristic amino acid motif of SLSKDLGFPGFA that has been reported to be the amino acid sequence of the active center in ripening and fruit specific ACS enzymes. This 1093 bp gene fragment was inserted in reverse orientation between the CaMV-35S promoter and the NOS terminator sequences in to the vector pBI101. Young apple leaves were transformed with *Agrobacterium tumefaciens* strain EHA101harboring the binary vector, and transformants selected on kanamycin containing medium. Shoot formation was initiated on shooting medium [3] and root formation on rooting medium [16]. Transgenic explants were propagated in glass jars and their ethylene production measured (Figure 1).

From the investigated explants one (TG 196-1) was an ethylene over producer, one line (TG 168-1) showed no significant change in ethylene production, while two other transgenic lines (TG 198-1 and TG 197-1) showed a markedly decreased ethylene production.

The explants were micrografted onto M7 rootstock, propagated in the greenhouse and after one year transplanted to the field. No morphological differences were visible between control and transgenic plants. Flowers were observed after three years in the field, and one plant produced two transgenic fruits. Flowering in the fourth year was moderate to heavy, and fruit production varied between 2 and 45 per transgenic tree. Fruit size varied from small to normal between individual lines of a transformant, and in most cases was similar to that of non-transgenic control fruits. The transformation lines, number of trees deriving from these and fruit count is shown in Table 2.

Transgenic fruits were harvested on September 8, 1999 and stored at 4°C. Samples (3 fruits from each transgenic tree) were removed from storage after one day, stored in glass jars under lid and ethylene production and other ripening parameters were measured (Table 3).

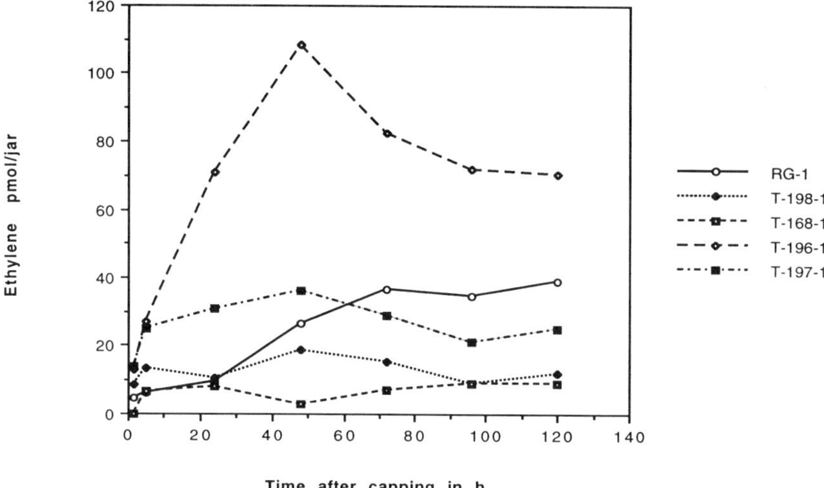

Figure 1: Ethylene production by transgenic explants. Transgenic explants were grown in glass jars on agar medium. The jars were capped, and ethylene was measured in intervals indicated in the figure. Control explants not containing the antisense construct (RG-1) are shown as solid line, transgenic plants of different transformation lines as broken lines.

Table 2: Number of transgenic trees and fruits within Royal Gala transformation lines.

Transformation line	Number of trees	Fruit count
TG 196	5	20
TG 197	5	26
TG 198	4	62
TG 213	7	68
TG 215	2	17
TG 218	8	130
TG 234	4	31
TG 235	4	60
TG 236	4	22
TG 237	1	23
TG 239	6	84
TG 240	2	0
TG 507	3	36
TG 508	3	0

For reasons of simplicity the values from fruits deriving from individual trees within a transgenic line were averaged. These averaged data show no significant decrease in ethylene production between control and the transgenic lines or in other ripening parameters. However, there were significant differences between the individual trees within a transgenic line that are not obvious from the averaged data (see statistical deviation).

Ethylene production was continuously monitored for 69 days after harvest. Fruits were removed from storage, placed in the jars, the jars sealed and allowed to come to room temperature. Ethylene was measured 24 hours after the jars were sealed. During the storage time ethylene production in the control plants rose to an average of 200 nmol/100g/h, while in the transgenic lines ethylene production was below 100 nmol/100g/h.

Table 3. Ripening parameters of Royal Gala fruits from transgenic lines.

Apple line	Ethylene* (nmol/100g h)	Soluble solids (Brix°)	pH	Force (N)
Control	2.36±0.38	14.7	3.93	128.2
TG 196	3.81±0.76	14.4	3.97	142.2
TG 197	2.60±0.64	15.1	4.04	164.4
TG 198	2.75±0.79	14.7	3.98	139.0
TG 213	2.94±0.39	17.7	3.88	131.5
TG 215	2.92±0.97	12.6	4.08	136.0
TG 218	3.04±0.43	16.0	3.97	132.8
TG 234	3.29±0.52	15.3	4.04	170.9
TG 235	4.08±0.92	17.2	3.95	121.0
TG 236	3.12±0.99	16.9	3.94	153.3
TG 237	6.03±0.00	-	-	-
TG 239	3.11±0.40	14.2	4.01	126.6
TG 507	5.55±1.18	17.3	3.99	136.2

* Ethylene calculated as average ± SE from 2-8 replications (transgenic trees).

On the per cent basis ethylene production in apples from the individual trees from the different transgenic lines was down regulated after 69 days of storage by 11 to 97%, with the majority of fruits falling in the 60-80% range.

To insure that the data we obtained here are not caused by other factors that were outside the scope of this investigation, the individual transgenic Royal Gala trees will have to be evaluated on a continual basis for a number of years, and ethylene production has to be correlated yet with ACS enzyme activity and mRNA production. However, the presently obtained data are good indicators that the strategy that was successfully used in tomatoes [10-13] for extending shelf life also works in apples.

Acknowledgement

This work was supported in part by the New York State Apple Research and Development Program, and the Cornell Center for Advanced Technology in Biotechnology, which is sponsored by the New York State Science and Technology Foundation and industrial partners.

References

[1]   R. Ayub et al., Expression of ACC oxidase antisense gene inhibits ripening of cantaloupe melon fruits. *BioTechnology* **14** (1996) 862-866.

[2]   I.M. Bartley, Exo-polygalacturonase of apple. *Phytochemistry* **12** (1978) 213-216.

[3]   A. De Bondt et al., *Agrobacterium*-mediated transformation of apple (*Malus x domestica* Borkh): An assessment of factors affecting regeneration of transgenic plants. *Plant Cell Reports* **15** (1996) 549-554.

[4]   J.G. Dong et al., CLoning of a cDNA encoding 1-aminocyclopropane-1-carboxylate synthase and expression of its mRNA in ripening apple fruit. *Planta* **185** (1991) 38-45.

[5]   J. Gray *et al.*, Molecular biology of fruit ripening and its manipulation with antisense genes. *Plant Mol. Biol.* **19** (1992) 69-87.

[6]   M. Harpster *et al.*, Characterization of a PCR fragment encoding 1-aminocyclopropane-1-carboxylate synthase in pepper (*Capsicum annuum*). *J. Plant Physiol.* **147** (1996) 661-664.

[7]   D.J. Huber, The role of cell wall hydrolases in fruit softening. *Hort. Rev.* **5** (1983) 169-219.

[8]   J.M. Leliévre *et al.*, Ethylene in fruit ripening. *Physiol. Plant.* **101** (1997) 727-739.

[9]   J.E. Lincoln *et al*, LE-ACS4, a fruit ripening and wound-induced 1-aminocyclopropane-1-carboxylate synthase gene of tomato (*Lycopersicon esculentum*). Expression in *Escherichia coli*, structural characterization, expression characteristics, and phylogenetic analysis. *J. Biol. Chem.* **268** (1993) 19422-19430.

[10]   P.W. Oeller *et al.*, Reversible inhibition of tomato fruit senescence by antisense RNA. *Science* **254** (1991) 437-439.

[11]   S. Picton *et al.*, Altered fruit ripening and leaf senescence in tomatoes expressing an antisense ethylene-forming enzyme transgene. *Plant J.* **3** (1993) 469-481.

[12]   A.J. Reed *et al.*, Delayed ripening tomato plants expressing the enzyme 1-aminocyclopropane-1-carboxylic acid deaminase. 1. Molecular characterization, enzyme expression, and fruit ripening traits. *J. Agric.Food Chem.* **43** (1995) 1954-1962.

[13]   A. Theologis *et al.*, Modification of fruit ripening by suppressing gene expression. *Plant Physiol.* **100** (1992) 549-551.

[14]   Q. Wu *et al.*, (1993) Endopolygalacturonase in apples (*Malus domestica*) and its expression during fruit ripening. *Plant Physiol.* **102** (1993) 219-225.

[15]   S.F. Yang and N.E. Hoffman, Ethylene biosynthesis and its regulation in higher plants. *Ann. Rev. Plant Physiol.* **35** (1984) 155-189.

[16]   J.-L. Yao *et al.*, Regeneration of transgenic plants from the commercial apple cultivar Royal Gala. *Plant Cell Reports* **14** (1995) 407-412.

*Use of Agriculturally Important
Genes in Biotechnology
G. Hrazdina (Ed.)
IOS Press, 2000*

Elements of Biotechnology Applied to Potato Breeding at IHAR Mlochów

E. Zimnoch-Guzowska, B. Flis, R. Lebecka, W. Marczewski, E. Mietkiewska,
J. Syller

*Plant Breeding and Acclimatization Institute (IHAR), Mlochów Research Center,
05-832 Rozalin, Poland*

Abstract. At the Mlochów Research Center of IHAR various techniques of
biotechnology are used in the following areas of potato breeding and breeding
research: (1) Increase of the genetic pool utilized in potato breeding by GMO
sources obtained by genetic transformation and somatic hybridization. Studies
on transgenic lines in the respect of transgen effects on their resistance to PVY
and PLRV and on variability of agronomic and morphological traits among
transgenic lines. (2) Studies on potato-pathogen interactions. Aphid
transmission of PSTVd encapsidated by PLRV particles. (3) Application of
the PCRto the diagnostics of TRV and PSTV in the potato. (4) Marker
assisted selection for PVS resistance in diploid and tetraploid breeding
program with an application of SCAR markers linked to the gene Ns. (5)
Mapping the potato genome. QTL mapping for tuber and leaf resistance to
Erwina sp., tuber resistance to *Phytophthora infestana*, resistance to PVS,
PVM and PLRV. Some of the above projects are carried out in the
collaboration with the Institute of Biochemistry and Biophysics of Polish
Academy of Sciences in Poland, Max Planck Institute for Breeding Research in
Germany and Cornell University in the USA.

1. Introduction

Potato belongs to the main agricultural crops in Poland. It is cultivated on 1.3 million
hectares, with total production ranging from 20 to 25 million tons yearly. These data have
placed Poland as the fourth largest potato producing area in the world. The potato crop in
this region is affected by many diseases and pests. Hence for breeding new cultivars,
breeding for resistance in Poland is an important activity in addition to selection for
quality.

For over 30 years the Mlochów Research Center, IHAR, has been involved in research
on potato genetics and breeding parental lines for new cultivars, which are outstanding in
resistance to main potato viruses, late blight and quality traits. The research is related to
the breeding program.

The research on enlarging the genetic pool for plant breeders by utilizing wild species
with conventional selection techniques or with transformation of existing cultivars is well
established. The Center is involved in work on mapping the potato genome, especially for
genes responsible for resistance to pathogens. The biology of potato pathogens, including
viruses, with special attention to their new pathotypes is being studied. Improvement of
diagnostics and selection methods include preliminary studies on marker assisted selection
and application of PCR to diagnostics of potato infection with pathogens. Some projects
that use the tools of biotechnology are presented in this paper.

2. Unconventional sources as the genetic pool utilized in potato breeding

The aim of the potato transformation was selective improvement of resistance to viruses in two Polish cultivars. This work was done in close collaboration with the Institute of Biochemistry and Biophysics of the Polish Academy of Science (IBB-PAN), where molecular parts of the projects were done. Studies on transgenic lines conducted at IHAR Mlochów were carried out to investigate the phenotypic expression of the transgen (evaluation of resistance level to PVY or to PLRV, respectively) and the variability of the agronomic and morphological traits among transgenic lines. Various parts of the viral genome were used for the constructs. In both projects highly resistant lines were selected. The lines highly resistant to PLRV were selected from lines of cv. Bzura transformed with the antisense CP PLRV gene [6], while lines of cv. Irga resistant to necrotic strain of PVY were transformed with sense and antisense constructs of the truncated PVY^N polymerase gene, or with non-translated region of PVY^N genome [2], or with CP gene of lettuce mosaic virus (LMV, which is also a potyvirus). The lines mentioned above appeared to be resistant to pathogens after mechanical or graft inoculation with the respective virus. Further details about the projects are given by others at the workshop.

Primary results of our studies on transgenic potato indicate that a transgen of viral origin can be a source of high resistance, similar to that conferred by the potato genes. The genetic transformation of the cultivar that is popular on the market was believed to be a method that can be used for the improvement of this cultivar in desired traits without any additional changes in its agronomic value. The assumption that the cultivar maintains an intact genome after the transformation was not correct, although the range of variability of the various agronomic traits after transformation was lower than that observed after sexual crossing of the parents. There remain several general questions which should be answered in the future: (i) how far can genetic transformation complement a conventional breeding progress, (ii) how to assess the risk of introducing tranformants into a large scale agriculture, (iii) how to select the best genes and the best plant genotypes to create transgenics, (iv) which transgenics will be accepted by the public.

The somatic fusion, the second unconventional method of increasing the genetic pool for breeding, is applied to genetically distant parental combinations of *Solanum* species. At IHAR, the work has begun on fusion of diploid interspecific potato hybrids, carrying combinations of desired characters and showing difficulties in sexual crossing. In collaboration with IBB-PAN, where somatic hybrids between *Solanum tuberosum* and *Solanum nigrum* were obtained, fusion hybrids are tested for resistance to *Phytophthora infestans,* as *S. nigrum* was used as a donor of resistance to late blight, with the intention to further transfer this resistance to the *S. tuberosum* germplasm.

3.　Studies on potato–pathogen interactions

The aphid transmission of potato spindle tuber viroid (PSTV) encapsidated by potato leafroll virus (PLRV) particles has been demonstrated. The PSTV is a quarantine potato pathogen, which has no aphid vector under natural conditions and is known to be transmitted by mechanical infection. This is also an important potato pathogen, as it belongs to rare pathogens, which are transmitted to the next sexual generation by true seeds. PLRV is one of the most important potato viruses, that is readily transmitted by some aphids, *Myzus persicae* Sulz. being the most efficient vector. It was shown that PSTV was transmitted by *M. persicae* to potato, tomato, *Physalis floridana* and *Datura stramonium* from potato and *P. floridana* plants doubly infected with PLRV and PSTV. The viroid was not transmitted by aphids from plants infected with PSTV alone. Heterologous encapsidation of PSTVd RNA by coat protein of PLRV is the most probable explanation of the viroid transmission by *M. persicae*. When PLRV preparations purified from doubly infected plants were treated with RNase, the viroid was protected against digestion with the enzyme and gave positive signals in dot blot hybridization [7].

4. Application of the RT-PCR techniques to the diagnostics of pathogens in potatoes

To simplify routine diagnostics of contamination of potato with some pathogens, a PCR approach was used. RT-PCR was introduced to detect the tobacco rattle virus (TRV), for which routine testing by ELISA is not efficient. The reason for this is because one of the TRV particles (RNA1), which does not produce coat protein, is not detected [5].

Until recently, the PSTV diagnostics has relied mainly on biological indexing and polyacrylamide gel electrophoresis [4]. Both methods are laborious and expensive. Nucleic acid hybridization techniques are also used for viroid detection, but they are not used widely in laboratories. So far RT-PCR techniques have not been used for viroid indexing in large scale certification [8]. Recently, RT-PCR techniques have been adopted for such certification. A simple method for RNA extraction without organic solvents has been introduced. It has been shown that RNA extracted by this method was useful for viroid detection from potato tubers as well as from sprouts and leaves in RT-PCR. Comparison of PSTVd detection by RT-PCR and dot-bot hybridization confirmed a high consistency of the results obtained by the use of these methods.

5. Marker-assisted selection in potato breeding program

Wild species of the genus *Solanum* have played an important role as a source of resistance in potato breeding. *Solanum tuberosum* ssp. *andigena* was found to be a source of the dominant gene *Ns*, which is responsible for the hypersensitive reaction of potato to potato virus S (PVS) infection [1]. PVS is considered to be one of the most important viruses and is found in nearly all countries, where potatoes are grown. The *Ns* gene was used in potato breeding programs, both at the diploid and tetraploid levels, at the Mlochów Research Center.

Using RAPD method and bulked segregant analysis four RAPD markers were identified, that were linked to the gene *Ns* in diploid potato clones [3]. The distances between the gene *Ns* and these markers ranged from 2.6 to 6.6 cM. SCAR primers are longer than RAPDs and initiate DNA amplification in a more specific way. The OPG17-450 RAPD marker has been cloned and sequenced. Using this sequence, specific PCR primers were designed and used for PCR amplification. We have found such a pair of SCAR primers which result in amplification on a single specific band linked with the *Ns* gene in diploid potatoes. However, only one of the RAPD markers, OPG17-450, is amplified incidentally for DNA isolated from tetraploid, PVS resistant clones or cultivars. The development of SCARs from this RAPD marker did not enlarge the list of PVS-resistant lines identified with the SCAR products. Diploid and tetraploid potatoes carry the *Ns* gene derived from different resistant clones, both having an *andigenum* origin. This may suggest that there are different alleles at the *Ns* locus in tetraploids.

The work on marker identification is extended for resistance against PVM, governed by genes for necrotic response from *S. megistracrolobum* or genes from *S. gourlayi,* and for resistance against PLRV derived from *S. tuberosum* and *S. chacoense.*

6. Mapping the potato genome

In collaboration with the Max Planck Institute the mapping was done for quantitative trait loci (QTL) having effect on potato resistance to *Erwinia carotovora* ssp. *atroseptica* in tubers and leaves. The chosen mapping population was a diploid F1 progeny from crossing a resistant parent with a susceptible one. A linkage map was constructed on AFLP and RFLP markers including three resistance-gene-like (RGL) markers. Genetic factors affecting resistance were located on all the twelve potato chromosomes. Ten putative QTLs for tuber resistance were identified on nine linkage groups. Two groups for major effects have been mapped on chromosomes I and VI. Seventeen putative QTLs for leaf resistance mapped to ten potato chromosomes. Six groups of QTL for leaf and tuber resistance occupied the same map positions. Four out of the seven QTLs linked to the RGL loci were mapped to genome segments known to containing factors for resistance to different pathogens of the *Solanaceae* [9].

The *Ns* is being mapped on the potato genetic map using an AFLP method. We have found a few AFLP fragments cosegregating with phenotypic data. They can be the candidates to continue the experiments to develop closely linked, simple PCR-based marker forms of the *Ns* locus that could be used as an easy selection tool for practical breeding.

In phenotypically distinguishable hybrid populations there are several desired traits, which will be mapped in the near future: (i) tuber resistance to *P. infestans,* (ii) chipping quality, (iii) and tuber flesh discoloration.

References

[1]　M.-L. Baerecke, Uberempfindlichkeit gegen das S-Virus der Kartoffel in einem bolivianischen Andigena –Klon. *Zuchter* **37** (1967) 281-286.

[2]　B. Flis *et al.*, Transgenic potato clones resistant to potato virus Y. Abstr. 10th EAPR Virology Section Meeting, Baden, Austria, 1998, 37-38.

[3]　W. Marczewski *et al.*, Identification of RAPD markers linked to the Ns locus in potato. *Plant Breeding* **117** (1998) 88-90.

[4]　W.H.H. Mosch *et al.*, Development of a standard method for detection of potato spindle tuber viroid in potato plants. *Neth. J. Plant. Pathol.* **88** (1982) 113-122.

[5]　T. Muchalski, Use of RT-PCR technique for detection of tobacco rattle virus in potato tubers. *Phytopathol. Pol.* **13** (1997) 31-37.

[6]　A. Palucha *et al.*, An antisense coat protein gene confers immunity to potato leafroll virus in a genetically engineered potato. *Eur. J. of Plant Path.* **104** (1998) 287-293.

[7]　J. Syller *et al.*, Transmission by aphids of potatospindle tuber viroid encapsidated by potato leafroll luteovirus particles. *Eur. J. of Plant Path.* **103** (1997) 285-289.

[8] Y.F. Wau Chow Wah and R. Symons, A high sensitivity of RT-PCR assay for the diagnosis of grapevine viroids in field and tissue culture samples. *J. Virol. Methods* **63** (1997) 57-69.

[9] E. Zimnoch-Guzowska *et al.*, QTL analysis of new sources of resistance to *Erwina carotovora* ssp. *atroseptica* in potato using AFLP, RFLP and resistance-gene-like markers. *Crop Science* (submitted 1999).

Mapping of the Wheat Genome for Genetic Diversity

A. Kubek
Slovak Agricultural University, Nitra, Slovak Republic

1. Introduction

Application of molecular genetics principles in plant breeding is based on genotype or genofond of population identification of the DNA level by means ``marker genes´´ which are also called candidate genes. These markers are usually the result of sudden changes caused by point mutations in genes which are playing a decesive role in metabolic processes, or they direct key hormones or cell membrane receptors (Hruban, 1996).

Development of molecular biology technics enabling pursue a polymorphism of DNA mutations on the molecular level and search or identified genes which are responsible for hereditary anomalies or metabolical disorders as well as for hereditary predisposition to create economic important traits.

Most economical important traits in different species are quantitative traits, it means they exhibit continuous variation resulting from the action of multiple genes that is modified by environment. The selection process of plant improvement is thought to act on frequency of different version of these genes. Determining the location and number of genes that condition quantitative traits and estimating the magnitude of individual gene effects have occupied quantitative geneticists since the begining of this century. However, a preliminary dissection of the genetic factors affecting quantitative characters was achieved only in experimental species of animals and very recently in some plant species.

Utilisation of DNA polymorphism has a great importance, before all, in such traits improvement which are measurable with difficulties, they have a low heritability, they are expressed in higher age or in such traits they have a negative correlation each to other, where effectiveness of traditional methods of selection is low.

From the increasing knowledge, different and sophisticated approaches have been established for polymorphism of DNA in coding but mainly in noncoding parts of genome. Principially, genome analysis can be divided into analysis of structure and function of distinct genes, detection and comparison of DNA variants in populations, and mapping of DNA markers.

Up to now, a great number of polymorphic differences in DNA sequences have been detected in plant populations and the availability of DNA probes as markers dramatically increased the potential to detect new and special selected DNA variants. In most cases the identification of DNA variants is done by the amplification of selected DNA fragments using PCR, followed by the action of special restriction enzymes which cut the fragment only if the base substitution is present or absent. In such cases are produced fragments of various length which is possible by means of electrophoresis to detect. A detected polymorphism is called RFLP. Thereby, the patterns depend directly on the genotypes of the individuals. Of course there exists other form of polymorphism where the basis of DNA variants can be caused by variable number of tandem repetitive elements (VNTR). Thus for the assay of DNA variants two different approaches are available: A base-pair alteration in the DNA sequence can change the restriction site for an enzyme and thereby initiate restriction site variants (RSVs) detectable by RFLPs, and tandem repeat alteration based on a PCR allele-specific assay (PASA).

Screening of base pairs variants has revealed that about one nucleotide site in 100 is potentially polymorphic and about one nucleotide in 500 will actually differ between two randomly chosen chromosomes. That means for a gene of 5 to 10 kbp as much as 50 to 100 variants can be detected on the level of DNA.

For genome analysis the mapping of DNA markers has become a main topic in research. Mapping requires landmarks against which genes can be located within genome. These landmarks or reference markers are in genetics usually termed markers.

Gene mapping can be divided in genetical and physical or cytogenetic mapping. Mapping of gene on the basis of recombination frequencies by family studies – genetic mapping – requires allelic variants in order to follow their segregation within groups of individuals. These groups must have defined linkage disequilibria. Physical mapping is performed on the basis of the chromosome structure itself. In this case, allelic variation is not necessary, but the association of a marker to a particular chromosome region should be defined. For genetic mapping, one would want to have all points of the genome within not much more that 20cM (centimorgans) of a mapped marker locus, in order to enable linkage relationships to be established with a resonable number of individuals.

The types of gene mapping are different in their levels of details and the extent to which the map provides and owerview to the mapped markers. At the ``high end'' of gene mapping, the chromosomal banding patterns observed by light microscopy or the in situ hybridisation allow an average chromosome to be subdivided into 10 to 20 regions, each of a lenght of 10 to 12 milion nucleotides. The resolution of the ``low end'' of the physical maps is obtained by restriction enzymes or DNA sequencing and is described by distance of single nucleotides. The large gap in resolution between sequencing and chromosome staining can be bridged by cleaving the DNA into large fragments and then analyse sequences that surround the cleavage sites. It is very important that DNA variants used as physical markers can also be followed genetically. Thus DNA variants link the genetical – measured in cM – and physical maps – measured in bp – at large numbers of genomic sites.

An important feature of mapping is the comparative consideration of the genetic material itself if the same genes or DNA sequences are identified in different species. In such situation the selection of marker genes within one species can use data of other species.

Research programmes in dealing with utilisation of genome analyses results in relation to economically important traits can be used at MAS (marked assisted selection). The first study oriented to this problem was made by Sax (1923). Since that time up to now were dicovered many such associations-marker x QTL (Kluge and Geldermann, 1983, Lande and Thompson, 1990, Weller and Fernando, 1991). Comparison of plant populations on the level of DNA sequences will mainly analyse the differences between and within populations and try to define genetic resources. Consequences of new biotechnical approaches in plant rearing means the estimation of success and risk for breeding programmes as well as for breeding organisations. In all study where is necessary to identify QTL or marker limited QTL, which could be utilized in target breeding programmes are important the famillies of plants which segregate according to economic important traits. They are called resource famillies (Roche et al. 1992). If the association of QTL x marker gene is proved than application method of statistical l model (MAS), can be highly ussefull by increasing of selection gain in quite short time than is able to reach by traditional selection (Gibbson et al. 1994).

You are looking at a Line Map from GrainGenes

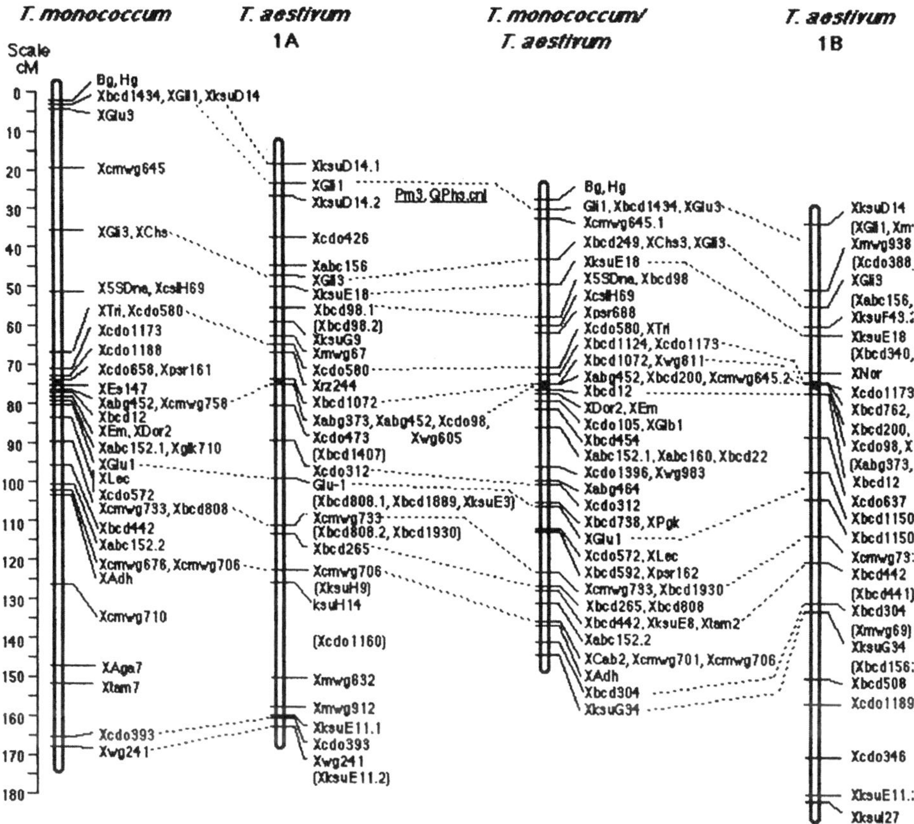

You are looking at Line Map from GrainGenes

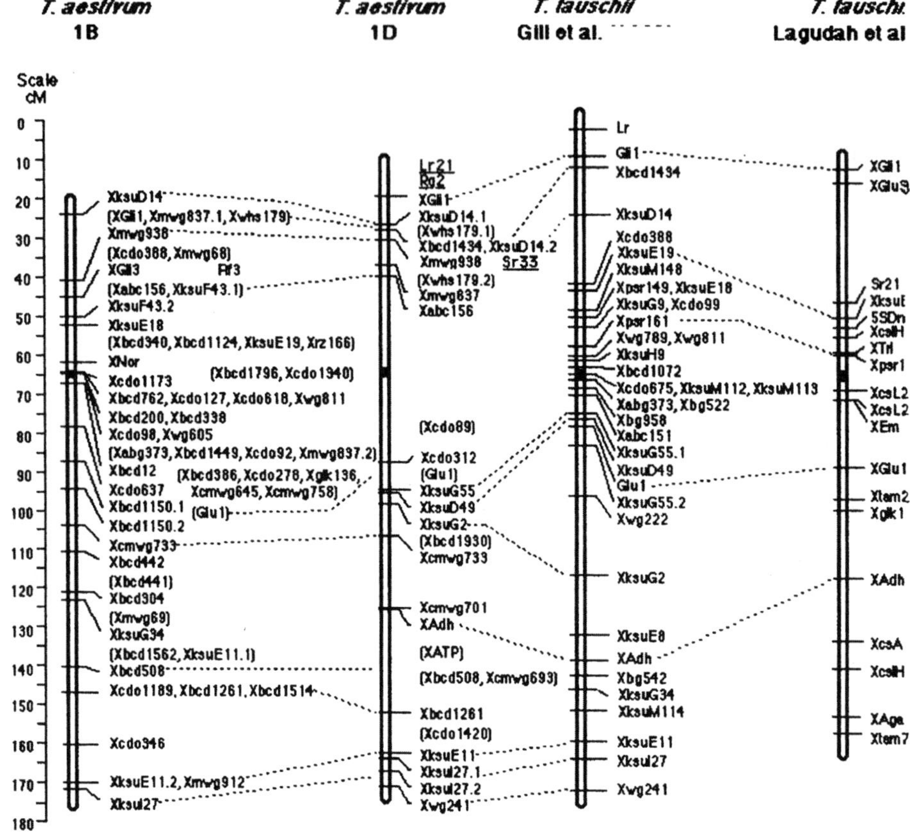

You are looking at Line Map from GrainGenes

Curtis and Lukaszewski   **Kota et al.**

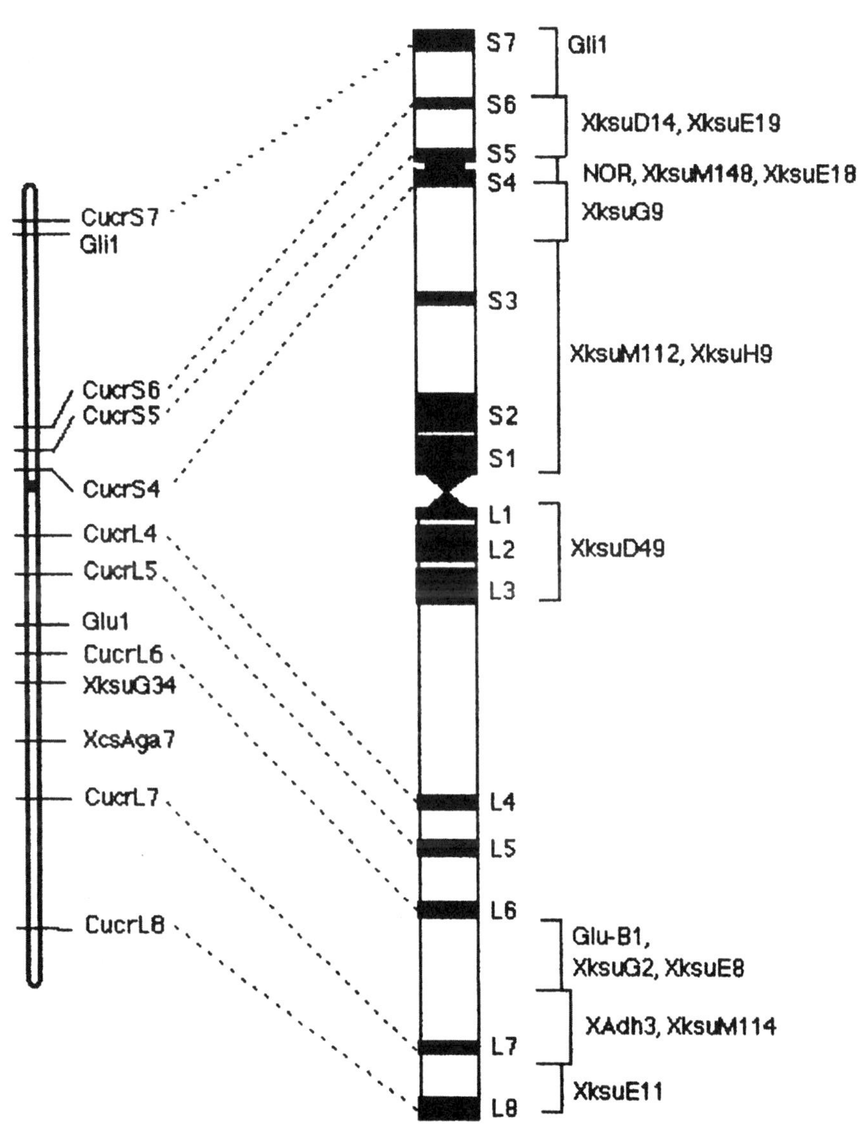

2. Implications

Development and use of genetic linkage maps for marker-assisted selection is aided by a database and information system maintaining marker genotypes and trait phenotypes on the plant species as well as DNA sequences of markers and genotyping protocols. Critical success factors for the effective implementation of genetic linkage maps are:

1. Maximizing flexibility through effective conceptual database modeling and normalization;
2. Maintenance of sufficient genotypic and phenotypic data within families to be able to determine transmission of chromosomal fragments between generations;
3. Achieving a genetic linkage map with 20 cM resolution;
4. Maintenance of DNA sequence data to aid in standardizing nomenclature.

Molecular data could change direction of genetic improvement more quickly than phenotypic data alone if records are available on all traits. In each case, marker data will be combined with performance data of the plants to make optimal progress.

QTL mapping is an idea whose time has come. The number of markers and traits has become so large for many species that joint analysis of all data derived from single experiment is necessary. Application of marker-assisted selection potentially can increase rates of genetic gain by up to 20 % but will require significant changes in breeding programs.

3. Biodiversity

Biological diversity is a variety of all forms of life that comprises ecosystems, plant and animal species, microorganisms, and genetic information they contain. It is constantly changing as result of evolution and since the time of a man has developed also of human interferences. Its conservation is a substance for human well-being since it provides food, medicines and industrial products and serve numerous ecological functions.

Despite our dependence on the limited resources of the Earth we are taking over natural resources and disturbing natural systems. Keeping this course we are at risk of survival. Loss of biological diversity has became a global problem reflected in many international documents and is waiting for a sound solution.

It had become clear that traditional approach based on the conservation of selected territories and species was not more effective enough to stop the course taken.

Therefore at the end of eighties the international community building upon existing conventions became a process of negotiating of new global treaty covering not only conservation of biodiversity but at the same time the issues of access to genetic resources, sustainable use of biodiversity, biotechnology's, and moreover building of partnership between the countries based on the idea of equitable sharing of benefits that biodiversity use brings to humankind.

Identification is the first step to determination of status of biological diversity components. It is necessary to identify components of biological diversity important for its conservation and sustainable use having regard to the generally accepted categories.

The comprehensive biological and ecological inventories are one of the most fundamental technology for conservation of biodiversity and the sustainable use of biological resources.

Among others also population biology characteristics of species can give us very often satisfactorily information on genetic variability without exact knowledge of species gene pool. From this point of view there are some strategic directions:

- priorities need to be set to address gaps in our knowledge on status of biodiversity components;
- high attention should be given to increasing knowledge's on vulnerable, threatened and endangered species and ecosystems, critical habitats, little-studied taxonomic groups, taxonomic groups of economic importance, areas of high diversity and areas where human development and disturbance are the most significant;
- link zoological, botanical and habitat, biotope or community inventories with soil, climate and other surveys;
- to finish Biotope Mapping and Wetland Mapping to gain comprehensive data base of valuable habitats and suggest measures for their protection, management and sustainable use.

The sources of the adverse impacts on biological diversity may have different nature, but those we would like to address are all of human origin. They *inter alia* include: pollution (heavy metals, halogen compounds, phenols, dioxins, complex-forming substances etc.), inappropriate management practices, global climate change, artificially introduce alien species or genetically modified organisms.What is necessary in biodiversity to:

- prevent the introduction of, control or eradicate those alien species which threaten ecosystems, habitats or species;
- to finish taxonomic study of invasive species including distribution;
- to continue in population biology study of invasive species;
- prevent unsustainable use of natural resources;
- control pollution's that have adverse impact on biodiversity;
- strengthen controlling the risk associated with the use and release of genetically modified organisms *inter alia* by establishing appropriate structures.

Gene pool of species represents an important part of biological diversity and the mankind widely benefits of it. Gene pool of particular cultivated species consists not only of modern and exploited cultivars and hybrids which have been primarily used in intensive agriculture, but also restricted cultivars, world assortment cultivars, land races and old cultivars, ecotypes of some cultivated species and their natural populations, which have been maintained and directly used by small-scale growers for different purposes.

Which are the strategic directions:

- improve inventories to determine the genetic diversity of domesticated and non-domesticated biological resources to maximize the conservation and economic use of genetic resources;
- gather of valuable genotypes of old cultivars and land races as well as ecotypes within natural populations and subsequently evaluate, documentate and multiplicate them for long-term maintenance;
- establish special repositories for long-term maintenance of valuable genotypes of vegetatively propagated species;
- utilize of establishing collections for both home and international breeding programs, for research and study activities, broadening of public knowledge, in alternative agriculture and landscape engineering:

The most utilized plant in agriculture from the economic point of wiev is wheat.

Comparative genome mapping of the wheat rye and barley and related wild species coordinated by ITMI (international triticale mapping initiative) which form a group of cereal geneticists they develop genetic stock, molecular probes and technologies for the purpose of producing linkage maps of genomes in the grass triticeae which includes the crops (wheat, barley, rye ets). These maps will facilitate the identification and development of genes for improvement of crop species, the identification and use of genes from wild species to improve crop species and the elucidation of genetic and evolutionary relationship among the species of Triticeae.

The linkage maps constructed according bread wheat genome libraries select clones to complete a 20 cM RFLP on the maps. Some of the laboratories working with clones will take to the acount at least 500 clones the nulitrasomic and dietelosomic aneuploid stock to the construction of maps to find the similarities between the wheat and barley genomes.

The way is directed mapping which determine genotypes of all backcross progeny by Southern blotting using RFLP loci spaced as close to 20 cM apart as possible to locate the genes affecting resistance and quantitative traits. Recently was identified RFLP markers linked to HESSIAN FLY resistance genes, ESI genes, PHI, genes responsible for Russian wheat aphid resistance etc.

Microsatellite markers are a PCR-based marker system which requires only minimal amounts of DNA and which can easily be automated. Microsatellite markers isolated from the wheat (Triticum aestivum) genome display a much higher variability than RFLP markers. In the last years, was isolated GA and GT microsatellite markers from the wheat genome. The isolation of functional microsatellite markers from wheat is severely hampered by its large genome size which results in many primer pairs that do not amplify the correct product or multiple fragments. At the moment, is isolated approximately 250 functional microsatellite markers from the wheat genome. Wheat microsatellites are in general genome specific in so far as they usually amplify only a single fragment from either the A or B or D genome. More than 175 microsatellites have been mapped onto a framework of the ITMI wheat mapping population. The markers appear to be more or less evenly distributed along the wheat chromosomes.

If it is possible to complete such genome maps of different wheat species, clones and sorts there is possible to find which gene on which loci is different and variable. The variability is the tool to keep biological diversity. As I propose to show some example of map of grain genome where is possible to see the common gene for more varieties but also the gene which are different in each of them. According to Neie it is for qualificated judgement of genetic diversity estimation necessary to analyse as much as possible genetic loci. The parameters of genetic diversity are than e.g. gene frequencies, level of homozygosity and heterozygosity, the absolute and effective number of alleles etc. The final judgement of genetic diversity will than result of complex ``reading of genome'' for example by the methods of DNA sequencing and constructing of genome maps for each chromosome of each variety of wheat.

References

Curtis and Lukaszewski, In: The structure and function of the expressed portion of the wheat genomes. The project, 1991.

Lepica, S., 1998: Mapping of the farm animals genome. Proceeding of XVIIIth Days of Genetics, Budejovice, 1998.

Gill *et al.*, 1993: Construction of chromosome arm-specific DNA libraries of wheat using flow cytometry: Plant and animal genome 4 (1993).

Kota et al, 1993: Plant and animal genome VI.

Korzun, V., Ganal, M., Roder, M.S., 1998: Genetic mapping of microsatellite markers in the wheat (Triticum aestivum) genome. Plant and Animal genome. V. Conference. San Diego CA, 1997, p. 57.

Kubek, A., Bulla, J., Dvorák, J., 1997: Genetic markers and their association with QTL. Proceeding of XVIIIth Days of Genetics. Budejovice, s. 21 – 26.

Thaller, G., 1998: Methodological procedures for mapping of quantitative loci. Proceeding of XVIIIth Days of Genetics, Budejovice.

Linkage Studies in *Pisum* and *Lupinus* Genomes

B. Wolko, W. K. Swiecicki, K. Kruszka, L. Irzykowska
*Institute of Plant Genetics, Polish Academy of Sciences,
Strzeszynska 34, 60-4 79 Poznan, Poland*

Abstract. The genetic map of *Pisum sativum* chromosomes has been developed over the past 50 years. Our contribution to the newest versions of the pea linkage map is 32 spontaneous and induced morphological mutants and 6 isozyme markers. We cooperate also with other scientists within the framework of the *Pisum* Linkage Map Committee on marker alignment verification, rearrangement of existing linkage groups as well as genetic terminology. The linkage research of *Lupinus* species is preliminary in character. Cross combinations of *L. angustifolius* and *L. albus* were selected for mapping studies. The genetic analysis in *L. angustifolius* has made it possible to map 4 isozymes and 60 RAPD markers in 17 groups. The linkage map of *L. albus* consists of 18 groups which contain 3 morphological traits, 3 allozymes and 49 DNA markers. The construction of the linkage map of *L. albus* resulted in the finding of RAPD markers which showed tight linkage (5.2 cM) to the agronomically important chemical trait of low alkaloid content. The obtained RAPD fragment was converted into SCAR to increase the specificity of the marker system. The specific primers were designed and used for screening of segregating plant material.

1. Introduction

The pea *(Pisum sativum)* and three species of the Old World lupins: white lupin *(Lupinus albus)*, narrow-leafed lupin *(L. angustifolius)*, yellow lupin *(L. luteus)* are the major representatives of grain legume crops in Central Europe. Their importance is due to high protein content in seeds, place in crop rotation and N-fixation as well as high yield of green mass. Our genetic studies were concerned with the wide variability of pea and lupin lines gathered in the Polish Gene Bank.

The phenomenon of linkage makes it possible to mark one gene with another if the two are located near each other on the chromosome. A practical application of it which breeders have used for years is to screen for commercially important traits by following a linked morphological character. The major drawback of this strategy has been the lack of appropriate markers tightly linked with commercially important genes. Additionally, the marker expression should not detract from the quality of the crop. Although molecular markers introduced recently to the linkage maps may not be as easy to score as some of the morphological mutants, they possess the advantage of neutrality, freedom from epistatic or pleiotropic interactions, and are expressed relatively early in plant development.

Linkage maps summarize our genetic knowledge about the species and can be a good reference point in searching for markers for monogenic and polygenic commercially important traits to increase the effectiveness of breeding programs. The recent linkage studies in pea have been concentrated on expanding and saturating the existing map. Although many of the linkage groups have been well documented, questions remain regarding the contents of some of them. The linkage studies of lupin species are actually of preliminary character. The variability of molecular markers which has been revealed as a result served as a basis for the construction of the first map of the *Lupinus* genome.

2. Gene mapping in *Pisum sativum*

Pea is an important model species not only for studies on the physiological, biochemical and molecular control of plant developmental genetics. It has also been a favorite organism for genetic investigations, the result of which is the chromosomal linkage map that is comparable with that of any other plant species [1]. From the over 500 loci identified in pea, a considerable number have been mapped. A linkage map based on classical morphological and physiological markers [2] has evolved during the last 50 years to the newest versions containing protein polymorphisms [3], RFLPs of known genes or cDNA sequences [4, 5] RAPID and AFLP markers [6, 7].

The localization of 38 loci has been our share in the construction of the pea linkage map (Table 1). Thirty two of these are spontaneous and induced mutants and 6 are isozyme loci. The mapped mutants are involved in changes in metabolism of chlorophyll , carotene and anthocyanins, modifications in morphology of leaves, seeds and plant architecture as well as necrotic changes (Table 2). Some of these mutants can be economically important, e.g. *Orc* - higher amount of carotene, *dnd, art2* - lodging resistance or *det* - *s*elfcompleting of vegetation. One of the isozyme loci, Prx-3, was localized in linkage group VI close to *sbm* gene conferring resistance to seed-borne mosaic virus and can be used as a marker for selection of breeding materials.

Table 1. The list of genes mapped by a research group in Poznan.

Linkage group	Gene no.	Gene symbol
I	6	*Orc, orl, Ans. Ast, And, py2*
II	6	*crd, Dia-3, Fum, calf, orp, pal*
III	6	*Dia-1, lum2, dnd, chi32, apu, brac*
V	10	*det, nlb, coch, nec, gn, lk, art2, lum3, rms5, len2*
VI	9	*bls, Acp-4, Gpi-c, rui, Gty, Prx-3, chi33, len3, smi*
VII	1	*def*

The cooperation with other scientists allowed us to participate in the verification of marker alignments and rearrangements of several linkage groups. Results of these linkage studies have provided a basis for the following versions of the pea linkage map. The introduction of new types of markers as isozymes and RFLPs has resulted in considerable rearrangements of some linkage groups. The large segment *r - tl - bt* from linkage group VII has been moved to linkage group V and segment *Skdh - def* has been transferred from linkage group II to VII [4]. In the following years the pea map has been extended and saturated with other molecular markers such as RAPDs, AFLPs and SCARs. It has resulted in the moving of *crd - His7* fragment from linkage group I to II [6]. The recent version of the consensus pea map has been combined from a set of markers generated from a single cross population and marker loci which did not segregate in the cross combination but whose position relative to markers segregating in this population is sufficiently well known to approximate their location on the primary map. The linkage segment *np -le* formerly assigned to linkage group IV is actually part of group III. The map spans nearly 800 cM with the saturation of about 1 marker per 2 cM [7].

Table 2. The classification of mapped genes according to mutation changes

Mutation changes	Gene symbol
Chloropyll metabolism	*chi32, chi33, py2, lum2, lum3*
Carotene metabolism	*Orc, orl, orp*
Anthocyanin metabolism	*Ans, Ast, And*
Leaf morphology modifications	*calf, crd, pal, apu, nlb, coch, bls, smi, py2, lum2, lum3*
Seed morphology modifications	*def, Gty*
Plant architecture	*dnd, brac, det, lk, art2, rms5, rui*
Necrosis	*nec, gn, len2, len3, bls*
Isozymes	*Dia-1, Dia-3, Fum, Acp-4, Gpi-c, Prx-3*

3. Linkage maps of lupin species

The majority of *Lupinus* species are wild and semi-domesticated. Only a few of them *(L. angustifolius, L. albus* and *L. luteus,* originating from the Old World and *L. mutabilis,* from the New World gene pool, are considered as crop species. As a result of intensive breeding efforts several favorable genes such as those for low alkaloid content, non-shattering pods, thermoneutrality, unbranching and soft seeds were introduced to cultivars. Consequently, lupin has become an essential part of temperate farming systems, as a protein source for animal and human consumption [8]. The exploitation of the wide variability of lupin forms to improve cultivars is limited by crossing barriers, especially among Old World species. An additional problem is associated with a low level of intraspecific variability and limited genetic knowledge of species [9].

We have started our research on the development of the lupin linkage maps several years ago with testing the range of isozyme polymorphism of European lupin crops and their wild relatives. On the basis of these analyses we selected the parental lines for cross combinations. The largest allozyme variation was observed within *L. angustifolius* (8-14 segregating allozymes). *L. albus* and *L. luteus* possessed a much lower level of polymorphism (3-4 segregating allozymes). However, in any cultivated or wild lupin species the allozyme analysis proved that it could be difficult to identify a sufficient number of polymorphic protein markers to develop a linkage map.

Due to above results we were interested in the recognition of the range of molecular variation in the tested plant material. Segregation studies were mainly done with RAPD markers due to the ease of the technique and in order to map random coding, as well as non-coding sequences. The results indicated that the range of DNA markers revealed would be sufficient to create a linkage map. Isozymes, commonly used in our laboratory were also used for mapping because of their codominant nature of inheritance. Crosses for linkage studies were chosen on the basis of two criteria: (1) high level of molecular polymorphism, and (2) the occurrence of differences in some morphological and chemical traits.

On the basis of the isozyme variation, the Emir x B540 cross combination of *L. angustifolius* was designated for detailed mapping studies. Forty four arbitrary primers (10-meres) were used for testing RAPDs variability in the F_2 population representing 48 plants. We obtained 105 polymorphic DNA fragments. For constructing of the linkage map only RAPDs and isozymes were used which showed an undisturbed monohybrid segregation. The computer program MAPMAKER version 3.0 was used for the multi-point analysis with LOD ≥ 3 [10]. Statistical analysis provided the map position of 60 RAPDs and 4 allozyme loci *(Aat-1, Gpi-2, 6Pgd-2, Est-3)*. The length of the generated linkage map is close to 987 cM and markers are distinguished into 17 groups (Figure 1).

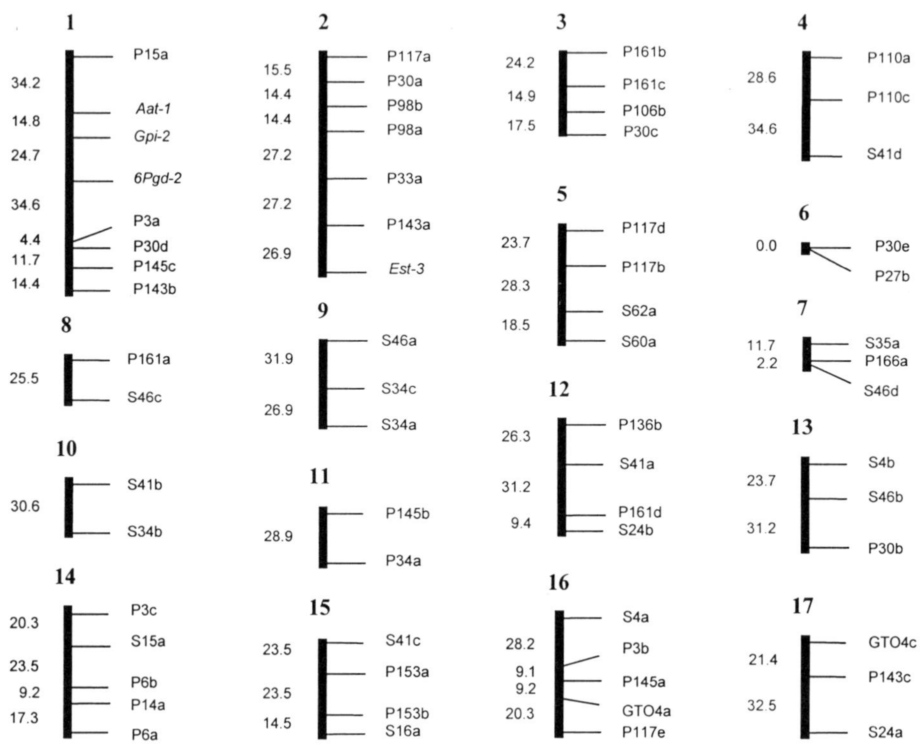

Figure 1. Linkage map for *Lupinus angustifolius*. Map distances are given in cM.

The same approach has been used for the construction of a linkage map of white lupin. The F$_2$ population of the *L. albus* var. *graecus* x Kalina cross combination was selected for mapping studies. This particular cross was chosen on the basis of two criteria: relatively high level of molecular polymorphism and the occurrence of differences in some morphological, physiological and chemical traits. The F$_2$ plants were tested for the segregation of flower color, early flowering, and fast growth of the main stem characters. After harvesting a seed coat color trait was examined and an alkaloid content was analyzed. A preliminary screening was conducted with the use of Dragendorff's reagent. To confirm above results, the total alkaloid content in leaf tissue was estimated by using of colorimetric method [11]. Only 4 isozyme loci were polymorphic in segregating plant material: *Mdh-1, Pgm-1, Est-2* and *Est-3*. To obtain RAPD markers, 120 oligonucleotide arbitrary primers were used. Nearly 200 polymorphic DNA fragments were amplified, from which 124 exhibited undisturbed monohybrid segregation.

All morphological and chemical traits, allozymes as well as RAPDs with undisturbed monohybrid segregation were analyzed in a multi-point analysis using the MAPMAKER program, with a LOD of 3. The constructed *Lupinus albus* linkage map contains two morphological traits (flower color - FLCOL and seed coat color - SDCOL) and one chemical character (alkaloid content - ALCON) as well as 3 isozyme loci *(Pgm-1, Mdh-1, Est-3)* and 49 RAPD markers (Figure 2). The length of the *L. albus* map is 801 cM with an average distance between markers close to 15 cM. This size is comparable to the *L. angustifolius* and maps of some other plant species. The resulting map is unique in *Lupinus,* since it covers linkages between molecular markers and some genes for morphological and chemical characters for the first time.

Different sets of isozyme loci and morphological traits as well as a cross specificity of RAPD markers make it impossible to compare both maps to search for similarity or rearrangements within linkage groups. Considering the polyploid origin of the lupin genome, the maps presented cover only a part of the genome, but can be used as a starting point for further genetic studies.

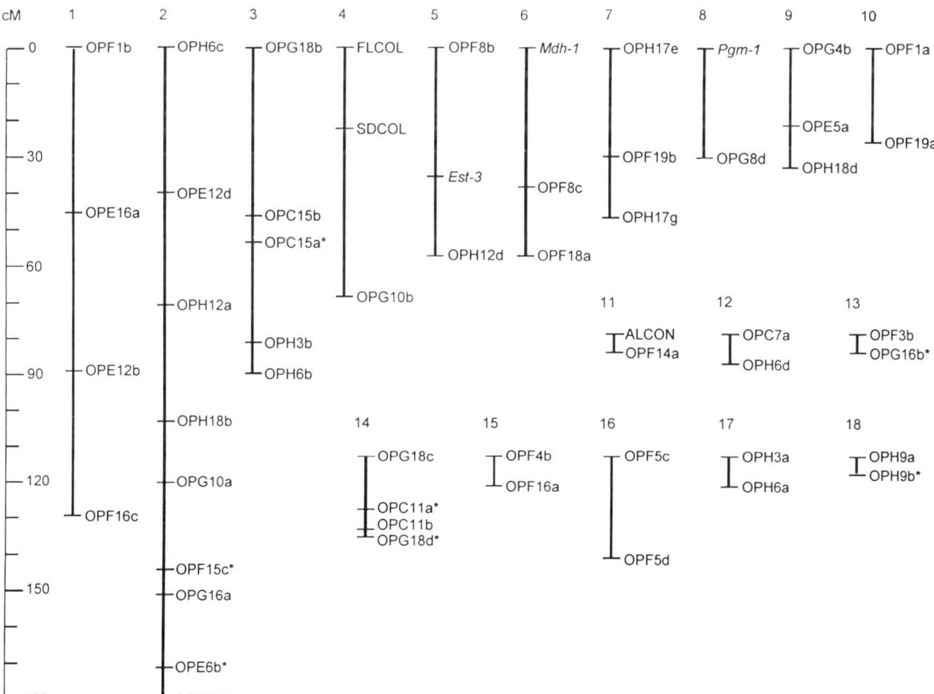

Figure 2. Linkage map of morphological, isozyme and RAPD markers for *Lupinus albus*.

4. SCAR marker for a low alkaloid content trait

Mapping studies of the white lupin reveal a close linkage (5.2 cM) between the OPF14a marker and the gene determining a low alkaloid content. The closely linked RAPD marker can play a significant role in breeding programs for low/high alkaloid form selection in *L. albus*. Because the RAPD marker system has exhibited a number of disadvantages, we decided to convert this particular DNA fragment into more specific and easier to use a SCAR marker.

Among products amplified by using the primer OPF14, the band containing a 231 bp DNA fragments was especially interesting to us. It was extracted from an agarose gel by using the extraction kit. To obtain DNA quantity suitable for ligation into the pCR®2.1 vector, the obtained product was reamplified. The recombined vector was used for transformation of *E. coli* competent cells. Then plasmid recombinants were selected by the Colony PCR method. Afterwards, the nucleotide sequence of cloned DNA fragment was established. The information of that sequence made it possible to designate a pair of specific primers and we examined their efficiency. We observed the presence of a band containing the amplification product (215 bp in length), when the template DNA from low alkaloid content plants was used. This band was absent when the DNA from high alkaloid plants was introduced (Figure 3).

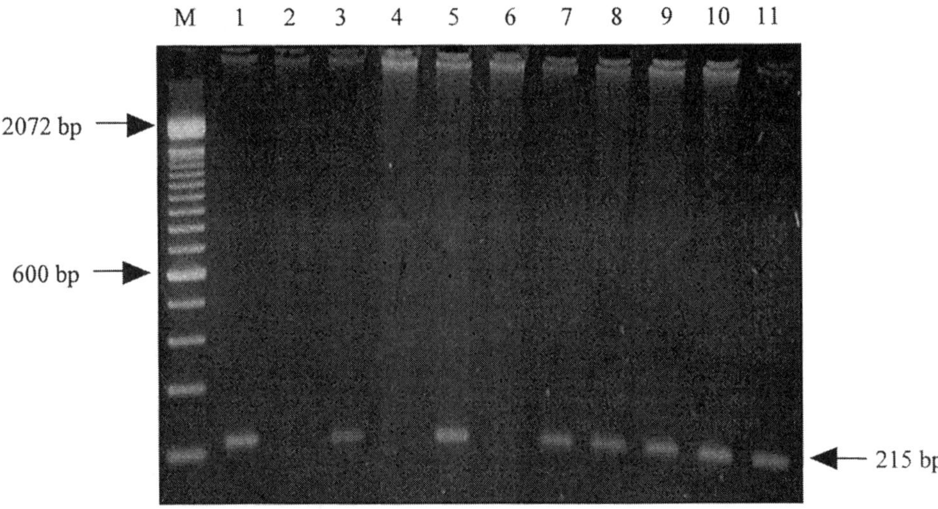

Figure 3. Amplification products obtained by using SCAR primers. Samples: 1, 3, 5, 7-11 - amplification with DNA from low alkaloid plants; 2, 4, 6 - amplification with DNA from high alkaloid plants; M - molecular weigh marker 100bp DNA Ladder; bp - base pairs

References

[1]　S. Blixt, The pea. In: R.C. King (ed.), Handbook of Genetics. Plenum Press, New York, 1975, pp. 181–221.

[2]　H. Lamprecht, The variation in linkage and course of crossing-over, *Agri. Hort. Genet.,* **6** (1948) 10-48.

[3]　N.F. Weeden, An isozyme linkage map for *Pisum sativum.* In: P.D. Hebblethwaite, M.C. Heath, T.C.Dawkins (ed.), The Pea Crop. Butterworths, London, 1985, pp. 55-66.

[4]　N.F. Weeden and B. Wolko, Linkage map for the garden pea *(Pisum sativum)* based on molecular markers. In: S. J. O'Brien (ed.), Genetic Maps (5 ih edition). Cold Spring Harbor Laboratory Press, 1990, pp. 6.106-6.112. N. F. Weeden *et al., Pisum* sativum, pea. In: S. J. O'Brien (ed.), Genetic Maps, Locus Maps of Complex Genomes; Book 6 - Plants (5 th edition). Cold Spring Harbor Laboratory Press, 1993, pp. 6.24-6.34.

[5]　N.F. Weeden *et al.,* The current pea linkage map. *Pisum Genetics* **28** (1996) 1-4.

[6]　N.F. Weeden *et al.,* A consensus linkage map for *Pisum sativum. Pisum Genetics* **30** (1998) 1-4.

[7]　W.A. Cowling *et al.,* Lupin Breeding. In: J.S. Gladstones, C. Atkins, J. Hamblin (eds), Lupins as crop plants - biology, production and utilization. CAB International, Oxon, New York, 1998, pp. 93-120.

[8]　J.S. Gladstones, Lupins as crop plants, *Field Crop Abstracts* **23** (1970) 123-148.

[9]　S.E. Lincoln *et al.,* Constructing Genetic Linkage Maps with MAPMAKER/EXP Version 3.0: A Tutorial and Reference Manual. A Whitehead Institute for Biomedical Research, Technical Report (3rd edition), 1993.

[10]　I. Reifer and S. Niziolek, Kolorymetryczna mikrometoda oznaczania alkaloidow w nasionach lubinu (eng. A colorimetric microdetermination of alkaloids in lupine seeds). *Acta Biochem Pol* **4** (1957) 165-180.

Use of Molecular Markers in Wheat Breeding for Disease Resistance

L. Prunhauser[1], G. Gyulai[2], M. Tar[2], M. Csösz[1], A. Mesterhazy[1], L. Heszky[2]

[1]Cereal Res. NP Co., Szeged, Hungary, [2]Dept. Genetics and Plant Breeding, Godollo Univ., Hungary

Abstract. Leaf rust is one of the most important leaf diseases of wheat in Hungary. Breeding for resistance is considered to be the most economical and environmentally safe method to reduce damages. About fifty resistance genes have been characterized against leaf rust. The traditional way of transferring one or more resistance genes to a single wheat cultivar relay on field and greenhouse screening with different races, which is a very laborious and time consuming process. In recent years DNA-based markers have shown great promise in lessening the time and expense for pyramiding resistance genes. So far 26 molecular markers have been reported which are closely linked to 14 different leaf rust resistance (*Lr*) genes. The most widely used marker system is based on RFLPs. Now, there is a moving from the expensive and technically more difficult RFLP marker system to the user friendly PCR-based markers. In this study a search for PCR markers linked to *Lr* genes was carried out by screening on *Thatcher*-based wheat leaf rust near isogenic lines (NILs). Four RAPD (*Lr20* and *Lr29*), SSR (*Lr18*) and ISSR (*Lr3,3bg*) markers were selected. The linkage analysis of these putative markers to *Lr* genes (F_2 segregation analysis) is in progress.

1. Introduction

Common wheat (*Triticum aestivum* L.) is seriously damaged by a variety of fungal diseases. In Hungary, the most dangerous diseases of wheat are: leaf rust, stem rust, fusarium head blight, and powdery mildew. Of these, leaf rust (*Puccinia recondita* f.sp. *tritici*) is often the most serious disease, which can reduce total yield by 1% for each 1% increase of infection (6)

Although many fungal diseases can be successfully controlled with fungicides, however, breeding for resistance is considered to be the most economical and environmentally suitable method to reduce crop damages. About fifty resistance genes (*Lr*) have been characterized against leaf rust. Large part of these genes were introduced from alien species to wheat. Wheat cultivars carrying resistance genes to leaf rust can avoid yield loss due to this disease. By developing near isogenic lines (NILs) for each resistance gene on the same genetic background, the efficiency of these genes could be evaluated separately. For *Lr* genes in wheat, *Thatcher* based NILs are available.

The traditional way of transferring one or more resistance genes to a single wheat cultivar relay on field and greenhouse screening with different races, which is a very laborious and time consuming process. In recent years DNA-based markers have shown great promise in lessening the time and expense for pyramiding resistance genes.

In this study we evaluated the effects of *Lr* genes on disease resistance and searched for PCR markers linked to *Lr* genes by screening *Thatcher*-based wheat leaf rust NILs.

2. Materials and Methods

Thatcher NILs were obtained from Dr. J. Kolmer, Canada (Win.). Natural infections of NILs in field conditionson were evaluated according to the Cobb's scale. Genomic DNA samples were extracted from seedlings of the *Thatcher* NILs by CTAB the method used by Dweikat et al. (1994). Amplification reactions were run in a volume of 25 µl by a CoyTempCycler. The reaction mixture contained 100 µM of each 100 deoxynucleotides; 2.0 mM $MgCl_2$; 0.5 U Taq-polymerase; 0.5-5.0 nM of primers, either simple sequence repeat (SSR), inter simple sequence repeat (ISSR) or random amplified polymorphic DNA (RAPD); 2.5 µl of 10 x thermophylic buffer (50 mM KCl, 10 mM TRIS-HCl); and 20 ng template DNA. The reaction mix was overlaid with a drop of mineral oil. For PCR amplification, 40 cycles were run with the following steps: (1) 20 sec at 94°C for DNA denaturation prior to a preheating of samples at 84 °C for 5 min; (2) 20 sec at 40°C (for RAPD 36°C) for primer annealing; (3) 90 sec at 72°C for DNA polymerase reaction; (4) a final extension at 72°C for 5 min. Amplifications were assayed by agarose (1.8% SeaKem LE, FMC) gel electrophoresis, stained with ethidium bromide.

3. Results and Discussion

Each year we planted the NILs to check the effects of different *Lr* genes under changing climatic conditions. In Table 1. the intensity of infection on NILs are shown in a 5 years long period. Out of the 41 *Lr* genes investigated 7 genes, especially the *Lr9*, could provide a high level of resistance annually under the climatic conditions of Hungary. Ten genes showed medium resistance. However, the rest of the genes did not give satisfactory defence responses. Pathogen populations can rapidly change, by this way many resistant wheat varieties/*Lr* genes lost their resistance against the pathogens. These *Lr* genes are not effective if used alone. Pyramiding more „weak" *Lr* genes into one elite line may still provide a high level of resistance. It seems that pyramiding of strong genes is even more effective.

To win this race, breeders must be faster than pathogens. A set of molecular markers linked to resistance genes are required to accelerate breeding programs. In Table 2. only those leaf rust genes are shown for which markers are available. To date 26 different type of molecular markers have been reported which are closely linked to 14 different *Lr* genes. The most widely used marker system is based on RFLPs. However, it is rather difficult to use RFLP markers in practical breeding programmes. Therefore a gradual shifting from the expensive and technically more difficult RFLP marker system to the user-friendly PCR-based markers is taking place presently.

Many resistance genes were introduced to wheat from alien species such as rye. The most famous translocation in wheat is the 1BL/1RS in which the long arm of wheat chromosome 1B is replaced by the short arm of the rye chromosome 1R. This translocation has been reported to have positive effects on yield and several resistance genes, e.g.: *Pm8*, *Sr31*, *Yr9* and *Lr26*. (Francis et al.) [5] found a RAPD marker specific for the rye chromosome. We tested this markers in our breeding program. Out of 35 wheat cultivars we have found 7 cultivars which carry the 1BL/1RS (20% of cultivars investigated, data not shown).

Table 1. Effectiveness of the leaf rust genes in adult stage (1995-1999, Szeged, Hungary)

| Gene | NILs | Leaf rust infection (%) and reaction types | | | | |
| | | Years | | | | |
		1995.	1996.	1997.	1998.	1999.
	Effective					
Lr9	Transfer/Tc*6	0	0	0	0	0
Lr19	Tc*7/Translocation 4-Agr. elong.	5R	0	0	tMR	0
Lr24	Tc*6/Agent	0	tR	5MR	tMR	0
Lr25	Tc*6/Transec	0	30R	0	tMR	0
Lr29	Tc*6/CS7D-Ag#11	10R	20R-MR	tR	30MR	20;R
Lr35	Tc*6/RL 5711	2R-MR	tMR	tR	0	30R-MR
Lr38	Tc*6/T7 Kohn	5R	0	tR	0	30R-MR
	Moderately effective					
Lr12	Exchange/Tc*6	20R-MR	80R-MR	40R	20MR	80;R-MR
Lr13	Tc*6/Frontana	40MS(MR)	80R-MR	30R-MR	40MR	80;R-MR
Lr17	Klein Lucero/Tc*6	5R	30R-MR	20R-MR	10MR	100;R-MR
Lr18	Tc*7/Africa 43	30R	100R-MR	20R-MR	10R	80;R-MR
Lr23	Lee 310/Tc*6	5R-MR	40MS	30R-MR	tMR	40;R-MR
Lr28	Tc*6/C-77-1	40R-MR	30MR	20MS-S	5R	40;R-MR
Lr32	Tc*6/3/Ae. squarrosa	30MR-MS	60MR	20R-MR	10MR	80;R-MR
Lr37	Tc*8/VPM	10R-MR(MS)	80MR	20R	5MR	60;R-MR
Lr44	Tc*6/T. spelt	30MS	80R-MR	50R-MR	tMS	70;R-MR-MS
LrW	Tc*6/V336	40MS(MR)	60MR	20R-MR	60MS	30;R-MR
	Ineffective					
Lr1	Tc*6/Centenario	25MR-MS	100S	40MR-MS	40MS	50R-MR
Lr2a	Tc*6/Webster	20MR	100S	60MR	40MS	70MR-MS
Lr2b	Tc*6/Carina	30MR-MS	100S	80MS	-	100MS
Lr2c	Tc*6/Loros	50MS-S	100S	100S	60MS	100MS
Lr3	Tc*6/Democrat	50S	100S	100S	80MS	100S
Lr3bg	Bage/Tc*8	30S	100S	100S	40MR	100S
Lr3ka	Tc*6/Aniversario	50S	100S	80MS-S	20MR	60MS-S
Lr10	Tc*6/Exchange	60S	100S	80S	30MR	100S
Lr11	Tc*2/Hussar	60S	100S	100S	30MR	100S
Lr14a	Selkirk/Tc*6	40MS-S	100S	80S	40MR	100S
Lr14b	Tc*6/Mario Escobar	15MR-MS	100S	80MS	10MR	80R-MR-MS
Lr15	Tc*6/W1483	60S	100S	100S	20MR	100S
Lr16	Tc*6/Exchange	10S	100S	100S	40MR	100S
Lr20	Tc*6/Jimmer	30MR-MS-S	100S	60R-MR-MS	10MR	80R-MR-MS
Lr21	Tc*6/RL 5406 T.C. X Ae. sq var. mey.	50MS-S	100S	60MR	10MR	100S
Lr22	Tc*6/RL 5404 T.C..XAe. sq. var. stran.	30MR	80S	20R-MR	tMR	60;R-MR
Lr26	Tc*6/St-1-25	50S	100S	80MS	5MR	100S
Lr30	Tc*6/Terenzio	30MR-MS-S	80S	60R-MR	5MR	80MR-MS-S
Lr33	Tc*6/PI 58548 (1+gene)	40S	100S	30MR	40MS	100S
Lr34	Tc*6/PI 58548 (2+gene)	20MS-S(MR)	80MS	20R-MR	20MR	60S
Lr38	Tc*6/TMR-514-12-24	40MR-MS	80S	50R-MR	5MR	50;R-MR-MS
Lr44	Tc*6/8404	40MR-MS-S	80MS	50MR-MS	tMS	80-R-MR-MS
LrB	Tc*6/Carina	40S	100S	80MS-S	60MS	100S
LrB	Tc*6/PI 268316	40S(MS)	80S	50MS	60MS	100S
Tc	Tc	60S	100S	80S	60MS	100S

Reaction types: 0=no visible uredia, ;=very resistant, R=resistant, MR=moderately resistant, MS=moderately susceptible, S=susceptible

SCAR primers designed from the sequence data of the polymorphic fragment of RAPD product provided a more specific and reliable marker for rye translocation in wheat. In addition the presence of single amplification products allow the direct detection of the marker in test tube by the addition of EtBr, instead of the use of gel electrophoresis.

All these results underline the importance of DNA markers in the characterization of resistance genes and their interactions toward the final goal of constructing durably resistant wheat cultivars via marker-controlled pyramiding of specific genes. The cultivation of leaf rust resistant wheat seems to be the most economical and environmentally benign way to reduce yield losses. Cultivars with the combination of major and minor *Lr* genes as a gene pyramiding was reported to posses durable resistance (8). Uniquely, the *Lr34* gene, the only effective adult plant resistance gene known for durability, combined with other resistance genes were introgressed in about 60% of CIMMYT-derived spring wheat cultivars (10). To be able to monitor these *Lr* genes in different breeding programs, the use of molecular markers linked to *Lr* genes is inevitable. Recently, about fifty *Lr* genes were identified and developed in *Thatcher*-based NILs (8).

Table 2. DNA markers for leaf rust resistance in wheat

Gene	Chr. loci	RFLP marker	PCR-based marker RAPD	Other	References
Lr1	5DL	RFLP		STS	[4]
Lr 9	6BS/6BL	RFLP	RAPD	STS	[13]
Lr10	1AS	RFLP		STS	[15]
Lr13	2BS	RFLP			[16]
Lr19	7DL		RAPD		[1]
Lr19	7DL	RFLP			[11]
Lr23	2BS	RFLP			[10]
Lr24	3DL/3BL	RFLP	RAPD	STS	[14]
Lr24	3DL/3BL		RAPD	SCAR	[2]
Lr25	4BS		RAPD	DGGE	[12]
Lr27	3BS	RFLP			[10]
Lr28	4AL		RAPD	STS	[9]
Lr29	7DS/7D Ag. tr.		RAPD	DGGE	[12]
Lr31	4BS	RFLP			[10]
Lr34	7DS	RFLP			[10]
Lr35	2B	RFLP			[16]

In our study putative markers were selected in the following NILs: *Lr3/3bg* gene by the ISSR primer (GACA)$_4$ CA; for the gene *Lr18* by the SSR-primer (ACTG)$_4$, as well as for the *Lr20* and *Lr29* gene by the RAPD primer OPJ 17 and OPY 10, respectively. Each of the polymorphic PCR fragments detected were less than 1000 bp in length (Table 3). It is obvious that the polymorphic PCR fragments observed were developed as a result of the change in the genomes caused by *Lr* gene insertion during NIL backcrossing, which generated new binding sites for the primers of RAPD, SSR, and ISSR. The final linkage analysis (F$_2$ segregation analysis) is in progress.

Table 3. New putative markers identified for *Lr* genes in wheat

Lr gene	Loci	NIL gene donor	Marker	Marker primer	PCR fragm.
Lr3	5DL	Tc*6/ Democrat	ISSR	(GACA)$_4$CA	~650 bp
Lr3 bg	5DL	Tc*8/ Bage	ISSR		
Lr18	5BL	Tc*7/ Africa43	SSR	(ACTG)$_4$	~950 bp
Lr20	7AL	Tc*6/ Jimmer	RAPD	OPJ17	~700 bp
Lr29	7DS	Tc*6/ CS7D-Ag#11	RAPD	OPY10	~1000 bp

There are several resistance genes identified for the main wheat diseases and pests (3, 7). However, few of them cause strong resistance if they are alone in a wheat cultivar. In most cases pyramiding of two or more resistance genes are needed for controling a given disease. It should be of major significance to develop wheat cultivars which are resistant for not only one disease, but also for the other main diseases. The complex resistant cultivars have a great advantage namely farmers need not treat them with costly and environmentally hazardous fungicides. In Hungary there are several cultivars which are resistant to one or more main diseases, but non of them show a complex resistance for all important diseases. By finding markers for most of the resistance genes, a marker assisted selection could provide a faster development of more effective and durable resistance in the field than the conventional breeding techniques used today.

Table 4. Number of the identified disease resistance genes and markers

Disease	No. of resistance genes	No. of resistance genes for which markers are identified
Leaf rust	~50	14
Stem rust	~70	2
Powdery mildew	~25	7
Septoria	4 - 8	0
Fusarium head blight	6 - 10	0

Acknowledgements

The project was supported in part by Hungarian Research Grants OTKA-T-026559, and by the SZPÖ/Gyulai-73/99. Authors wish to thank, Kinga Hornok, and József Marticsek for contributing to the experiments.

References

[1]   E. Autrique *et al.*, Molecular markers for four leaf rust resistance genes introgressed into wheat from wild relatives. *Genome* **38** (1995) 75-83.
[2]   F. Dedryver *et al.*, Molecular markers linked to the leaf rust resistance gene *Lr24* in different wheat cultivars. *Genome* **39** (1996) 830-835.
[3]   C. Feuillet and B. Keller, Molecular aspects of biotic resistance in wheat. In: Proceedings of the 9th International Wheat Genetics Symposium (Ed. by A E Slinkard). Univ. Extension Press. Univ. of Saskatchewan, 1998, p. 171-174.
[4]   C. Feuillet, *et al.*, Genetic and physical characterization of the *Lr1* leaf rust resistance locus in wheat (*Triticum aestivum* L.). *Mol Gen Genet* **248** (1995) 553-562.

[5]　H.A. Francis *et al.*, Conversion of a RAPD-generated PCR product, containing a novel dispersed repetitive element, into a fast and robust assay for the presence of rye chromatin in wheat. *Theor Appl Genet* **90** (1995) 636-642.

[6]　M.A. Khan *et al.*, Quantitative relationship between leaf rust and wheat yield in Mississippi. *Plant Disease* **81** (1997) 769-772.

[7]　P. Langridge and K. Chalmers, Techniques for marker development. In: Proceedings of the 9th International Wheat Genetics Symposium (Ed. by A E Slinkard). Univ. Extension Press. Univ. of Saskatchewan, 1998, p. 107-117.

[8]　R.A. McIntosh *et al.*, Wheat rust. An atlas of resistance genes. CSIRO Australia, Melbourne and Kluwer Academic Publisher, Dordrecht, Boston, London, 1995.

[9]　S. Naik *et al.*, Identification of STS marker linked to the *Aegilops speltoides*-derived leaf rust resistance gene *Lr28* in wheat. *Theor Appl Genet* **97** (1998) 535-540.

[10]　J.C. Nelson *et al.*, Mapping genes conferring and suppressing leaf rust resistance in wheat. *Crop Sci.* **37** (1997) 1928-1935.

[11]　R. Prins and G. F. Marais, An extended deletion map of the *Lr19* translocation and modified forms. *Euphytica* **103** (1998) 95-102.

[12]　J.D. Procunier et.al., PCR-based RAPD/DGGE markers linked to leaf rust resistance genes *Lr29* and *Lr25* in wheat (*Triticum aestivum* L.). *J Gent Breeding* **49** (1995) 87-92.

[13]　G. Schachermayr *et al.*, Identification and localization of molecular markers linked to the *Lr9* leaf rust resistance gene of wheat. *Theor Appl Genet* **88** (1994) 110-115.

[14]　G. Schachermayr *et al.*, Identification of molecular markers linked to the Agropyron elongatum-derived leaf rust resistance gene *Lr24* in wheat. *Theor Appl Genet* **90** (1995) 982-990.

[15]　G. Schachermayr *et al.*, Molecular markers for the detection of the wheat leaf rust resistance gene *Lr10* in diverse genetic background. *Mol Breeding* **3** (1997) 65-74.

[16]　R. Seyfarth *et al.*, Development and characterization of molecular markers for the adult leaf rust resistance genes *Lr13* and *Lr35*. In: Proceedings of the 9th International Wheat Genetics Symposium (Ed. by A E Slinkard). Univ. Extension Press. Univ. of Saskatchewan, 1998, p. 154-155

*Use of Agriculturally Important
Genes in Biotechnology
G. Hrazdina (Ed.)
IOS Press, 2000*

Identification of Molecular Markers Linked to Gene Resistance and Marker Assisted Introduction in Barley and Wheat

J. Kraic, E. Gregová, M. Závodná, [1]S. Sliková, [1]V. Sudyová
Research Institute of Plant Production, Departament of Cell and Molecular Biology,
[1]*Departament of Cytology and Genetics, Bratislavská cesta 122, 92168 Piestany, Slovakia*

Abstract. Two years ago we started a research program for marker assisted introduction (MAI) of specific resistance genes against leaf rust in winter wheat and against viruses (BaYMM, BaMMV) in winter barley. The aims of this program were to test applicability of the MAI principle in breeding practice for resistance against important phytopathogens, and to create wheat and barley lines usable in our breeding programs. Gene *Lr19* causing resistance against leaf rust in wheat and gene *ym4* causing resistance against both viruses were introgressed from donor genotypes into acceptor genotypes by the MAI method. Both programs seem to be succesfull and usable for production of lines possessing the desired genes. F_2 and F_3 progenies, respectively were obtained and individual plants possessed molecular markers linked to the desired resistance genes.

1. Introduction

Incorporation of effective race-specific resistance genes and their pyramiding are possible ways to create biological material with high or higher level of resistance against important phytopathogens. Improvement of wheat and barley for resistance to pathogens includes at present time application of the marker assisted introduction (MAI) principle. This consist of classical crossing of selected gene donor and gene acceptor genotypes, simultaneously controlling the gene transfer into offspring and selection of individuals by molecular markers tightly linked to the desired genes. This approach overcomes the problems of the absence of specific pathogen isolates and selection of lines possessing a combination of two or more specific alleles in resistance breeding.

Leaf rust caused by *Puccinia recondita* f. sp. *tritici* is one of the most important diseases on wheat. Resistance breeding against leaf rust is based on many resistance genes which have been identified and introgressed into wheat from wild relatives (*Lr* genes). The resistance gene *Lr19*, that is effective worldwide, was translocated into wheat from *Agropyron elongatum* [1]. Isoenzyme marker - endopeptidase null allele *Ep-D1c*, was identified and recombination distance of 0.33 cM between *Ep-D1c* and *Lr19* was calculated [2]. The resistance gene *Lr19* is interesting from the point of view of resistance against leaf rust in future years. None or limited virulence has been found in Europe with the other perspective leaf rust resistance gene - *Lr24*. Nevertheless, due to wide use of this gene in wheat breeding in United States and Canada, this gene should be used preferable in combination with other *Lr* genes. RFLP, RAPD, and STS molecular markers linked to the *Lr24* gene have been also identified and characterized previously [3].

The barley mild mosaic virus (BaMMV) and barley yellow mosaic virus (BaYMV) are responsible for diseases of winter barley in central Europe also. The only way to protect barley against these viruses is cultivation of cultivars possessing effective resistance genes. The single recessive gene *ym4* confers immunity to almost all European winter barleys. This was located on the long arm of chromosome 3 [4] from 2.4 _ 1.6 cM away from an RFLP marker [5]. Based on this information a co-dominant STS marker for the gene was developed [6].

2. Materials and Methods

The first step was to select donor and acceptor genotypes for genes *Lr19* and *Lr24*. Wheat cultivars Agrus and Sunnan were used as donors of the gene *Lr19*, the near isogenic line Thatcher/Lr24 as a donor of *Lr24* gene, and the cultivars Sofia (possess *Lr26* and *Lr3* genes), Simona (withouth *Lr* genes), and Lívia (*Lr26* gene) were used as acceptors. All donor cultivars were spring and all acceptor cultivars were winter wheats. Presence of the *Lr19* gene in donors and in their offspring was detected by the presence of the null endopeptidase allele *Ep-D1c*. Isozyme markers were detected by isoelectrofocusing in polyacrylamide gels using ampholyte (pH 4.2-4.9). Proteins were extracted from mature embryos and from young leaves. Specific staining of endopeptidases were performed with Fast Black K.

In the barley/MAI program we selected the donor cultivars of the *ym4* gene resistance against BaYMM and BaMMV (Target, Romanze) and the acceptor cultivars (Torrent, Kamil, Luxor, Copia, KM-104). In this program we used a DNA marker - a codominant STS marker [6] that derived from the RFLP marker MWG838 [7]. DNA was extracted from young leaves by alkali treatment. Amplification products were digested with the *Rsa*I restriction enzyme and separated in agarose gels.

3. Results and Discussion

We decided to incorporate effective resistance genes against leaf rust-genes *Lr19* and *Lr24* individually into our wheat cultivars by the MAI methode. We have also tried to incorporate both these genes into one cultivar to pyramid two genes. At first we transferred the gene *Lr19* to create F_3 individuals homozygous for this gene and subsequently to pyramid gene *Lr24*. F_2 offsprings were obtained from all combinations of parents. Segregation of *Ep-D1* loci in F_2 seedlings significantly deviated from the 3:1 ratio. Individuals homozygous in *Ep-D1c* null allele were selected and self-pollinated to create F_3 progenies. Lines with winter habitus homozygous in *Ep-D1c* will be selected in next steps. Selected homozygous F_3 individuals will be crossed with the donor of *Lr24* gene and tagged with the STS marker that is tightly linked to the *Lr24* gene.

Donor and acceptor genotypes and F_1 offsprings in the barley/MAI program were analyzed by the STS marker linked with *ym4* gene. STS primer pairs amplified a DNA fragment which, after digestion with *Rsa*I, differentiated both parents and showed the presence of both markers in the F_1 offspring. F_2 progeny is being selected at presently and homozygous individuals will be selfed to obtain F_3 and F_4 populations.

In conclusion, the endopeptidase marker system for the *Lr19* gene is not ideal in comparison with a more unambiguous DNA marker system, mainly in distinguishing between homozygous and heterozygous individuals in *Ep-D1* loci. Moreover the *Ep-D1c* is the null allele, however, this marker is usable in the hands of an experienced molecular breeder. Both program seem to be succesful and usable for production of lines having the desired genes. F_2 and F_3 progenies, respectively were obtained and individual plants having molecular markers linked with the desired resistance genes were selected. These biological materials will be included into our wheat and barley breeding programmes.

References

[1]   D. Sharma and D. R. Knott, The transfer of leaf-rust resistance from *Agropyron* to *Triticum* by irradiation, *Canadian Journal of Genetics and Cytology* **8** (1966) 137-143.
[2]   M. Winzeler, *et al.*, Endopeptidase polymorphism and linkage of the *Ep-D1c* null allele with the *Lr19* leaf-rust-resistance gene in hexaploid wheat, *Plant Breeding* **114** (1995) 24-28.
[3]   G. M. Schachermayr, *et al.*, Identification of molecular markers linked to the *Agropyron elongatum*-derived leaf rust resistance gene *Lr24* in wheat. *Theoretical and Applied Genetics* **90** (1995) 982-990.
[4]   R. Kaiser and W. Friedt, Gene for resistance to barley mild mosaic virus in German winter barley located on chromosome 3L, *Plant Breeding* **108** (1992) 169-172.
[5]   A. Graner and E. Bauer, RFLP mapping of the *ym4* virus resistance gene in barley, *Theoretical and Applied Genetics* **86** (1993) 689-693.
[6]   S. Tuvesson, *et al.*, Molecular breeding for the BaMMV/BaYMV resistance gene *ym4* in winter barley, *Plant Breeding* **117** (1998) 19-22.
[7]   E. Bauer and A. Graner, Basic and applied aspects of the genetic analysis of the *ym4* virus resistance locus in barley, *Agronomie* **15** (1993) 469-473.

Identification and Mapping of Molecular Markers for Scab Resistance Genes in Apple (A progress report)

M. Hemmat[1], F. S.Cheng[1], N. F. Weeden[1], S. K. Brown[1], H. S. Aldwinckle[2] , and J. Fu[1]
*[1]Department of Horticultural Sciences and [2]Department of Plant Pathology
Cornell University, New York State Agricultural Experiment Station, Geneva, NY
14456 USA*

Abstract: Apple scab, *Venturia inaequalis* Cke. (Wint), is one of the most serious fungal disease of apple in the world. Genetically controlled resistance exist for at least five genes(V_f, V_m, V_r, V_b, and V_{bj}) that each confer resistance to apple scab. Primers of arbitrary sequence were screened by bulked segregant analysis to identify markers close to and flanking each gene. We have identified several molecular markers that are linked to a number of the scab genes and mapped their position on the standard linkage map of apple (Rome Beauty x White Angel). We have developed a BAC library for Russian seedling R12740-7A. We obtained over 40,000 clones which will represent coverage of more than 5 times the apple genome. Our goal is to develop bases for marker assisted cloning of the scab genes and further pyramiding of scab resistance genes to confer resistance in commercially important scab susceptible apple cultivars.

1. Introduction

Apple is the most important deciduous tree fruit crop in the United States as well as in many other temperate regions throughout the world. Until recently, the North American market has been dominated by relatively small number of varieties. However, several new varieties with enhanced quality characteristics have been introduced and this created demand for improved varieties. In addition, considerable pressure such as increasing governmental regulations, diminishing availability of pesticides, and development of resistance to pesticides in certain *Venturia* populations is forcing the industry to demand disease resistant varieties that do not require several sprays per year. One of the most damaging disease of apple is scab caused by a fungal pathogene called *Venturia inaequalis*. Production of scab-free apples requires 10-12 sprays per year in the regions where conditions are suitable during the growing season for the fungus to grow and damage the susceptible trees. Introducing resistance to apple scab to good quality otherwise susceptible varieties has been one of the goals of the apple breeders. Genetically controlled resistance exists for several of the major fungal and bacterial pathogens of apple. Included among the resistance genes are at least five (*Vf*, *Vr*, *Vm*, *Vb*, and *Vbj*) dominant genes that each confer resistance to one or more strains of *Venturia inaequalis* [1,2]. The most important gene conferring resistance to apple scab is *Vf*, derived from *Malus floribunda* [3] that confers resistance to nearly all races of scab. A report by Parisi et al. in 1996 [4], suggests that this resistance may be broken down in certain areas in Europe [4]. Although the resistance of *M. floribunda* 821 itself remains effective. There are no reports of this breakdown of *Vf* in North America. Therefore, this gene together with other scab genes remains useful for pyramiding scab resistance, and to assure the availability of scab-free apple cultivars for commercial production.

2. Materials and Methods

One of our immediate objectives was to identify tightly linked markers for the major scab resistance genes in different apple populations. We have studied populations from five different crosses. Each population was carrying one or two sources of scab resistance. Scab inoculations, using spores from a mixture of five races of *V. inaequalis* as described by Manganaris et al.[5], were carried on when the seedlings were at the two-leaf stage. The seedlings were incubated for 48 h at 15-20°C at 95% relative humidity then transferred to the greenhouse and maintained at 20°C for two weeks, until disease symptoms were expressed. Seedlings that did not show symptoms were reinoculated and scored after two weeks. Bulked Segregant Analysis was used for identifying markers for scab resistance in each population. DNA was extracted from young unfolding leaves using CTAB buffer. DNA extraction and amplification was performed as described by Hemmat et al., 1994 [6]. DNA was amplified in a PCT100 thermocycler (MJ Research, Inc., Watertown, Mass.). pCRII (TA Cloning Kit, Invitrogen, San Diego, Calif.) was used for cloning of the identified RAPD markers. Sequencing of the markers was performed at Cornell University, New York State Center for Advanced Technologies.

A population from cross between 'Golden Delicious' x 'Prima' was used for generating a linkage map for the *Vf* region. We identified several RAPD markers (Table 1), and developed a linkage map for *Vf* region (Fig. 1).

Table 1. List of the newly identified and previously reported primers that were used in the construction of the linkage map of the *Vf* region.

RAPD primer	Sequence	Reference	
B398	5'CAGTGCTCTT-3'	Hemmat, et.al., 1998	[12]
B505	5'-CCCTTTACAC-3'	Hemmat et.al., 1998	[12]
OPAL07	5'-CCGTCCATCC-3'	Tartarini 1996	[11]
OPK16	5'-GAGCGTCGAA-3'	Hemmat et.al., 1998	[12]
OPA15	5'-TTCCGAACCC-3'	Durham and Korban 1994	[8]
OPAM19	5'-CCAGGTCTTC-3'	Tartarini 1996	[11]
OPC08	5'-TGGACCGGTG-3'	Tartarini 1996	[11]
OPC09	5'-CTCACCGTCC-3'	Tartarini 1996	[11]
OPD20	5'-ACCCGGTGAC-3'	Yang and Kruger 1994	[7]
OPM18	5'-CCAGGTCTTC-3'	Koller et.al., 1994	[10]
OPU01	5'-ACGGACGTCA-3'	Koller et.al., 1994	[10]
P198	5'-GTGTTTGCGG-3'	Hemmat et.al., 1998	[12]
S29	5'-CCAGACAAGC-3'	Hemmat et.al., 1998	[12]
$S5_{2500}$	5'-CCGGCTCTTG-3'	Hemmat et.al., 1998	[12]
$S5_{1450}$ forward	5'-ACA GGT CGA TTA CTT GCG	Hemmat et.al., 1998	[12]
$S5_{1450}$ reverse	5'-GGC TCT TGC AGT TGG GAA	Hemmat et.al., 1998	[12]
$U1_{400}$ forward	5'-GTS AAG CAA GCA CTT CAA CG	Gianfraneschi et al., 1996	[9]
$U1_{400}$ reverse	5'-GTA AAA TAG ATG GTG GGG TAGC	Gianfraneschi et al., 1996	[9]

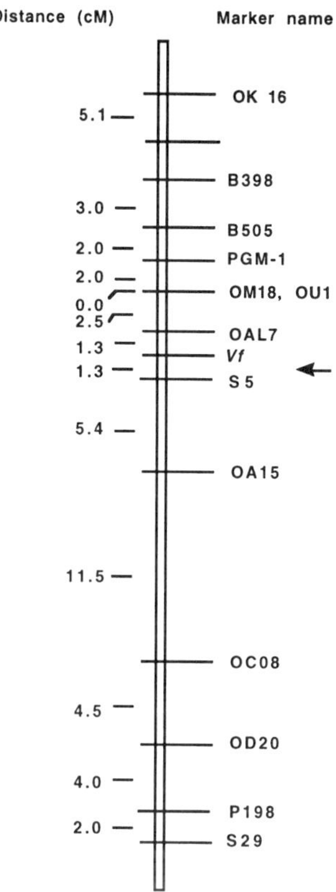

Fig.1. Genetic linkage map of the *Vf* region from progeny of 'Golden Delicious' x 'Prima' using previously published and newly identified RAPD markers.

3. Results

The map consist of several previously reported RAPD markers from different laboratories [7-11] and five new markers identified in our laboratory [12]. We determined the most tightly linked markers that flank the *Vf* gene (Fig. 1). The closest marker, $S5_{2500}$, was cloned and sequenced. Based on sequence data we designed two specific primers for the polymorphic region of the sequence and developed an STS marker ($S5_{1450}$) for *Vf*. A multiplexed assay for simultaneous amplification of the two markers flanking the gene was used [12]. This assay is being used routinely in our breeding program for screening seedling populations as early as the first true leaf stage. We placed the *Vf* gene on linkage group 8 of the standard linkage map of apple [6].

Another important source of scab resistance is from the Russian seedling R12740-7A (*Vr*) that was reported to have two major genes, one of which is resistant to all known races of *V. inaequalis* [13]. Apple cultivars carrying the *Vr* gene developed stellate (SN) upon scab infection [14]. These stellate necrotic lesions are used now generally in detecting *Vr* resistance. The second dominant gene in R127-7A has been designated as *Vx* and shows a pit like reaction in the resistant plants. We used a population generated from cross between 'Empire' x Russian seedling R12740-7A for our study. We identified several markers for *Vr* (Table 2). OB18 (STS) and CH02B10(SSR) are the two tight flanking markers identified for *Vr*. We mapped the *Vr* markers on linkage group 11 of the standard apple map.

Several markers were identified for *Vx* (Table 2). We tagged *Vx*, which often produces a pit type lesion similar to *Vm*, with a 1300bp RAPD fragment. The fragment was then cloned and sequenced and two specific primers were generated to produce an STS marker (CS22) for *Vx*. This marker was not linked to any marker identified for *Vf* or *Vm*. We demonstrated for the first time at the DNA level that there are two dominant genes, *Vf* and *Vx* that confer resistance to scab in Russian seedling R12740-7A. Subsequent testing with CS22 indicated a possibility that *Vx* and *Va* are similar or identical. The position of the *Vx* marker (CS22) is yet to be identified on the linkage map. These tightly linked markers could be useful for screening of the BAC library developed from the Russian Seedling in our laboratory [14].

Table 2: Newly identified DNA markers for different scab resistance genes.

Scab gene	Identified markers
Vr	BP415, BC562, OB 18 (STS), CH02B10 (SSR)
Vb	BC220, BC301, BC346, BC439, BC543, OPAM19
Vm	OB12 (STS)
Vx	BC895x2, PGAJF28, CS22 (STS), S6, B149

We generated a population from 'Empire' x NY74828-12 for marker identification linked to *Vm* [16]. A fragment generated from OB12 showed close linkage to *Vm*. Using sequence data, two specific primers were designed to produce a STS marker (OB12) for *Vm*. This marker was used in the survey of *Malus* germplasm that was obtained from USDA, Plant Genetic Resources Unit(PGRU) at Geneva, NY., to test for the presence of the *Vm* gene. OB12 is useful for possible pyramiding scab resistance genes in cultivated apples.

About 400 primers were used to screen resistant and and susceptible individuals from a cross generated from 'Empire' x Hansan's Baccata #2 for *Vb*. We identified six markers (BC220, BC301, BC346, BC439, BC543 and OPAM19) that were linked to *Vb*. BC220 was the closest marker to the *Vb* gene and the primer consistantly amplified a band about 700 bp in every accession in which the pedigree indicated that *Vb* was present.

We will take advantage of the map of 39 Simple Sequence Repeat (SSR) markers on the apple linkage map (Hemmat et al.in prepration) for quick identification of the corresponding linkage groups in our new populations segregating for scab genes and hope to map *Vx*,*Vm* and *Vb* markers using the mapping populations prepared for next spring.

4. BAC library

For most plant species, a practical method available for isolating genes of unknown sequence or product is map based cloning. This method has been shown to be feasible in several species with relatively small genomes [16,17]. Map based cloning requires 3 molecular tools before the method can be applied:

(1) a large- insert genomic library

(2) a set of DNA markers tightly linked to the gene of the interest

(3) a transformation system to permit confirmation that the DNA sequence isolated carry the desired gene.

We have constructed two BAC library's of apple [15] from Russian seedling R 12740-7A, using the methods described by Zhang et al., 1999 [19]. We used two restriction enzymes (*Eco*R1 and *Hind* III) in preparing the library in order to increase likelihood of obtaining a random sample of genomic fragments. About 40,000 clones have been obtained from both libraries. The average insert size of the clones was about 110 Kb. This library represents a coverage of more than 5 times the apple genome. Pooled plasmid DNA from BAC colonies have been isolated. Upon completion of the library, we will be using the identified markers tightly linked to both *Vr* and *Vx* genes in a marker assisted cloning strategy. We will screen the genomic library constructed from the Russian seedling for DNA clones containing the marker sequences. Our next logical steps in this system is the cloning of the resistant genes for transformations.

References

[1]   E.B. Williams and J. Kuc, Resistance in *Malus* to *Venturia inaequalis*. *Ann. Rev. Phytopathol.* 7 (1969) 223-246.

[2]   A.R. Biggs, Apple scab. pp.6-9. In: A.L. Jones and H.S. Aldwinckle (eds).Compendium of Apple and Pear Diseases. ASP) American Phytopathology Society Press. MN, USA, 1990.

[3]   L.F. Hough, A survay of the scab resistance of the foliage on seedlings in selected apple progenies. *Proc. Amer. Soc. Hort. Sci.* 44 (1944) 260-271.

[4]   L. Parisi and Y. Lespinasse. Pathogenicity of *Venturia inaequalis* strains of race 6 on apple clones (*Malus* sp.). *Plant Dis.* 80 (1996) 1179-1183.

[5]   A.G. Manganaris *et al.*, Isozyme locus *Pgm-1* is tightly linked to a gene (*Vf*) for scab resistance in apple. *J. Amer. Soc. Hort. Sci.* 94 (1994) 1286-1288.

[6]   M. Hemmat *et al.*, Molecular marker linkage map for apple. *J. Heredity* 85 (1994) 4-11.

[7]   H.Y. Yang and J. Kruger. Identification of an RAPD marker linked to the *Vf* gene for scab resistance in apples. *Plant Breeding* 112 (1994) 323-329.

[8]   R.E. Durham and S.S. Korban,. Evidence of gene introgression in apple using RAPD markers. *Euphytica* 79 (1994) 109-114.

[9]   L. Gianfranceschi *et al.*, Molecular selection in apple for resistance to scab caused by *Venturia inaequalis*. *Theor. Appl. Genet.* 93 (1996) 199-204.

[10]   B.L. Koller *et al.*, DNA markers linked to *Malus floribunda* 821 scab resistance. *Plant Mol. Biol.* 26 (1994) 597-602.

[11]   S. Tartarini, RAPD markers linked to the *Vf* gene for scab resistance in apple. *Theor. Appl. Genet.* 92 (1996) 803-810.

[12]   M. Hemmat *et al.*, Molecular markers for scab resistance (*Vf*) region in apple. *J. Amer. Soc. Hort. Sci.* 123 (1998) 992-996.

[13]   D.F. Dayton *et al.*, Apple scab resistance from R12740-7A, a Russian apple. *Proc. Amer. Soc. Hort. Sci.* 62 (1953) 334-340.

[14]   H.S. Aldwinckle *et al.*, Early determination of genotypes for apple scab resistance by forced flowering of test cross progenies. *Euphytica* 25 (1976) 185-191.

[15]   J. Fu *et al.*, Development of a BAC library from Russian seedling R12740-7A, an apple selection with two genes conferring resistance to scab. Abstract. Plant Genome VII San Diego, Jan. 17-22, 1999.

[16]   F.S. Cheng, *et al.*, Development of a DNA marker for *Vm*, a gene conferring resistance to apple scab. *Genome.* 41 (1998) 208-214.

[17]   V.B. Arnodel *et al.*, Map based cloning of a gene controlling omega-3 fatty acid destruction in *Arabidopsis*. *Science* 258 (1992) 1353-1355.

[18]   G.B. Martin *et al.*, Map-based cloning of a protein kinase gene conferring disease resistance in tomato. *Science* 262 (1993) 1432-1436.

[19]   H-B Zhang *et al.*, Preparation of megabase-size DNA from plant nuclei. *Plant J.* 7 (1995) 175-184.

Use of Agriculturally Important
Genes in Biotechnology
G. Hrazdina (Ed.)
IOS Press, 2000

Construction of an Improved Linkage Map of
Diploid Alfalfa *(Medicago sativa)* and Mapping Symbiotic Genes Conditioning Ineffective and Non-nodulation Phenotype

G. B. Kiss, P. Kalo, A. Kereszt, S. Mihacea and G. Endre

Institute of Genetics, Biological Research Center, 6701 Szeged, P. 0. Box 521, Szeged, Hungary

Abstract. The improved genetic map of diploid (2n=2x=16) alfalfa has been developed by analyzing the inheritance of more than 1500 genetic markers on 137 plant individuals of an F2, population. The genetic map of alfalfa in its present form contains 8 morphological, 5 SSCP, 10 isozyme, 26 seed protein, more than 400 RFLP, and more than 700 RAPD and 200 specific PCR markers in 8 linkage groups. The genetic map covers 754 centimorgans with an average marker density of 0.8/cM. The correlation between the physical and genetic distance is about 1000- 1300 kilobase pairs per centimorgan. The location of an ineffective *(in6)* and a non-nodulation *(nnl* mutation were determined in diploid and tetraploid alfalfa populations, respectively. BAC clones originating from a *M. truncatula* BAC library were isolated with the help of closely linked markers. Chromosomal walking approach is taken to identify and isolate the genes conditioning the ineffective and the non-nodulation phenotype.

1. Construction of the improved genetic map of alfalfa

An improved genetic map of diploid (2n=2x=16) alfalfa has been developed by analyzing the inheritance of more than 1500 genetic markers on 137 individuals of an F_2 population. This mapping population derived from a self-pollinated F1, hybrid individual of the cross *Medicago sativa ssp. quasifalcata x Medicago sativa ssp. coerulea,* and was the same, which had been used for the construction of our basic alfalfa genetic map [1]. The genetic analyses were performed by using classical and special maximum-likelihood equations [2], related computer programs [3] and color mapping [4].

The constructed detailed linkage map of alfalfa [5] contains more than 1500 markers of which eight were morphological, one resistance, 5 SSCP, more than 400 RFLP and more than 700 RAPD markers, and 200 specific PCR markers, as well as 10 isozyme and 26 seed protein markers. The genetic markers span 754 centimorgan genetic distance and the correlation between the physical and genetic distance is 1000-1300 kb per cM on average, taking the DNA content of the haploid genome as 750-1000 Mbp [6, 7]. The map position of several nodule specific genes, 140 genes with known function or sequence, was determined. The genetic mapping of two multicopy genes, localized earlier by cytological experiments, allowed us to correlate LG 6 (ribosomal RNA genes) with chromosome No. 8 carrying the Nucleolus Organizing Region detected by cytology [8, 9], and LG 4 carrying the B-tubulin genes with the small metacentric or another submetacentric chromosome to which these genes hybridized in *situ* [10].

In the improved linkage map [5], the linkage relationships of some markers on linkage groups 6, 7, and 8 are different from the previously published one [1]. The cause of this discrepancy was that the genetic linkage of markers displaying distorted segregation (characterized by an overwhelming number of heterozygous individuals) had artificially linked genetic regions that turned out to be unlinked. To overcome the disadvantageous influence of the excess number of heterozygous genotypes; the recombination fractions, we used recently described maximum-likelihood formulas [2] and colormapping [4] which allowed us to exclude the misleading linkages and to estimate the genetic distances more precisely.

2. An effort towards the map based cloning of a gene conditioning non-nodulation trait in tetraploid alfalfa

Our group intends to isolate plant genes impaired in the symbiotic nitrogen fixation process by map-based cloning. Two mutations conditioning a non-nodulation and an ineffective symbiosis are under intensive investigations.

The MN-1008 mutant conditioning non-nodulation phenotype was isolated from tetraploid *Medicago sativa* cultivar, and from early segregation studies it was proposed that two recessive mutations were responsible for this phenotype [11]. This mutant displayed no root hair deformation, no root hair curling or cortical cell division [12], however variability in membrane depolarisation after *Nod* factor treatment was detected [13]. In addition, this mutant was resistant to vesicular-arbuscular mycorrhiza [14] but spontaneous nodules appear on the roots [15]. According to the above properties it is thought that the perception of the rhizobial signal molecules or the first steps of their subsequent signal transduction pathway are impaired in this mutant. Consequently, it seems likely that this mutant can be an important key to understand some of the early events in nodule formation elicited by rhizobial *Nod* factors.

In order to isolate the *nod* gene behind the phenotype first we had to determine its location on the linkage map of alfalfa. For mapping this trait, tetraploid F_2 populations segregating the non-nodulation phenotype were produced after crossing the NN-1008 plant with a nodulating M. *sativa (cv.* Nagyszenasi). Self-mating of the resulting F_2, plants generated segregating F_1 populations. Selecting appropriate individuals and based on the Bulked Segregant Analysis [16] we identified RAPD markers co-segregating with the mutation. With the help of the diploid mapping population these RAPD markers were mapped on linkage group five (LG5) of the *Medicago* genetic map [17]. This region spanned more than 10 cM genetic distance on LG5, therefore more molecular markers had to be searched for to saturate the region and find more tightly linked markers as starting points for the chromosomal walking experiments. The diploid mapping population and the Bulked Segregant Analysis were used to look for additional molecular (mostly RAPD) markers between RFLP markers U71 and U492 in LG5 [5], where the non-nodulating mutation linked to. Testing more than 500 different 10-mer oligonucleotide primers in PCR experiments resulted in 43 new RAPD markers mapped in this region. Some of these selected markers were also mapped in the tetraploid *Nod*-segregating populations (marker transfer).

At the same time two of the tetraploid F_2 segregating populations (progenies of NAB and NBW F_1 plants) were extended in order to determine the segregation ratio of the non-nodulating phenotype as well as the recombination frequencies in the region more precisely. The plant nodulation test, and the segregation data allowed us to conclude that: (i) many plants with low viability died before reaching maturity; (ii) the evaluation of the nodulating phenotype of the plants 6 weeks after the infection gives ambiguous results, therefore the non-nod characters should be tested later as well; and (iii) the segregation ratios of the non-nod phenotype (detected non-*nod* plants in the surviving population) in different F_2 populations are closer to the theoretical segregation ratio characteristic for a single gene mutation.

For fine mapping of the *nod* gene, the F_2 segregating population of the NAB F_1 plant has been chosen (with more than 2500 individuals) and selected markers from the diploid mapping population were tested. The first aim of the genotyping work was to detect polymorphism for all of the four alleles, namely the two from the non-nodulating parent NM-1008 and two from the wild type *M. sativa* plant, respectively. For this purpose different techniques were used, RFLP hybridizations, specific PCR amplifications (SCARs from RAPDs), and SSCP. Several markers for which genotype could be determined for the appropriate alleles were used to screen all individuals of the NAB F_2 population and plants carrying recombinant chromosomes for this region were selected for further analyses. Besides ordering the molecular markers the non-nodulating phenotype itself could be mapped more precisely with the increased number of individuals in the population. Having tested more than 2500 plants for recombination and finding markers tightly linked to the *nod* gene, isolation of BAC clones and the physical mapping was started.

For physical mapping and chromosomal walking experiments, the BAC library of *M. truncatula* were used [18]. Before starting the recombinant DNA work, however, the microsynteny in this region between *M.* sativa and *M. truncatula* was tested using some *M. sativa* markers. By mapping these markers in *M. truncatula* it was demonstrated that the location and order of these markers were the same in both plant species. Markers proved to be tightly linked to the *nod* gene were used as hybridization probes to screen the library. Positive clones were identified and analyzed further. The isolated end-fragments of the BAC clones allowed us to identify overlapping clones and thus to build a contig in the region.

We are in the process of sequencing this contig which may lead to identify candidate genes. This gene than can be used to complement the mutation in transformation experiments. We also consider the possibility that no candidate gene will be found. In this case systematic complementation experiments can help to identify the *nod* gene.

3. Saturation of the region containing the *fix*1 mutant gene

The diploid ineffective alfalfa mutant, the Fix1 (McIN6) develops nodules but is incapable to utilize the atmospheric nitrogen in the presence of *Rhizobium meliloti*. The Fix1 individual plant was crossed with the yellow-flowered *M. sativa* ssp. *quasifalcata* k93 and the generated F_1 plants were self-mated to produce segregation populations. RFLP and PCR markers were used to determine the map position of the fix 1 (in_6) mutation close to the RFLP marker U69 in linkage group 7. RAPD markers were identified using 240 10 base-long oligonucleotide primers to saturate the region in the vicinity of the fix1 (in_6) mutation by the so-called Bulked Segregant Analysis method [16]. Thirteen

RAPD markers were mapped in this region and mapping of other RFLP markers revealed that the *fix*1 mutation located between U69 and L83 RFLP markers which were mapped in 4.6 cM genetic distance from each other. The aim of our work is to determine the map position of the *fix*1 mutation as accurate as possible, therefore the screening for tightly linked markers to the *fix*1 locus will be continued.

Acknowledgements

This work was supported partly by grants OTKA (Hungarian Scientific Research Fund) T025467, F030408, Dr. Janos Bastyai Holczer Foundation, Volkswagen Stiftung Grant No. 1/72 244, and the European Commission (EuDicotMap contracts PL96-2170).

References

[1] G.B. Kiss *et al.*, Construction of a basic genetic map for alfalfa using RFLP, RAPD, isozyme and morphological markers. *Mol Gen Genet* **238** (1993) 129-137.

[2] M. Lorieux *et al.*, Maximum-likelihood models for mapping genetic markers showing segregation distortion. 2. F2 populations. *Theor Appl Genet* **90** (1995) 81-89.

[3] S. Lincoln *et al.*, Constructing genetic maps with Mapmaker/EXP 3.0 Whitehead Institute Technical Report. 3rd edition, 1992.

[4] G.B. Kiss *et al.* Colormapping: a non-mathematical procedure for genetic mapping. *Acta Biologica Hungarica* **49** (1998) 47-64.

[5] P. Kalo *et al.*, Construction of an improved linkage map of diploid alfalfa *(Medicago sativa). Theor Appl Genet* (in press), (2000)

[6] K. Arumuganathan K and E.D. Earle, Nuclear DNA content of some important plant species. *Plant Mol Biol Rep* **9** (1991) 208-219.

[7] I. Winicov *et al.* (1988) Characterization of the alfalfa *(Medicago sativa)* genome by DNA reassociation. *Plant Mol Biol* **10** (1988) 369-371.

[8] G.R. Bauchan and T.A. Campbell, Use of an image analysis system to karyotype diploid alfalfa *(Medicago sativa* L.). *The Journal of Heredity* **85** (1994) 18-22.

[9] E. Falistocco *et al.* Karyotype and C-banding pattern of mitotic chromosomes in alfalfa, *Medicago sativa* L. *Plant Breeding J.* **4** (1995) 451-453.

[10] D.A. Schaff *et al.*, In situ hybridization of B-tubulin to alfalfa chromosomes. *The Journal of Heredity* **81** (1990) 479-483.

[11] .A. Peterson and D.K. Barnes, Inheritance of ineffective nodulation and non-nodulation traits in alfalfa. *Crop Sci* **21** (1981) 611-616.

[12] M.E. Dudley and S. Long, A non-nodulating alfalfa mutant displays neither root hair curling nor early cell division in response to *Rhizobium meliloti.* Plant Cell **1** (1989) 65-72.

[13] H.H. Felle *et al.*, Nod signal-induced plasma membrane potential changes in alfalfa root hairs are differentially sensitive to structural modifications of the lipochitooligosaccharide. *Plant Journal* **7** (1995) 939-947.

[14] S.M. Bradbury *et al.*, (1991) Interaction between three alfalfa nodulation genotypes and two Gloinus species. *New Phytol* **119** (1991) 115-120.

[15] G. Caetano-Annollés *et al.*, Nodule morphogenesis in the absence of *Rhizohium.* New Horizons in Nitrogen Fixation, Kluwer Academic Publishers, Dordrecht, The Netherlands, 1993, pp. 297-302.

[16] R.W. Michelmore *et al.*, Identification of markers linked to disease-resistance genes by bulked segregant analysis: a rapid method to detect markers in specific genomic regions by using segregating populations. *Proc Natl Acad Sci* USA **88** (1991) 9828-9832

[17] G. Endre *et al.*, Genetic analysis of *Medicago sativa* nodulation genes. Biological Nitrogen Fixation for the 21st Century, Kluwer Academic Publishers, Dordrecht, The Netherlands, 1997, pp. 3 15-3 16.

[18] C. Nam *et al.*, Construction of a bacterial artificial chromosome library of *Medicago truncatula* and identification of clones containing ethylene responsive genes. *Theor Appl Genet* **98** (1999) 638-646.

Use of Agriculturally Important
Genes in Biotechnology
G. Hrazdina (Ed.)
IOS Press, 2000

Application of Chromosomal Map and Gene Probes of *Arabidopsis* in Studies on *Brassica* Genomes

D. Babula, M. Kaczmarek, P. Ziólkowski, J. Sadowski
Institute of Plant Genetics, Polish Acad. of Sciences, Poznan, Poland

Abstract. The plant model system, *Arabidopsis thaliana* provides an opportunity for systematic identification and isolation of genes on a large scale, and for reaching a comprehensive understanding of various aspects of plant biology. For example, where synteny can be established with the model genome, gene identification and studies on chromosomal organization in crop species may be greatly facilitated. We have undertaken systematic comparative mapping between *A. thaliana* (n=5) and *Brassica oleracea* (n=9). *Arabidopsis* probes comprising of two gene complexes mapped to chromosomes 3 and 4 were selected. These were hybridised to the *B. oleracea* mapping population in such a way identifying regions of the genome corresponding to the targeted *A. thaliana* chromosomal regions. Each region of the *A. thaliana* genome investigated was found to be homologous to 2-3 distinct regions in the *B. oleracea* genome. Some of the genes used in the study however, mapped to 1-5 regions in the *B. oleracea* genome. They were found as single genes or in a form of partial complexes assembled with some of other genes from *A. thaliana*-like conserved complexes. In another line of our studies we have mapped to existing *B. oleracea* RFLP map over 100 loci corresponding to cDNA probes from the *A. thaliana* genome sequencing program. Our results indicate some extent of conservation of chromosome colinearity at the micro and macro levels. This suggests that the DNA sequence data from *A. thaliana* will be of direct use in the identification and isolation of genes in *Brassica* crop species. The studies confirmed earlier observations in our laboratory and that of others that there are many intragenomic duplications in the *Brassica* genomes reflecting their polyploidal origin.

1. Introduction

It has become increasingly clear that plant genome research is needed as the basis to reach new level of efficiency in the application of genetics and breeding to crop plants. Future progress in plant biology also depends fundamentally on plant genome research. The current *A. thaliana* and rice genome projects have proven to be of immediate value to research in biology. What's more, recent research on cereal genomes has revealed that, because of extensive synteny and colinearity of chromosomes of related species, new information on the genome of one cereal species can be used to understand the genomes and predict the location of genes on chromosomes of other cereals. Genome research applied to cereals as a group of related species thereby can increase the effectiveness of breeding improved cereal species, which includes wheat, corn, and rice, the world's most important sources of food.

Less chromosomal colinearity has been observed in other dicotyledonous crop plants like rapeseed and a variety of leafy vegetables, among others than in the model plant *A. thaliana* [1, 2, 3]. However, long chromosomal segments covering up to 20-30 cM are expected to be collinear in *Brassica* and *Arabidopsis* [4]. Considering the difference in physical size between *A. thaliana* (100-120 Mbp) and *Brassica* diploid species (500-650

Mbp) [5], *A. thaliana* can serve not only as a source of many gene probes, but it can simplify the identification and isolation of genes important to breeding the *Brassica* species. In general, coding gene sequences are conserved among genera and even among families. Therefore, comparative mapping could allow the transfer of information on chromosome structure and gene organization among related species. Furthermore, this information serves to gain insight into their evolution. Chromosomal aberrations followed by the number of minor mutations together with environmental selection have created the architecture of genomes and their functional integrity. Many of these changes can be readily detected by comparative micro-synteny and macro-synteny studies. Because of the absence of the *Brassica* ancestral species, which are likely to be extinct, *A. thaliana* serves as a temporary model for comparative studies. However, the crucifers *A. thaliana* and *Brassica* are classified taxonomically in different tribes (Sisymbrieae and Brassiceae) unlikely a direct ancestral conclusions between these two genera making. Earlier comparative mapping of *Arabidopsis* and *Brassica* species identified conserved islands in the chromosome segments tested [1, 2, 3, 4]. Some of these segments conserving the gene repertoire and order may represent orthologous segments between these species. It has been suggested recently that each of the *Brassica* genomes contains three complete copies of the *A. thaliana* genome [6]. Research conducted by our group in the Institute of Plant Genetics aims at testing micro- as well as macro-colinearity of *Brassica* and *Arabidopsis* genomes based on well described gene probes of *A. thaliana*. Two kinds of maps evaluating the micro- and macro-collinearity of these genomes are in preparation:

- in terms of micro-colinearity studies, we selected two *A. thaliana* segments from chromosome 3 and 4 covering 17 and 6 genes, respectively. Our earlier comparative genetic analysis showed that the gene complex from *A. thaliana* chromosome 4 is structurally conserved in *B. nigra* [7]. Coding sequences corresponding to these two homologous chromosomal segments in *A. thaliana* and *B. oleracea* were used as molecular probes in Southern hybridization to determine the level of conservation and possible structural changes distinguishing these two crucifers.

- in terms of macro-colinearity, we are completing the construction of a RFLP map for *B. oleracea* based on over 300 gene probes from the *A. thaliana* genome.

2. Mapping procedure

In order to conduct RFLP analysis and mapping of *B. oleracea* genome, nuclear DNA was isolated from young leaves of F2 plants resulting from crossing kale x cauliflower. Isolation of DNA was performed by CsCl-gradient ultracentrifugation [8]. DNA samples were digested with 17 restriction endonucleases (*Alu* I, *Bam* HI, *Eco* RI, *Eco* RV, *Eco* 72 I, *Hae* III, *Hin* 6 I, *Hin*d III, *Hin*f I, *Kpn* I, *Mbo* I, *Pst* I, *Pvu* II, *Xba* I, *Xho* I, *Rsa* I, *Taq* I) in order to identify polymorphism of hybridizing restriction fragments. Following electrophoresis in 0.8% agarose gel, DNA was transferred to nylon membranes. Gene probes from *A. thaliana* (ESTs) were labeled with DIG-system (Boehringer) and non-radioactive hybridization was performed as was described earlier [9]. The program MAPMAKER version 3 [10] was used to establish linkage groups.

3. Results

3.1. Micro-colinearity

We have investigated the arrangement of two gene complexes in the *B. oleracea* genome. Their organization on *A. thaliana* chromosomes have already been determined in detail. One of the two *A. thaliana* gene complexes consists of 6 genes: *ABI1, RPS2, CK1, NAP, X9,* and *X14. RPS2* gene codes for resistance to *Pseudomonas syringae* [11, 12, 13], a topic of obvious interest and potential economic significance.

Alltogether the *A. thaliana* sequences were mapped on four linkage groups of *B. oleracea* (not shown). The *A. thaliana* multiple gene linkage arrangement from chromosome 4 was almost absolutely conserved in a single linkage group of *B. oleracea* (Fig. 1), although not all members (*X14*) of the complex could be followed up due to lack of polymorphism. Tightly linked *ABI1, RPS2, CK1, NAP,* and *X9* were mapped on the *B. oleracea* C4 linkage group. One major rearrangement differentiating *B. oleracea* and *A. thaliana* complex is segmental inversion which concerns the *NAP* and *CK1* genes. Similar as in *A. thaliana*, the *B. oleracea RPS2* homolog was present in single copy. Additional copies for some genes were present in three other linkage groups. Some of the members of gene complexes are tandemly duplicated and exist as small gene families. Most abundant among six genes in the *B. oleracea* genome was *CK1* coding for caseine kinase. Four *CK1* loci at different *B. oleracea* chromosomes were mapped.

3.2. Macro-collinearity

In the present studies, the *B. oleracea* a linkage map was developed by set of *A. thaliana* gene probes to study the macro-colinearity between the two genomes with different chromosome number. A large number of gene probes from *A. thaliana* are to be used to launch structural analysis of corresponding regions of the *Brassica* genomes. RFLP analysis permitted us to find 250 potential chromosomal markers by application of 180 cDNA probes. So far, we have mapped 190 corresponding loci on nine chromosomes of *B. oleracea*. Comparison between the two crucifers studied disclosed numerous genomic regions of conserved gene organization. Genes from the *A. thaliana* genome were found to be homologous mostly to 2-4 distinct loci in *B. oleracea* genome. Some of the genes used in our studies however, mapped to 1-5 regions in the *B. oleracea* genome. It was not possible to determine the chromosomal locations of all existing loci corresponding to *A. thaliana* gene probes for the lack of RFLP for part of the loci. Interestingly, we identified a group of *B. oleracea* loci (20%) displaying lack of RFLP even after digestions with 17 restriction endonucleases.

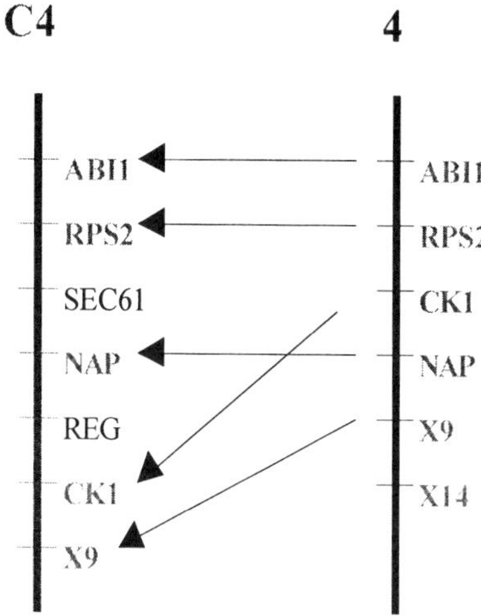

Fig. 1 Chromosomal location of conserved gene complex in *B. oleracea* and *A. thaliana*

The present work served to map a considerable number of loci to the corresponding linkage groups. All linkage groups were enriched with many loci with sequence homology to genes of known functions. When duplicated sequences are frequent, as in the *Brassica* genomes, the assignment of orthologous loci in the genomes compared is only tentative. The reason is that it is unclear which restriction fragments correspond to the same loci in different species. In order to diminish this problem we consider as orthologous 1. single-copy loci and 2. reiterated loci where linkage to neighboring loci is conserved among *B. oleracea* and *A. thaliana*. In the future, chromosomal maps saturated with these loci may be useful in gene tagging or cloning.

3.3. Intragenomic organization of Brassica species

Single gene probes revealed that each *B. oleracea* linkage group carries multiple loci. The distances between duplicated loci ranged from 0.0 cM to over 80 cM.

A refined genetic map of duplicated loci will be helpful in identification and discrimination between active genes and pseudogenes.

4. Discussion

These studies have shown that certain regions of the *Brassica* complex genomes and a simple *A. thaliana* genome are partially conserved in terms of gene organization. The presence of multiple copies of many genes in *Brassica* further proves the duplicated nature of the *Brassica* genomes [14, 15]. It is not clear however, whether multiple copies have originated in *Brassica* from hybridization of two or even more ancestral genomes. We confirmed earlier reports, that duplicated gene blocks were exposed to extensive rearrangements [16]. The *A. thaliana* genome is likely to have undergone some of the rearrangements found in the *Brassica* genomes. It seems obvious that the *A. thaliana* genome should serve as a frame of reference, but it would be an inappropriate model for an ancestral genome.

The increased copy number of various genes may be one of the basic reasons for the lack of progress in modifying the amounts of some gene products using mutagenesis approach followed by selection. It has been recently shown in a case of an attempted modification of oleic and linolenic acid level using this approach. Increased number of gene copies might also complicate the modification of other genes using molecular techniques. While most of *B. napus'* genes are highly duplicated, the number and chromosomal distribution of functional gene copies remains unknown. Their number may be very likely much smaller than the number of copies identified because some of the identified sequences may represent silenced genes or pseudogenes. It will be useful in the future to identify and analyse all loci of a gene of interest. It will help to determine the homology level between the loci and which of them code for functional products. For example, research on fatty acid profile modification with help of biotechnological methods such as antisense inhibition has shown that this information may be helpful. In this line of studies [17] it was found that gene copy factors could take part in the ability of the antisense RNA to inhibit desaturation.

There is overwhelming evidence that the diploid and amphidiploid *Brassica* species are extremely plastic. A non-homologous chromosome recombination has almost certainly played a role in this process [16]. Existing colinear segments in *Brassica* species and *A. thaliana* indicate that multiple copies of most genes within the species of *Brassica* genomes are the effect of reiteration of longer chromosomal segments rather than individual gene duplications. The genetic data suggest that wide identification of colinear regions will enable the *A. thaliana* molecular resources to be applied for the exploration, understanding and modification of the *Brassica* genome.

Acknowledgements

We thank Ms. Barbara Zbaszyniak for technical assistance. The work was supported by grant No. 225 PO6A 016 11 from the State Committee for Scientific Research to J.S.

References

[1]   S.P.Kowalski *et al.*, Comparative Mapping of *Arabidopsis thaliana* and *Brassica oleracea* Chromosomes Reveals Islands of Conserved Organization. *Genetics* **138** (1994) 499-510.

[2]   J. Sadowski *et al.*, Genetic and Physical Mapping of an *Arabidopsis* Gene Complex in *Brassica* Genomes. *Cruciferae Newsletter* **16** (1994) 47-48.

[3]   J. Sadowski *et al.*, Genetic and Physical Mapping in *Brassica* Diploid of a Gene Cluster Defined in *Arabidopsis thaliana*, *Mol. Gen. Genet.* **251** (1996) 298-306.

[4]   U. Lagercrantz and D. Lydiate. Comparative Genome Mapping in *Brassica*, *Genetics* **144** (1996) 1903-1910.

[5]   K. Arumuganathan and E.D. Earle, Nuclear DNA Content of some Important Plant Species. *Plant Molec. Rpt.* **9** (1991) 208-219.

[6]   M.D. Gale and K.M. Devos, Plant Comparative Genetics after 10 Years. *Science* **282** (1998) 656-658.

[7]   J. Sadowski and C.F. Quiros. Organization of an *Arabidopsis thaliana* Gene Cluster on Chromosome 4 Including the RPS2 Gene, in the *Brassica nigra* Genome, *Theor. Appl. Genet.* **96** (1998) 468-474.

[8]   J. Sambrook *et al.*, Molecular Cloning: A Laboratory Manual, 2nd ed. Cold Spring Harbour Laboratory Press, Cold Spring Harbour, NY, 1989.

[9]   F. Perez, Genomic Structural Differentiation in *Solanum*: Comparative Mapping of the A- and E-Genomes. *Theor. Appl. Genet.* **98** (1999) 1183-1193.

[10] E.S. Lander *et al.*, MAPMAKER: an Interactive Computer Package for Constructing Primary Genetic Linkage Maps of Experimental and Natural Populations. *Genomics* **1** (1987) 174-181.

[11] M. Mindrinos, *et al.*, The *A. thaliana* Disease Resistance Gene RPS2 Encodes a Protein Containing a Nucleotide-Binding Site and Leucine-Rich Repeats. *Cell* **78** (1994) 1089-1099.

[12] A.F. Bent *et al.*, RPS2 of *Arabidopsis thaliana*: a Leucine-Rich Repeat Class of Disease Resistance Genes. *Science* **265** (1994) 1856-1860.

[13] A.L. Caicedo *et al.*, Diversity and Molecular Evolution of the Rps2 Resistance Gene in *Arabidopsis thaliana*, *Proc. Natl. Acad. Sci. U.S.A.* **96** (1999) 302-306

[14] S.F. Kianian and C.F. Quiros. Generation of a *Brassica oleracea* Composite RFLP Map: Linkage Arrangements among Various Populations and Evolutionary Implications, *Theor. Appl. Genet.* **84** (1992) 544-554.

[15] S.F. Kianian and C.F. Quiros, RFLP Map of *B. oleracea* Based on Four Crosses. *Theor. Appl. Genet.* **84** (1992) 544-554.

[16] J. Hu *et al.*, Linkage Group Alignement from Four Independent *Brassica oleracea* RFLP Maps. *Genome* **41** (1998) 226-235.

[17] J. Kohno-Murase *et al.*, Improvement in the Quality of Seed Storage Proteins by Transformation of *Brassica napus* with an Antisense Gene for Cruciferin. *Theor. Appl. Genet.* **91** (1995) 627-631.

Use of Agriculturally Important Genes in Biotechnology
G. Hrazdina (Ed.)
IOS Press, 2000

Micromanipulation of Gametic Cells of Cereals

B. Barnabás, Zs. Pónya, P. Finy, M. Solymoss, I. Timár, A. Fehér*, D. Dudits*

Agricultural Research Institute of the Hungarian Academy of Sciences, Martonváscir
and
**Biological Research Center of the Hungarian Academy of Sciences, Szeged, Hungary*

Abstract. Sexual reproduction plays a key role in the propagation of flowering plants. Fertilisation takes place in the embryo sac, which is usually deeply encased in the ovule tissue. The intimate mechanisms of double fertilisation have still not been entirely discovered mainly due to the fact that the female gametophyte is inaccessible for direct micromanipulation. Recent advances in plant cell and molecular biology have brought new, powerful technologies to investigate and manipulate the reproductive cells of angiosperms including cereal crop plants. Experimental approach based on various microtechnics involving *in vitro* fertilisation at the single cell level and microinjection aimed at dissecting the cellular and subcellular events preceeding and occurring during and after fertilisation now is available in our laboratory. Our recent achievements in the field of wheat egg cell micromanipulation are the presened in this paper.

1. Introduction

In vitro fertilisation techniques via electrofusion of isolated gametes of maize have been developed [1] and then fertile plants can be regenerated from artificially produced zygotes [2]. This permits the examination at the molecular level of the early developmental processes in gametic fusion and the events occurring immediately after sperm-egg contact. However, there still remains a considerable lag in the field of plant embryology as compared to animal or human embryology.

Thus a number of important questions remain to be answered such as how and when cell polarity is established in the egg cell, what factors trigger the first asymmetric division of the zygote and how the fate of the zygote unfolds that reaches the stage of proembryo formation and what genes govern these mechanisms.

Some of these questions can only be addressed if systems of gamete isolation and fusion are associated with foreign gene delivery into the gametes/zygotes of higher plants. When it comes to attempting to introduce foreign DNA into the germ cells (the number of which is severely limited), the method of choice can be microinjection, which allows the targeting cellular compartments (especially the nuclear region) more easily than other transformation techniques. However, the problem of cell immobilisation has to be circumvented in adopting microinjection of DNA to single plant cells/protoplasts.

Hitherto only one paper [3] has reported the successful microinjection of maize zygotes. However, no experimental data are available on DNA transfection of female gametes of higher plants by microinjection. Isolation of highly functional female gametes seems to be essential for implementing successful micromanipulation (e.g. DNA microinjection). Therefore, transmission electron microscopic investigations were designed to clarify the structural changes taking place during the life-span of wheat egg cells. Wheat egg cells in various developmental stages have been used as targets for DNA microinjection.

2. Special Morphological Features of Wheat Egg Cell Maturation

Our work on wheat (*Triticum aestivum L.*) represents the commencement of a series of investigations on the functionality of the female gametophyte and on its suitability for various micromanipulations.

Cultivated wheat is a protogynous plant in which the female gametophyte (embryo sac) develops three-four days before anthesis [4]. Biological studies carried out on flowering of higher plants in the 60's in connection with the development of hybrid wheat [5] already indicated the long-lasting receptivity of the wheat pistil. However, detailed observations were not made on the differentiation of the female gametophyte. Our examinations were made on egg cell protoplasts isolated mechanically from pistils of various stages of the spring wheat cultivar Chinese Spring fixed in agarose. After fixation and the embedding procedure semi and ultrathin sections were cut from protoplasts for light and transmission electron microscopic studies. Examinations of the cell structure of wheat egg cell protoplasts isolated from young (3 days prior to anthesis) and over-aged (12 days after anthesis) caryopses confirmed that wheat caryopses had a long life span. The nucleus of the egg cell, which became spherical during the isolation of the protoplasts, was centrally located. The characteristic cell organelles (mitochondria, amyloplasts, lipid bodies) surrounded the nucleus while the numerous, relatively large vacuoles were in peripheral position (Fig. 1.).

The maturation of the egg cell was accompanied by a number of fundamental morphological changes, one indication of which was the large increase in size. Six days after anthesis the initial volume of the egg cell doubled, after which it gradually decreased in volume during ageing. The changes in the volume of the protoplasts were reflected by an increase of the nuclei and the nucleoli. The ageing of the egg cell was accompanied by degradation of the chromatin reserves of the nucleus while the accumulation of reserved nutrients could be observed in the cytoplasm. There was a decline in the degree of vacuolisation, with the vacuoles becoming smaller and more uniformly distributed. In egg cells isolated 2 weeks after anthesis the characteristic symptoms of programmed cell death (apoptosis) could be observed [6].

3. A novel method of microinjecting egg cells of wheat

An efficient technique of microinjection has been established for the introduction of exogeneous DNA into egg cells of wheat. Egg protoplasts were isolated mechanically and

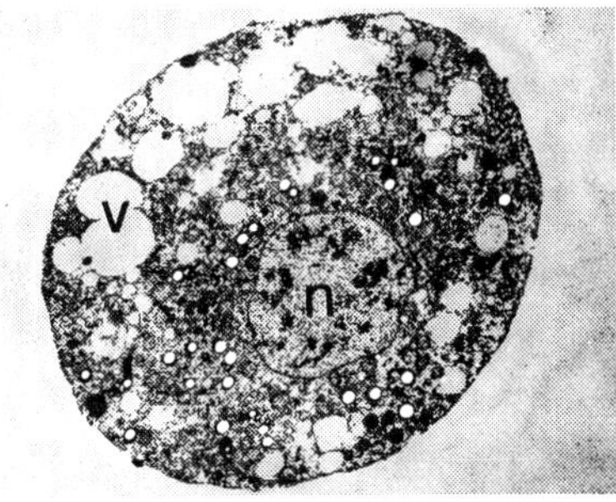

Fig. 1. Transmission electrom micrograph of a young egg cell of wheat isolated 3 days after emasculation (n, nucleus; v, vacuole).

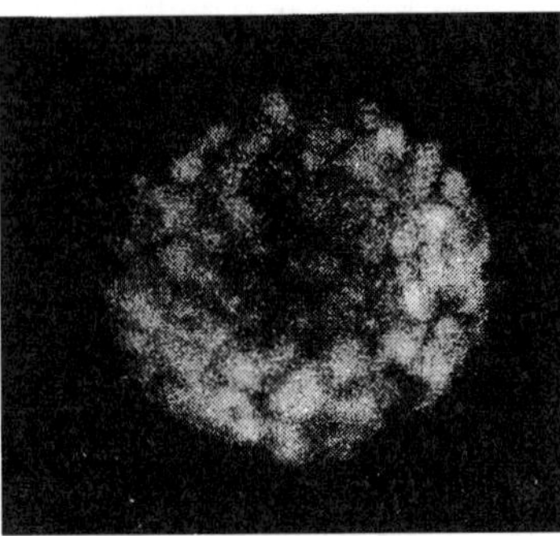

Fig 2. Transient expression of GFP reporter gene in a wheat egg cell.

than exposed to high frequency AC (alternating current) field for immobilisation. Capitalising on this novel approach, on average 15-20 egg cells could be microinjected per hour. Using GFP as a reporter system in transgene expression essays the analysed egg cells gave a positive response (Fig.2). The transformation rate seems to be greatly influenced by the time of egg cell isolation. Mature egg protoplasts isolated five days following emasculation, showed the highest frequency (73.9%) of transient gene expression (Table 1).

Table 1. Transient expression rate of isolated egg cells of wheat after microinjection of GFP reporter gene.

Timing of microinjection*	No. of microinjected egg cells	Frequency of transformation (%)
3	20	55.00
5	23	73.91
6	17	58.82
12	10	20.00
15	9	11.11

*Days after emasculation

The microinjection procedure presented here, coupled with the electrofusion system, makes it possible to attempt to transform cereals by using the sexual route and may contribute to filling the gap between sexual and somatic hybridisation. As the gametic fusion is brought about by a high frequency AC-field, the application of which obviates the need for the presence of specific gamete-recognition mechanisms, intergeneric and interspecific hybrids can be created. Furthermore, it can help in elucidating the molecular mechanisms underlying the early steps of embryogenesis.

References

[1]   E. Kranz *et al.*, In vitro fertilisation of single, isolated gametes of maize mediated by electrofusion, *Sex. Plant Reprod* 4, (1991) 12-16.

[2]   E. Kranz and H. Lörz, *In vitro* fertilisation with isolated single gametes results in zygotic embryogenesis and fertile maize plants, *Plant Cell* 5, (1993) 739-746.

[3]   N. Leduc *et al.*, Isolated maize zygotes mimic in vivo embryogenic development and express microinjected genes when cultured *in vitro*, *Developmental Biology* 177 (1996) 190-203.

[4]   I. Timar *et al.*, Comparative studies on the male and female gametophyte development in three different Triticum species, *Plant Science* 126 (1997) 97-104.

[5]   S. Rajki and E. Cicer, Pollination of the winter wheat Bankuti 1201 in different developmental stages of the pistil, *Növénytermelés* 10 (1961) 335-345.

[6]　Zs. Pónya *et al.*, Morphological characterisation of wheat (*Triticum aestivum L.*) egg cell protoplasts isolated from immature and overaged caryopses, *Sex. Plant Reprod* 11 (1999) 357-359.

[7]　Zs. Pónya *et al.*, Optimisation of introducing foreign genes into egg cells and zygotes of wheat (*Triticum aestivum L.*) via microinjection, *Protoplasma* 1999 (in press).

*Use of Agriculturally Important
Genes in Biotechnology
G. Hrazdina (Ed.)
IOS Press, 2000*

Identification of Storage Protein Genes as Marker for Breadmaking Quality of Wheat

Z. Gálová, H. Smolková
Slovak Agricultural University, Nitra, Slovak Republic

Abstract. The HMW subunits of glutenin, considered to be important components of bread-making quality of wheat, are encoded by genes at three genetically unlinked loci: Glu-A1, Glu-B1 and Glu-D1. These loci are located close to the centromere on the long arm of the homologous chromosomes 1A, 1B and 1D. The most important subunits are 5+10 and 2+12, which are expressed from the Glu-D1 loci. The relationship between the HMW glutenin subunit allels and the bread-making quality of different wheat varieties was investigated. Two cultivars of *Triticum durum* Desf., 18 landraces of *Triticum aestivum L.* and 30 varieties of *Triticum spelta L.* were analysed for their HMW glutenin subunits by SDS-PAGE. From the electrophoreograms the individual HMW glutenin subunits were determined and the Glu–score was calculated. The most common banding patterns in commercial wheat varieties were subunits "null" from the locus Glu-A1 (83%), subunits 7+9 from the locus Glu-B1 (83%) and subunits 5+10 (67%), 2+10 (33%), located on the locus Glu-D1. Secaline block Gld 1B3, i.e. marker of bad baking quality, but also that of resistence to stem rust of grain, was identified in gliadin spectra of the five varieties. The results of 30 winter spelt wheat grain cultivars showed a high frequency of occurence of HMW glutenin subunits with the composition of 1, 6+8, 2+12 and 6 Glu-score values, associated with medium bread-making quality. Two cultivars of *Triticum durum* Desf. were analysed and HMW subunits 0 at the Glu-A1 locus and 7+8 at the Glu-B1 locus with Glu-score 4 were determined. The results of wheat grain analysis confirmed the usefullness of Glu-score application for the prediction of bread-making quality of wheat flour.

1. Introduction

The relation between the HMW- glutenin subunit alleles and the bread-making quality of different wheat varieties was studied. Two cultivars of *Triticum durum* Desf., 18 varieties of *Triticum aestivum* L. and 30 cultivars of *Triticum spelta* L. were analyzed for their HMW glutenin subunits by SDS polyacrylamide gel electrophoresis. The standard reference method ISTA was used for identification and characterization of wheat varieties. From the electrophoretic separations individual HMW glutenin subunits were determined and the Glu-score was calculated. The verified correlations between bread-making quality and specific HMW subunits of glutenin can be useful to wheat breeders when using SDS-PAGE of as a screening test for the prediction of bread-making quality of wheat.

The glutenin proteins of wheat, particularly the highly viscous glutenin fraction (HMW) strongly influence the breadmaking properties of flours. Genetics studies of the HMW glutenin subunits have shown that they are controlled at three loci, recently designated Glu-Al, Glu-Bl and Glu-D1, located close to the centromere on the long arm of chromosomes 1A, 1B and 1D respectively. Alleles controlling different subunit bands, or band combinations, occur at all three loci. A recent catalogue of these alleles identified three alleles at the Glu-A1 locus, eleven at the Glu-B1 locus and six at the Glu-D1 locus.

This paper reports the allelic composition at each of the three loci controlling HMW glutenin subunits of different wheat varieties.

2. Material and methods

The relation between the HMW glutenin subunit allels and the bread - making quality of different wheat varieties was studied. The winter wheats included 18 common wheat varieties (*Triticum aestivum* L.), 30 winter spelt wheat (*Triticum spelta* L.) and 2 durum wheat cultivars (*Triticum durum* Desf.) which were obtained from the Gene Banks of the Slovak and Czech Republics.

Glutenins were extracted from individually ground kernels and separated on 10% polyacrylamide gels in the presence of SDS according to the method ISTA [5]. The gels were stained with Coomassie Brilliant Blue solution and destained with de-ionized distilled water. The stained bands were qualified by scanning densitometry (LD-01 Instrument). From the electrophoretic separation the individual HMW glutenin subunits were determined and the Glu - score was calculated.

The HMW subunits of glutenin were designated according to the numbering system of Payne & Lawrence [1]. The SDS sedimentation test was determined by Seditester LS 03.

3. Results

The storage proteins of hexaploid wheat are important. Because of the unique cohesive-elastic properties they bestow on doughs made from wheat flours. They are important for the textural characteristics of the bread quality. The HMW subunits of glutenin are considered to be the most important components with the respect to the baking quality. Correlations have been established between particular HMW glutenin subunits and bread-making quality of wheat [2, 4].

The most common banding patterns in 18 common wheat varieties (Table 1) were subunits "zero" from the locus Glu-Al (83%), subunits 7+9 from the locus Glu-B1 (83%) and subunits 5+10 (67%), 2+10 (33%) located at the locus Glu-D1. Secaline block GLD 1B3 (*), marker of bread baking quality, and a marker of resistance to stem rust of grain, was identified in the gliadin components of the five varieties - Fundulea 29, Iris, Livia, Sana and SK 3756-176. From the electrophoretic separations, the individual HMW glutenin subunits were determined and the so-called Glu- score was calculated. The highest value of the Glu-score was found in the cultivars Ilona, Vlada, Regia (9), and the lowest score (5) in the varieties Iris, Sana, Vala and SK 3756-1-76. Positive correlations were established between particular HMW glutenin subunits and SDS sedimentation test.

Two cultivars of *Triticum durum* Desf. were analysed and HMW subunits 0 at the Glu-A I locus and subunits 7+8 at the Glu-13 I locus with Glu-score 4 were determined (Table 2). Five electrophoretical profile groups with a different HMW glutenin subunit composition were determined in the 30 winter spelt wheat cultivars that were analyzed (Table 3).

The first electrophoretic profile with 1, 6+8, 2+12 HMW glutenin subunit composition and 6 Glu-score value was found frequent and was present in 63,33% of the analyzed winter spelt wheat cultivars (19 accessions).

Table 1. Variation in the individual HMW glutenin subunit alleles in winter wheat varieties.

Cultivar test	HMW glutenin subunits			Glu-score	SDS (cm^3)
	Glu-A1	Glu-B1	Glu-D1		
Astella	0	7+9	5+10	7	57
Barbara	0	7+9	5+10	7	50
Blava	0	7+0	5+1-	7	60
Fundulea 29*	0	7+9	5+10	7	43
Ilona	2*	7+9	2+12	5	48
Iris*	0	7+9	2+12	5	48
Kosutka	0	7+9	5+10	7	64
Livia*	0	7+9	5+10	7	35
Maris Marksman	0	7+8	2+12	6	40
Rada	0	7+9	5+10	7	47
Sana*	0	7+9	2+12	5	32
SK3756-76*	0	6+8	2+12	5	46
Solida	0	7+9	5+10	7	52
Torysa	0	7+8	2+12	6	54
Vala	0	7+9	2+12	5	56
Viginta	0	7+9	5+10	7	70
Vlada	1	7+9	5+10	9	79
Regia	1	7+9	5+10	9	70
average				6.8	53.6
min.				5	32
max.				9	79
v(%)				19.2	23.3

Table 2. Composition of HMW subunits of glutenin in durum wheat.

Cultivar	HMW glutenin subunits			Glu-score
	Glu-A1	Glu-B1	Glu-D1	
Soldur	0	7+8	-	4
Vendur	0	7+8	-	4

On the basis of the value for the Glu-score these cultivars can be classified as medium to low bread-making quality and they seem to be useful for common flour additions for increasing their nutritive and sensorial parameters. At two cultivars (Hercule, Steiner Roter Laupheimer) 6+8, 2+12 HMW glutenin subunit composition was determined and 4 Glu-score value was calculated what indicates cultivars with weak bread-making quality. On the other hand, cultivars Ardenne and Rouquin can be judged as good bread-making varieties because of its 1, 6+8, 5+10 HMW glutenin subunit composition and 8 Glu-score value. Two electrophoretic profiles were found only by one cultivar. The parameters for cultivar Baulander Spelz are 1, 13+16, 2+12 HMW glutenin subunits and an 8 value for Glu-score (good breadmaking quality); that of cultivar Renval 1, 20, 2+12 and 6 (medium bread-making quality). By 4 cultivars (H 92-20, H 92-27, KR 489-11 - 15, Lueg and LW 12) several electrophoretical profiles groups with different HMW glutenin subunit composition were determined and for this reason the value for Glu-score could not be calculated. Variation of HMW glutenin subunits in 118 spanish landraces was studied by Rodriguez-Quijano et al., [3], who confirmed the non homogeneity of a few of them. They found a new allele, designated 2.3 and assigned to Glu-DI locus in three landraces too.

Beside that, they found some new combinations of subunits and the couple of subunits 13+16 with good bread-making quality appeared in their collection in a higher frequency than in comparative common wheat one.

Table 3. Composition of HMW glutenin subunits in winter spelt wheat cultivars.

Cultivar	Origin	HMW glutenin subunits	Glu-score
Altgold	CHE	1, 6+8, 2+12	6
Ardemme	DEU	1. 6+8, 5+10	8
Baetting Niederuill	POL	1, 6+8, 2+12	6
Bauländer Spelz	DEU	1, 13+16, 2+12	8
Burgdorf 1	DEU	1, 6+8, 2+12	6
Fuggers Babenhauser	GER	1, 6+8 2+12	6
H 92-20 (P. Kunz)	CHE	more biotypes	--[b]
H 92-27 (p. Kunz)	CHE	more biotypes	--[b]
Hercule	BEL	6+8, 2+12	4
Kipperhaus weiBer Spelz	DEU	1, 6+8, 2+12	6
KR 489-11-15 (P. Kunz)	CHE	more biotypes	--[b]
Lueng	CHE	more biotypes	--[b]
LW 12 (Nuertingen)	DEU	more biotypes	--[b]
LW 13 (Nuertingen)	DEU	1, 6+8, 2+12	6
Oberkulmer Rotkorn	CHE	1, 6+8, 2+12	6
Oberkulmer Schwarzer	CHE	1, 6+8, 2+12	6
Ostro	CHE	1, 6+8, 2+12	6
Ostro Schwarzer	CHE	1, 6+8, 2+12	6
Renval	USA	1, 20, 2+12	6
Roter Kolbendinkel	DEU	1, 6+8, 2+12	6
Rotthweiler Fruehkorn	GER	1, 6+8, 2+12	6
Rouquin	BEL	1, 6+8, 5+10	8
Schwabenkorn	DEU	1, 6+8, 2+12	6
Spalda (Landrance 1-96)	DEU	1, 6+8, 2+12	6
Steiner Roter Laupheimer	AUT	6+8, 2+12	4
T. spelta (Uhrineves)	--[a]	1, 6+8, 2+12	6
TRI 2400 (Gatersleben)	GER	1, 6+8, 2+12	6
Waggershauser Kolbendinkel	GER	1, 6+8, 2+12	6
Weißer Kolbenspelz	DEU	1, 6+8, 2+12	6
Weißer winter Grandendinkel	POL	1, 6+8, 2+12	6

a - not prefaced, b - not calculated

4. Conclusion

The results of the wheat grain analysis confirmed the usefullness of Glu - score application for the prediction of bread - making quality of wheat flour. Correlations between bread - making quality and specific HMW subunits of glutenin can be useful to wheat breeders, using SDS - PAGE as a screening test for bread - making quality of wheat.

References

[1]   Payne, P. I., and Lawrence, G. J., Catalogue of allels for the complex gene loci, Glu-A1, Glu-B1 and Glu-D1 which code for high- molecular- weight subunits of glutenin in hexaploid wheat. *Cereal Res. Comm.*, 11, 1983, pp. 29-35.
[2]   Payne,P.I., Nightingale, M.A., Krattiger, A.F., Holt, L.M., The relationship between the HMW glutenin subunit composition and the bread-making quality of British-grown wheat varieties. *J. Sci. Food. Agric.*, 40, 1987, pp. 51-65.

[3]   Rodriguez-Quijano, M., Vázquez, J. F., and Carrillo, J. M., Variation of high molecular weight glutenin subunits in Spanish landraces of Triticum aestivum ssp. Vulgare and ssp. Spelta.*J. Genet. Breed. 44*, 2, 1990, pp. 121-126.

[4]   Wrigley, C. W., Identification of cereal varieties by gel electrophoresis of the grain proteins. In: Linkskens, H.F., Jackson, J.F.: *Seed analysis*. Springer - Verlag, Berlin, Heilderberg, 1992, pp. 17-41.

Crop Improvement
by
Transgenic Technology

G. V. Horváth[1,2], A. Oberschall[1,2], M. Deák[1,2], L. Sass[1], I. Vass[1], B. Barna[3], Z. Király[3], É. Hideg[1], A. Fehér[1,2] and D. Dudits[1,2,4]

[1]*Institute of Plant Biology, Biological Research Center, Hungarian Academy of Sciences, Szeged, Hungary;* [2]*Biotechnology Institute of Bay Zoltán Foundation for Applied Research, Szeged, Hungary;* [3]*Institute of Plant Protection of the Hungarian Academy of Sciences, Budapest, Hungary,* [4]Corresponding author.

ABSTRACT: Rapid accumulation of reactive oxygen species (ROS) and their toxic reaction products with lipids and proteins significantly contributes to the damage of crop plants under biotic and abiotic stresses. We have identified several stress activated alfalfa genes, including the gene of the alfalfa ferritin and a novel NADPH-dependent aldose/aldehyde reductase enzyme. Transgenic tobacco plants that synthesize alfalfa ferritin in vegetative tissues-either in its processed form in the chloroplast or in the cytoplasmic nonprocessed form retained photosynthetic function upon free radical toxicity generated by paraquat treatment and exhibited tolerance to necrotic damage caused by viral and fungal infections. We propose that ferritin protects plant cells from oxidative damage by sequestering intracellular iron involved in generation of the reactive hydroxyl radicals through a Fenton reaction. Our preliminary results with the other stress-inducable alfalfa gene (a NADPH-dependent aldo-keto reductase) indicate, that the encoded enzyme may play a role in the stress response of the plant cells. These studies reveal new pathways in plants that can contribute to the increased stress resistance with a potential use in crop improvement.

1. Introduction

Under abiotic and biotic stress conditions rapid accumulation of reactive oxygen species (ROS) and their toxic reaction products contributes significantly to the damage of crop species [1-4]. Production of the most harmful ROS, the hydroxyl radical (OH*), depends on the presence of free iron in the living cells [5]. Hydrogen peroxide (H_2O_2) can undergo the Fenton reaction, which gives rise to hydroxyl radicals in the presence of free Fe^{2+}. The iron-mediated Fenton oxidants can cause damage to all classes of biologically important macromolecules [6[. Because intracellular iron catalyses oxidative reactions, the control of the concentration of free iron could be a potential way to reduce oxidative damage. Most of the non-metabolic iron is sequestered intracellularly in ferritin. This iron-storage protein is widely distributed in living organisms from bacteria to mammals [7]. The plant ferritins, which show significant amino acid sequence homology and structural similarity to their mammalian counterparts are localised preferentially in the chloroplast, their synthesis is transcriptionally controlled in response to iron and influenced by the plant stress hormone, abscisic acid [8]. Under normal growth conditions, the amount of ferritin is low in vegetative organs; it accumulates in seeds during embryo maturation. Involvement of ferritin in the oxidative stress response is supported by experiments with mammalian cells that show stimulation of ferritin synthesis during oxidative damage [9-11].

The postulated role of the alfalfa ferritin in the cellular defence mechanisms during oxidative stress encouraged us to assess the protection afforded transgenic tobacco plants expressing the this gene after oxidative stress.

2. Results

2.1. *Overproduction of ferritin promotes paraquat and iron resistance in transgenic plants*

The deduced protein sequence of the full-length alfalfa ferritin gene (*MsFer*, EMBLNEW accession number: X97059) comprises 251 amino acids and shares sequence homology with ferritins of different origins; 39-49% identity with the human H and the horse L ferritin and 89% identity with pea ferritin [12]. The first 53 amino acids in the alfalfa ferritin correspond to a chloroplast transit peptide, in this region the pea and alfalfa proteins share only 47% of amino acid identity.

The full-length *MsFer* cDNA was cloned into an *Agrobacterium* transformation vector with either the viral CaMV 35S or Rubisco small subunit gene promoter and at least 10 independent transformants were identified and selfed. Northern blot analysis of kanamycin-resistant R1 progeny showed that transformants from both promoter combinations accumulated significant amounts of ferritin mRNA in their leaves and stems. Subcellular location of the alfalfa ferritin in transgenic tobacco plants was analysed using antibodies raised against the glutathione-S-transferase-ferritin fusion protein. These antibodies recognised the chloroplast localised processed form of ferritin lacking the leader sequence that exhibits 22.5 kDa molecular mass in transformants carrying the Rubisco vector construct. The FLAG-tagged alfalfa ferritin cDNA was expressed by the CaMV 35S promoter. Because of the effect of the charged FLAG motif on the chloroplast transport of the tagged ferritin, in these transformants the non-processed 30 kDa ferritin was detected by the antiferritin antibodies only in the total cellular extracts. These analyses demonstrate the ectopic synthesis of the mature ferritin in the chloroplasts and accumulation of a precursor form in the transformed tobacco plants.

Transgenic tobacco plants producing the alfalfa ferritin were tested for the resistance against oxidative damage by applying the herbicide paraquat (Pq). The electrons produced during the photosynthetic electron transport reduce the Pq, and form free radicals [13]. Cells containing high intracellular level of superoxide dismutase are more resistant to Pq, supporting the idea that the Pq is an inducer of oxidative stress [14]. Total leaves from control and transformed plants were exposed to 10 μM Pq. Functional damage was monitored by measuring continuously the light-induced chlorophyll fluorescence yield changes. Both kind of transformed lines exhibited considerable tolerance to the damage caused by paraquat treatment, while the control SR1 leaves completely lost their photosynthetic function.

The damage caused by the Pq-treatment was also reflected by the degree of bleaching of the leaf tissues. When we treated leaf discs with water and 10 or 20 μM Pq for 3 days, the total chlorophyll content decreased differently in the various clones. These measurements showed however that the loss of this pigment was significantly reduced in the transformants in comparison to the control plants. Similar results were obtained using extremely high dose of iron (500 μM Fe(III)-EDTA) in leaf disc experiments [15].

2.2. Reduced symptoms after infection with diffent necrotrophic pathogens in ferritin producing tobacco transformants.

ROS are produced by plants upon recognition of invading pathogens and have been suggested to be involved in signal transduction [16], limitation of pathogen ingress [17-19], and induction of plant tissue necrotization [4,20-22]. Thus, elevation of antioxidant capacity of plants should increase their tolerance to cell and tissue necrosis caused by pathogens. Given the key role of iron in generation of damaging ROS, we have tested whether the previously demonstrated antioxidant effect of ferritin can exert a protective function against plant pathogens that cause necrotic symptoms. Seven-week-old control (SR1) and transgenic tobacco plants grown in the greenhouse were inoculated with tobacco necrosis virus (TNV). The number of necrotic lesions was significantly reduced in transformants expressing the alfalfa ferritin cDNA under the control of both promoters regardless of the cellular localization of the ectopically synthesized protein. Based on the diameter of the necrotic leaf area all transformed plants exhibited increased tolerance to necrotization caused by *Alternaria alternata* and *Botrytis cinerea* [15].

Transgenic plants overproducing ferritin and grown in soil under greenhouse conditions did not show any visible alteration in morphology or growth rate. As iron is an essential component of photosynthetic pigments, it is important that neither the fluorescence studies nor the chlorophyll content measurements indicated any detrimental effect on photosynthetic function.

3. Discussion

Our results demonstrate that in vegetative tissues ectopic synthesis of the iron-binding and storage protein, ferritin, increases the tolerance of tobacco plants toward both abiotic and biotic agents that are expected to cause damage by the production of ROS, particularly OH* through the Fenton reaction: $Fe^{2+} + H_2O_2 \rightarrow Fe^{3+} + OH^- + OH^*$. Because ROS are produced by the herbicide Pq, excess iron and as a result of plant-pathogen interactions, we challanged with paraquat the tobacco transformants synthesizing ferritin, iron and necrotrophic pathogens. Transformed plants show Pq and iron resistance and reduced necrosis after viral or fungal infections. The protective function of ferritin against oxidative stress may result from an increased iron sequestering capacity in cells of the transformants. Although further work is needed to identify the molecular mechanisms responsible for the observed suppression of tissue necrotization, our results are in accordance with those of Zer et al. [23], who demonstrated that desferroxamine, the highly specific iron-chelator can reduce Pq toxicity in pea. Recent results of Van Wuytswinkel et al., showing that both cytoplasm and plastid targeted ferritin can decrease the paraquat dependent reactive aldehyde formation in transgenic tobacco also supported our results [24].

Ferritin-overproducing transformants also provided a new way to test the potential role of iron-activated ROS in pathogenesis. ROS production shows a characteristic profile during plant-pathogen interaction [25]. Since the oxidative burst is expected to be a key component during necrotization, we challenged the transformants with necrotrophic pathogens and demonstrated that transformed plants were more resistant than the wild type. Further experiments are in progress to clarify, whether these phenomena are due to reduced formation of ROS, enhanced oxidative capacity in the transgenic plants, or both.

Based on the detection of alfalfa ferritin in total cellular and in purified chloroplast protein extracts of transgenic tobacco leaves, we can conclude, that the alfalfa ferritin was properly processed in transformants carrying the Rubisco vector constructs, and after elimination of the leader sequence, this protein is accumulated in the chloroplast. In contrast, the transformants carrying the CaMV 35S vector construct exhibited 30 kDa ferritin molecules in the cytoplasm. During the construction of this expression vector the FLAG tag was introduced into the 5'-end of the leader sequence. Considering the size of the synthesized protein, we have to postulate that the addition of the charged amino acids of the FLAG epitope prohibited the correct processing of ferritin. Nevertheless, we could not recognize essential differences between the two classes of tobacco transformants in responses to paraquat, iron excess or pathogens. Therefore, we propose that both forms of this iron-binding protein can reduce cellular damage.

Plant development and the efficiency of crop production are highly dependent on iron-containing photosynthetic proteins. It is important, therefore, that ferritin overproduction did not change the photosynthetic activity or chlorophyll content in the transformed plants.

Recently, we isolated the cDNA of a NADPH-dependent aldo-keto reductase gene from alfalfa. The involvement of this gene in a wide range of stress responses was demonstrated by elevated accumulation of the mRNA in cultured alfalfa cells exposed to osmotic shock caused by polyethylene glycol (PEG, 15%) solution, heavy metal toxicity (250 μM $CdCl_2$) and the stress hormone abscisic acid (ABA, 75 μM), or a directly oxidative stress inducing agent (H_2O_2, 10 mM). In mammalian cells the different members of the aldo-keto reductase superfamily can catalyse the production of sorbitol from D-glucose, so they can serve as osmoregulators [26], and they play a role in detoxification processes [27]. Up to now plant aldose reductase homologue genes were cloned only from monocot species, in cultured bromegrass cells elevated level of the gene expression is associated with the induction of freezing tolerance [28], the expression of wild oat aldose reductase homologue can be significant in maintaining seed dormancy or longevity [29]. The best characterized barley aldose reductase gene is expressed during the desiccation phase of embryogenesis and in mature embryos [30]. Recent results of Guillén et al. [31] suggest the role of the members of this enzyme family in specific detoxification processes in plants. Further experiments are in progress to evaluate the role of the alfalfa aldo-keto reductase enzyme in stress response using transgenic tobacco plants.

The improved stress tolerance observed in alfalfa ferritin producing transgenic tobacco plants provides a basis for further application of this approach in crop improvement.

Acknowledgement

This work was supported by the Hungarian National Research Fund (OTKA, T 030262), Hungary.

References

[1]   A.H. Prince et al., Plant under drought-stress generate activated oxygen. *Free Rad Res Commun* 8: (1989) 61-66.

[2]   C.H. Foyer et al., Protection against oxygen radicals: an important defence mechanism studied in transgenic plants. *Plant Cell Environ* 17: (1994) 507-523.

[3]   K.E. Hammond-Kosack and J.D.G. Jones, Resistance gene-dependent plant defense responses. *Plant Cell* 8: (1996) 1773-1791.

[4]   C. Lamb and R.A. Dixon, The oxidative burst in plant disease resistance. *Annu Rev Plant Physiol Plant Mol Biol* 48: (1997) 251-275.

[5]   B. Halliwell and J.M.C. Gutteridge, Oxygen free radicals and iron in relation to biology and medicine: some problems and concepts. *Arch Biochem Biophys* 246: (1986) 501-514.

[6]   E.S. Henle and S. Linn Formation, prevention, and repair of DNA damage by iron/hydrogene peroxide. *J Biol Chem* 272: (1997) 19095-19098.

[7]   E.C. Theil, Ferritin: structure, gene regulation, and cellular function in animals, plants, and microorganisms. *Ann Rev Biochem* 56: (1987) 289-315.

[8]   S. Lobréaux et al., Abscisic acid is involved in the iron-induced synthesis of maize ferritin. *EMBO J* 12: (1993).651-657.

[9]   Gy. Balla et al., Ferritin: A cytoprotective antioxidant strategy of endothelium. *J Biol Chem* 267: (1992) 18148-18153.

[10] G.F. Vile and R.M. Tyrrell, Oxidative stress resulting from ultraviolet A irradiation of human skin fibroblasts leads to a heme oxygenase-dependent increase in ferritin. *J Biol Chem.* 268: (1993) 14678-14681.

[11] G. Cairo et al., Induction of ferritin synthesis by oxidative stress. Transcriptional and post-translational regulation by expansion of the free iron pool. *J Biol Chem* 270: (1995) 700-703.

[12] S. Lobréaux et al., Amino acid sequence and predicted three-dimensional structure of pea seed (*Pisum sativum)* ferritin. *Biochem J* 288: (1992) 931-939.

[13] C.E. Babbs et al., Lethal hydroxyl radical production in paraquat treated plants. *Plant Physiol* 90: (1989) 1267-1270.

[14] A. Perl et al., Enhanced oxidative stress defense in transgenic potato expressing tomato Cu, Zn superoxide dismutases. *Theor Appl Genet* 85: (1993) 568-576.

[15]   M. Deák et al., Plants ectopically expressing the iron-binding protein, ferritin, are tolerant to oxidative damage and pathogens. *Nature Biotech* 17: (1999) 192-196.

[16]   P. Low and J. Merida, The oxidative burst in plant defence: Function and signal transduction. *Physiol Plant* 96: (1996) 533-542.

[17]   M. Peng and J. Kuc: Peroxidase-generated hydrogen peroxide as a source of antifungal activity *in vitro* and on tobacco leaf disks. *Phytopathol.* 82: (1992) 696-699.

[18]   M.F. Ouf et al., The effect of superoxide anion on germination and infectivity of wheat stem rust (*Puccinia graminis* Pers. f. sp. *tritici* Eriks. and Henn.) uredospores. *Cereals Res Commun* 21: (1993) 31-37.

[19]   Z. Király et al., Effect of oxygen free radicals on plant pathogenic bacteria and fungi and on some plant deseases. In: Oxygen Free Radicals and Scavengers in the Natural Sciences (Gy. Mózsik, I. Emerit, J. Fehér, B. Matkovics, Á. Vincze, eds.) Akadémiai Kiadó, Budapest pp 9-19 (1993).

[20]   E.F. Elstner: Oxygen activation and oxygen toxicity. *Ann Rev Plant Physiol* 33: (1982) 73-96.

[21]   M.W. Sutherland, The generation of oxygen free radical during host plant response to infection. *Physiol Molec Plant Pathol* 39: (1991) 79-93.

[22]   E.F. Elstner and W. Osswald, Mechanism of oxygen activation during plant stress. In: Oxygen and Environmental Stress in Plants. (R.M.M. Crawford, G.A.F. Hendry and B.A. Goodman eds), The Royal Society of Edinburgh, 102B, Edinburgh, pp. 131-154 (1994).

[23]   H. Zer et al., The protective effect of desferrioxamine on paraquat-treated pea (*Pisum sativum*).*Physiol Plantarum* 92: (1994) 437-442.

[24]   O. Van Wuytswinkel et al., Iron homeostasis alteration in transgenic tobacco over expressing ferritin. *Plant J* 17: (1999) 93-97.

[25]   C.J. Baker and E.W. Orlandi Active oxygen in plant pathogenesis. *Ann Rev Phytopathol* 33: (1995) 299-321.

[26]   S.M. Bagnasco et al., Induction of aldose reductase and sorbitol in renal inner medullary cells by elevated extracellular NaCl. *Proc Natl Acad Sci USA* 84: (1987) 1718-1720.

[27]   D.L. Vander Jagt et al., Substrate specificity of human aldose reductase. Identification of 4-hydroxynonenal as an endogenous substrate. *Biochem Biophys Acta* 1249: (1995) 117-126.

[28]   S.P. Lee and T.H.H. Chen, Expression of an aldose reductase-related gene during the induction of freezing-tolerance in bromegrass cell suspension cultures. *J Plant Physiol* 142: (1993) 749-753.

[29]   B. Li and M.E. Foley, Cloning and characterization of differentially expressed genes in imbibed dormant and after ripened *Avena fatua* embryos. *Plant Mol Biol* 29: (1995) 823-831.

[30]   D. Bartels et al., An ABA and GA modulated gene expressed in the barley embryo encodes an aldose reductase related protein. *EMBO J* 10: (1991) 1037-1043.

[31]   P. Guillén et al., A novel NADPH-dependent aldehyde reductase from *Vigna radiata* confers resistance to the grapevine fungal toxin eutypine. *Plant J* 16: (1998) 335-343.

*Use of Agriculturally Important
Genes in Biotechnology*
G. Hrazdina (Ed.)
IOS Press, 2000

Transformation of Carnation Agrobacterium-mediated Transformation of Carnation with Antisense 1-aminocyclopropane-1-carboxylate Synthase (ACS) Gene

E. Kiss[1], A. Veres[1], Á. Varga[1], Z. Galli[1], N. Nagy[1], L. Heszky[1], E. Tóth[2], G. Hrazdina[3]

1 Department of Genetics and Plant Breeding, Gödöllő University of Agricultural Sciences, Hungary, Gödöllő, Páter K. u. 1. H-2103
2 Óbuda Horticultural Laboratory, Budapest, Hungary
3 Department of Food Science and Technology, Cornell University, Geneva Campus, Geneva, NY, 14456 USA

Abstract: ACS (1-aminocyclopropane-carboxylate synthase) is a key enzyme of ethylene biosynthesis. Ethylene influences different physiological and biochemical processes in plant from germination 'till maturation, including floral senescence and fruit ripening. Down-regulation of ethylene production via antisense transformation proved to be an effective tool in delaying fruit maturation. ACS - a member of a multigene family of relatively high DNA sequence homology - has already been isolated and sequenced from many different plant species e.g. tomato, apple, rice, avocado, winter squash, zucchini, wheat, carnation. In our carnation transformation experiments an apple-derived cDNA clone (1.1 kb MdACS-2 fragment) was inserted via subcloning steps into pBI121 binary vector (Clontech) in antisense orientation. This plasmid was introduced into two different *Agrobacterium* strains (LBA 4404 and EHA 105) and carnation varieties (Chabaud Pink, Dianthus chinensis, Improved White Sim, 'Bíbor'/Óbuda) were transformed using leaf and petal explants. Regenerating shoots were selected on kanamycin containing MS medium in two steps. Surviving plants were analysed at molecular level: (i) isolated genomic DNA was analysed by PCR using the 35S promoter, neomycin-phosphotransferase and transgene-specific primers, and (ii) hybridised with digoxigenin labelled probes. Transgenic regenerants were potted in glasshouse in order to observe phenotypic effect of the heterologous antisense ACS transformation on the intact plants e.g. flowering and flower senescence.

1. Introduction

Carnation is one of the leading varieties in the global flower trade, therefore genetic improvement of carnation in respect of several ethylene dependant traits such as flowering time, petal senescence characteristics or vase life can have beneficial economic effect in addition to scientific significance. ACS (1-aminocyclopropane- carboxylate synthase) is the key enzyme of ethylene biosynthesis. Ethylene - this "small hormone with many functions" [15] - influences different physiological and biochemical processes in plant from germination till maturation, including floral senescence and fruit ripening. Down-regulation of ethylene production *via* antisense transformation proved to be an effective tool in delaying fruit ripening in tomato [10]. ACS gene - a member of a multigene family of relatively high DNA sequence homology - has already been isolated and sequenced from many different plant species [14] e.g. tomato, apple, rice, avocado, winter squash, zucchini, wheat, carnation [6,11]. In our transformation experiments with carnation an apple-derived cDNA clone was used in antisense orientation [8,13] to down-regulate the ethylene biosynthesis [2].

2. Materials and Methods

Carnations tested in regeneration experiments are as follows: 'Nice' white carnation (*Dianthus caryophyllus*, Vetömag, Hungary), Chabaud pink, (*Dianthus caryophyllus* Denmark*), Dianthus chinensis* (Simon, Hungary), Chabaud mixed colours (*Dianthus caryophyllus*, Royal Sluis), Marie Chabaud yellow (*Dianthus caryophyllus*, Denmark), 'Bíbor'/Purple (*Dianthus caryophyllus,* Óbuda Horticultural Laboratory Hungary) Improved White Sim (*Dianthus caryophyllus,* Óbuda Horticultural Laboratory Hungary). For shoot regeneration MS [9] basal medium supplemented with 1 mg/l benzyladenin (BA) and 0.2 mg /l naphtalene acetic acid (NAA) was used both in case of leaf and petal explants. Leaves were harvested from in vitro plantlets, germinated on hormone free MS medium after desinfection of the seeds (1 min in 70 % ethanol, 20 min shaking in 1:3 diluted commercial sodium hypochlorite solution, 3x rinse in autoclaved distilled water). Petals were collected from buds and desinfected in the same way as the seeds after using a first washing step with steril water containing 10 % TWEEN 20. The scheme of transformation from gene isolation *via* plasmid construction, *Agrobacterium* transformation to regenerated plant can be seen in Figure 1. The 1.1 kb fragment of ACS cDNA was isolated from McIntosh apple cultivar and inserted *via* subcloning steps into pBI 121 binary vector (Clontech) in reverse orientation [8]. This plasmid was introduced into two different *Agrobacterium tumefaciens* strains, LBA 4404 and EHA 105. Only the well-regenerating *Dianthus chinensis*, Chabaud Pink, *Improved White Sim, Bíbor* were transformed using leaf [1,7] and petal explants [1, 16]. Regenerating shoots were selected on kanamycin containing MS medium in two steps. Green shoots developing on medium supplemented with 50 mg/l kanamycin were transferred to medium of higher kanamycin concentration (75-100-125-150 mg/l). Surviving plants were analysed at molecular level: (i) isolated genomic DNA [4] was subjected to PCR analysis using 35S CaMV specific promoter and neomycin-phosphotransferase (*npt II*) specific primers, and (ii) hybridised with digoxigenin (DIG) labelled probes according to the manufaturer's protocol [3]. The probe was the Pst I fragment of the pBI 121 plasmid (Clontech). Transgenic regenerants were potted in glasshouse in order to make observations of phenotypic effect of the heterologous antisense ACS transformation on the intact plants e.g. flowering, flower senescence.

3. Results and Discussion

Leaves of responsive cultivars harvested from three week old in vitro seedlings put on shoot regeneration medium [1] produced the first shoots after two weeks on the basal part of the leaves. Petal explants regenerated shoots also within three weeks.

For transformation four varieties were applied: *Dianthus chinensis, Bíbor, Improved White Sim* and *Chabaud pink* with two *Agrobacterium tumefaciens* strains EHA 105, LBA 4404. Table 1. shows the results of shoot regeneration from transformed leaf explants comparing Van Altvorst et al.'[1] and Hoersch et al.' S [7] infection method (Figure 2).

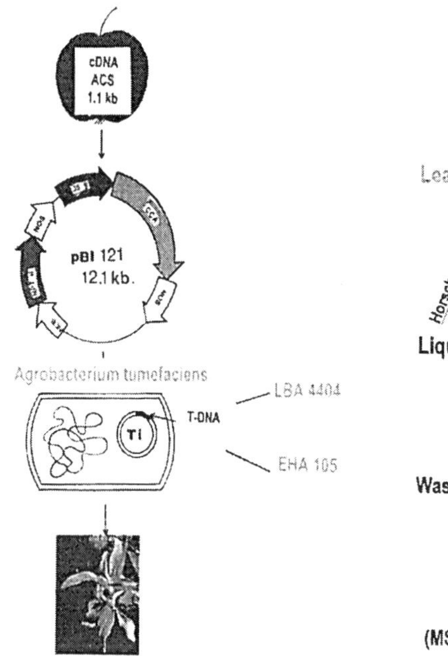

Figure 1. Scheme of transformation from gene isolation *via* plasmid construction, *Agrobacterium* transformation to regenerated transgenic plant.

Figure 2. Steps of *Agrobacterium* transformation of leaf and petal explants. Petals and leaves were infected according to van Altvorst et al.' method (1995) [1], in case of leaves Horsch et al. (1988) [7] protocol also was tried.

Table 1. Comparison of shoot regeneration percentage from leaf explants on MS medium supplemented with 1 mg/l BA+0.2 mg/l NAA+250 mg/l cefotaxime+50 mg/l kanamycine after transformation with Van Altvorst et al.' [1] and Hoersch et al.' [7] procedure.

Variety/cultivar	Van Altvorst et al. 1995 %	Hoersh et al. 1988
Chabaud pink	17	7
Improved white Sim	25	7
Dianthus chinensis	40	8
Bíbor (Purple)	48	13

Shoot regeneration from *Agrobacterium* infected leaves according to Van Altvorst et al. [1] protocol proved to be more efficient in case of each variety, therefore further transformations are being carried out according to this method.

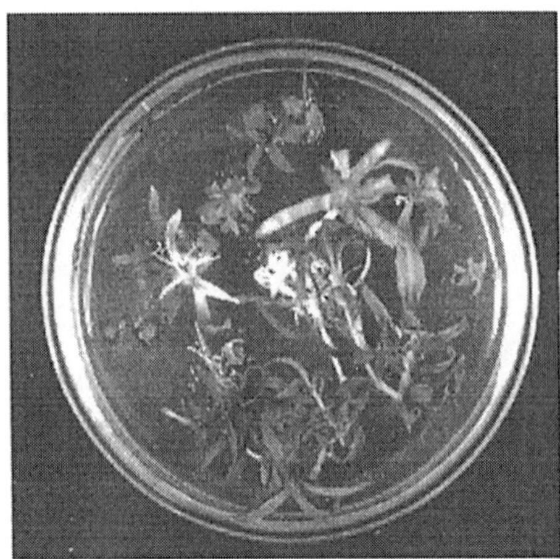

Figure 3. Green and white shoots of *Dianthus chinensis* developing on 50 mg/ml kanamycine containing regeneration medium.

In the two-step selection of transformants the green shoots developed on 50 mg/l kanamycin were transferred to 75-100-125-150 mg/l kanamycin containing regeneration media resulting in greening shoots even at 150 mg/l kanamycin concentration (Figure 3).

PCR and Southern hybridisation analysis of these green shoots served further proofs of transformation. The PCR amplification results of Dianthus chinensis with npt II specific primers (5'-CTGAATGAACTGCAGGACGAG-3' and 5'-GCCAACGCTATGTCCTGATAG C-3' determined a 503 bp fragment size, which can be seen on Figure 4. both in case of transformation carried out with EHA 105 and LBA 4404 *Agrobacterium* strains.

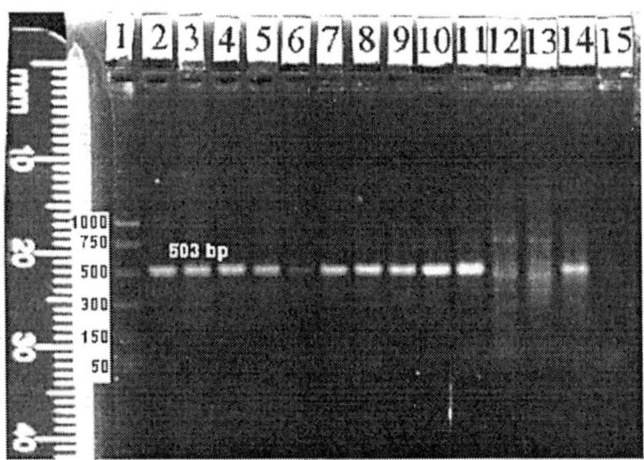

Figure 4. Results of PCR with *npt II* specific primers. The expected band of 503 bp can clearly be observed in lanes 2-5, 7-11 and 14, they are faint in lanes 6, 12-13 corrresponding the putative transformants of *Dianthus chinensis,* but it is missing from lane 15, representing the control non-transformed plant DNA. Lane 1: DNA molecular weight marker, Promega PCR marker .

The integration of npt II gene was confirmed with blotting of PCR amplification products and hybridised with DIG-labeled PstI fragment of pBI 121 (Figure 5).

503 bp → 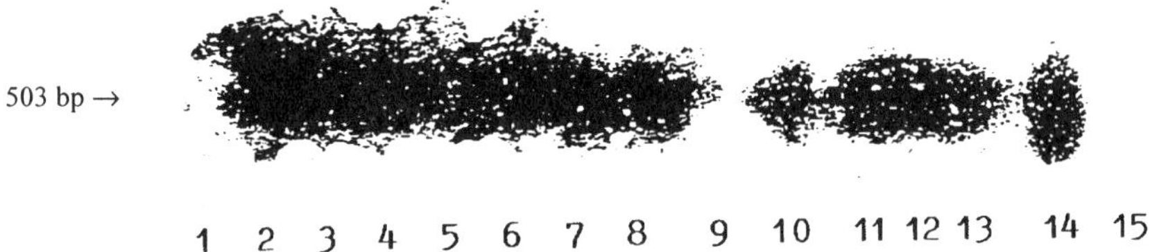

Figure 5. Southern hybridisation with digoxigenin labeled nptII probe. 1: DNA molecular weight marker, 2: putative transgenic Dianthus chinensis DNA samples, 15: non-transformed control plant DNA.

After the molecular analysis the plants were analysed for ethylene production, and despite of high individual differences both in the control and the transformant group, the average ethylene production of the antisense ACS transformant *Bíbor* plants were about by 40 % lower compared to the control plants (data not shown). *In vitro* shoot regeneration capacity of leaf explants of transformants were checked also, since there are data in the literature [5,12], that decreased ethylene concentration enhances shoot regeneration in plant tissue cultures. According to the first results the shoot regeneration of *Bíbor* plants increased by 10 % even on hormone free MS medium (Figure 6).

Figure 6. Comparison of shoot regeneration from transformed (antisense) and non-transformed (control) leaf

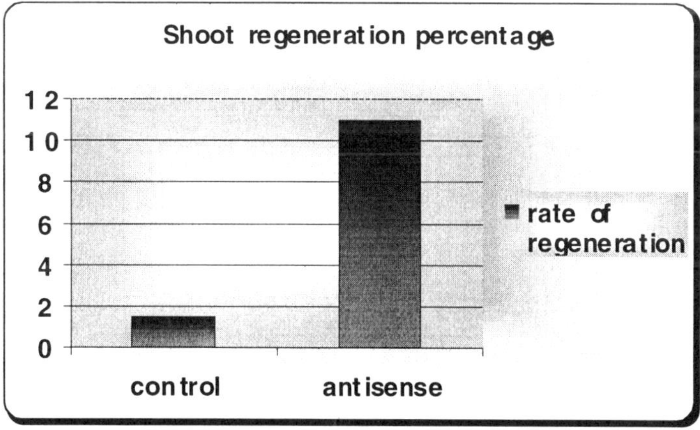

explants of Bíbor carnations on hormone free MS medium.

Figure 7. Growing transgenic *Dianthus chinensis* plants in pots before flowering.

Transformant and non-transformant carnations were potted (Figure 7) in a glasshouse in Óbuda Horticultural Laboratory and are being observed to determine what phenotypic changes can be the consequences of integration of antisense ACS cDNA.

Acknowledgement

The research is supported by National Science Foundation (OTKA TO 26502) and by the Ministry of Education (FKFP 0494).

References

[1]   A.C. van Altvorst *et al.*, Transgenic carnations obtained by Agrobacterium tumefaciens-mediated transformation of leaf explants. *Transgenic Research* **4** (1995) 105-113.

[2]   R. van Blokland *et al.*, Modulation of gene expression in plant by antisense genes. In: Antisense Research and Applications.ed: S.T.Crooke, B.Lebleu. CRC Press. Boca Raton, Ann Arbor, London, Tokyo, 1993, 126-143.

[3]   Boehringer-Mannheim: Digoxigenin Labeling Protocol.

[4]   S.L. Dellaporta *et al.*, A plant DNA minipreparation version II. *Plant Molecular Biology Reporter* **1** (1983) 19-21.

[5]   Pau Eng-Chong and JEE Lee, Ethylene regulating shoot regenerability in vitro. *Rice Biotechnology Quarterly* **21** (1994) 22-23.

[6]   A. Have and E.J. Woltering, Ethylene biosynthetic genes are differentially expressed during carnation (Dianthus caryophyllus) flower senescence. *Plant Molecular Biology* **34** (1997) 89-97.

[7]   R.B. Horsch *et al.*, Leaf disc transformastion. Plant Molecular Biology Manual A5: 1-9. Kluwer Academic Publishers, Dordrecht, 1988.

[8]   E. Kiss *et al.*, Down-regulation of ethylene biosynthesis in apples: cloning and sequencing of partial ACC-synthase gene in McIntosh. Proceedings of a Symposium on. Use of Induced Mutation and Molecular Techniques for Crop Improvement. Vienna, 1995, 562-564.

[9]   T. Murashige and F. Skoog , A revised medium for rapid growth and bioassays with tobacco tissue cultures. *Physiologia Plantarum* **15** (1962) 473-497.

[10] P.W. Oeller *et al.*, Reversible inhibition of tomato fruit senescence by antisense RNA. *Science* **254** (1991) 437-439.

[11] K.Y. Park *et al.*, Molecular cloning of an 1-aminocyclopropane-1-carboxylate synthase from senescing carnation flower petals. Plant Molecular, *Biology* **181** (1992) 377-386.

[12] L. Purnhauser *et al.*, Stimulation of shoot regeneration in *Triticum aestivum* and *Nicotiana plumbaginifolia* Viv tissue cultures using the ethylene inhibitor $AgNO_3$. *Plant Cell Reports* **6** (1987) 1-4.

[13] C.L. Rosenfield *et al.* Md-ACS-2 and Md-ACS-3: Two new 1-aminocyclopropane-1-carboxylate-synthases in ripening apple fruit. Plant Gene Register PGR96-122. *Plant Physiology* **112** (1996) 1735

[14] A. Theologis, *et al.*, Use of a tomato mutant constructed with reverse genetics to study fruit ripening, a complex developmental process. *Developmental Genetics* **14** (1993) 282-295.

[15] D. Van der Straeten *et al.*, Ethylene: a small hormone with many functions. Faculty of Agricultural and Applied *Biological Sciences*, Gent 27-29. September 1995.

[16] M. Nakano *et al.*, Adventious shoot regeneration from cultured petal explants of carnation. *Plant Cell, Tissue and Organ Culture* **36** (1994) 15-19

Use of Agriculturally Important
Genes in Biotechnology
G. Hrazdina (Ed.)
IOS Press, 2000

Field Experiments with Transgenic Maize in Hungary in 1993-94

S. Mórocz[1], S. Omirulleh[2], G. Donn[3], B. Szarka[1], M. Ladányi[1], L. Albrecht, D. Dudits[2]
[1]Cereal Research NP Co., Szeged, Hungary
[2]Biological Research Centre of the Hungarian Academy of Sciences, Szeged, Hungary
[3]Hoechst Schering AgrEvo Gmbh, Laboratory of Plant Cell Biology, Frankfurt a. M.,
Germany
[4]KSZE, Szekszárd, Hungary

Abstract. To test the laboratory and greenhouse results as well as to promote the introduction of the transgenic crops into our agriculture a field experiment was performed with transgenic maize offspring. The maize offspring were obtained by out-crossing plants heterozygotic for PATGUS MTE1 (*phosphinohtricin acetyltransferase β-glucuronidase* maize transformation event no.1) under strict regulations such as time and space isolation from other maize growing area or destruction of the field material. A one hectare area was planted with mix-variety recessive sugary endosperm (su/su) maize population, that contained a 2 x 2 m square in the center where the normal endosperm (Su/Su) genetically transformed and non-transformed plants were grown in a 1:1 ratio. Before flowering, the center square was sprayed with the glufosinate-containing herbicide Finale®, commercially available, to kill the non-transformed plants. At an appropriate ripening stage, when the kernels with the dominant normal endosperm (Su/su su) could be distinguished, ear samples were collected from four square meter areas in eight major directions of the compass. In each direction the sample squares were positioned at a distance of 2, 4, 8, 16 and 32 meter from the center square, respectively. The average pollen spread from the center square expressed in the percentage of out-crossed (normal endosperm) kernels in the recipient sugary population was 0.14 and 0.23 in 1993 and 1994, respectively. There was a decreasing tendency in the spreading with increasing distance from the centre, and in some directions presumably owing to the prevailing winds there were higher values. Inconsistent variations, however, were also observed, probably due to the activity of insects, birds, wind-storms etc. The low spreading percentage is, however, sufficient for the transfer of the gene to the neighbouring maize fields.

1. Introduction

The availability of the HE89-based maize protoplast culture system [1,2] supported the stable genetic transformation of maize *via* gene delivery into single cells and the production of transgenic maize populations [3,10] as well as the first field tests in Hungary [11, 12]. The aim of the field experiments were: (1) testing the expression of the transgene under field conditions, (2) studying the possible risks for the environment and (3) initiating the evaluation of glufosinate-ammonium resistance as an environmentally friendly weed control alternative offered by modern biotechnology in the maize growing areas.

2. Materials and methods

A control protoplast derived plant and a totipotent transgenic clone (PATGUS MTE1) transformed with a chimerical plasmid construction containing the *pMGP1, chimeric promoter -GUS fusion and PAT gene* [6] were cross-pollinated with the same hybrid plant. Following testing of the difference between the control and segregating heterozygotic PATGUS MTE1 T_2 offspring, the latter material was used for transgenic field trials.

A one hectare area was planted with a mixed-variety recessive sugary endosperm (su/su) maize population that contained a 2 x 2 m square in the center, where the normal endosperm (Su/Su) genetically transformed and non-transformed plants were grown in a 1:1 ratio. Before flowering the middle square was sprayed with Finale®, the commercially available herbicide containing glufosinate to kill the non-transformed plants. At an appropriate ripening stage when the kernels with dominant normal endosperm (Su/su su) could be distinguished ear samples were collected from four square meter areas in eight major directions of the compass. In each direction the sample squares were positioned at a distance of 2, 4, 8, 16 and 32 meters from the central square, respectively. The average pollen spread from the center square was expressed in the percentage of out-crossed (normal endosperm) kernels in the recipient sugary population.

The experiments were carried out in two successive years (1993-94) under the farming conditions of Corn Production System at Szekszárd, Trans-Danubia, Hungary.

As these experiments preceded the Hungarian law on biotechnological activities, efforts were made to avoid any escape of the transgene. The location of the experiment was at least at an isolation distance (1500m) from other maize growing areas, and additional natural barriers (forest, canal, hill) also existed. In addition to spatial isolation, the experimental plot was planted 2-3 weeks after the usual planting date. The field material was destroyed at the end of the sample collection.

3. Results

The phosphinotricin acetyltransferase (PAT) gene in PATGUS MTE1 T_2 progeny was expressed under the tissue culture conditions when the semi active substance (DL-glufosinate-ammonium, abbr. DL-pptNH$_4$) was added at 1mM concentration to N6M solidified medium [1], and could be distinguished from the non-resistant controls (Fig.1).

Approximately half of the plants that survived showed a mendelian segregation pattern after spraying the middle square with Finale, the commercial formulation of DL-pptNH$_4$ at the recommended dosage. The anticipated number of maize plants (with high probability the non-transformed segregants) and the weeds (mainly flowering *Convolvulus* sp.) completely withered in ten days in the treated central plot, as did the plants grown in a sprayed area both from the recipient wild-type maize population and the non-cultivated border area overgrown with grass mainly (Fig.2).

Fig.1. Germination test of 5 seeds each of the control non-transformed (from left 1st and 2nd culture vessels) and 1:1 segregating heterozygotic PATGUS transformed (3rd and 4th) maize in the absence (1st and 3rd) and the presence (2nd and 4th) of 1mM DL-pptNH$_4$ as selective agent.

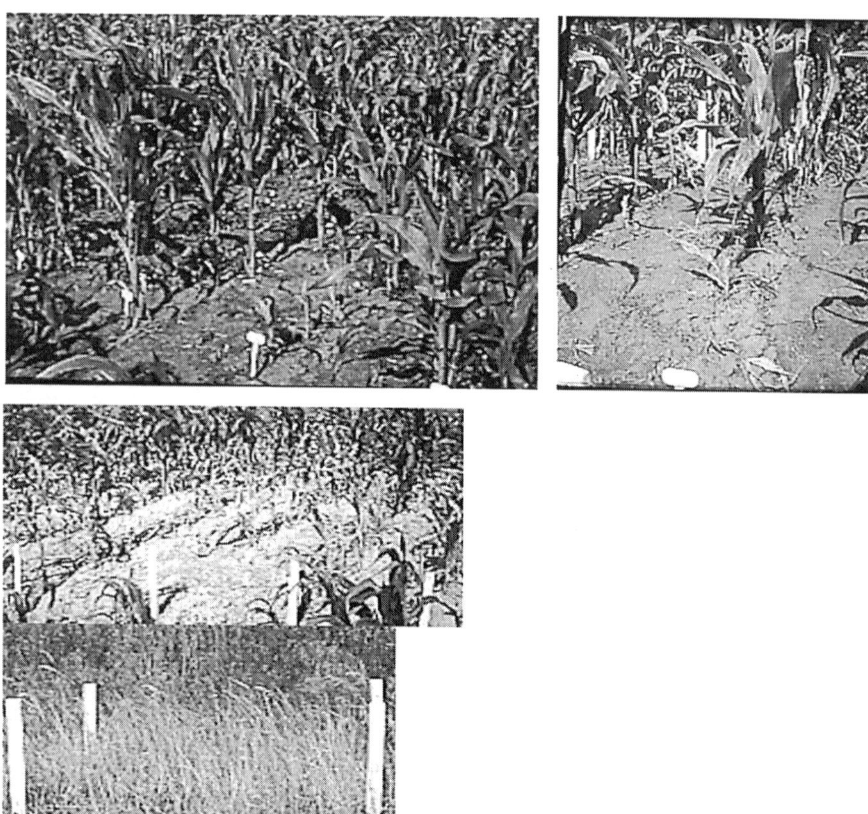

Figure 2. The effect of spray treatment with DL-pptNH$_4$-containing herbicide (Finale) on maize as well as on monocot and dicot weedy species. Upper pictures were taken from the central square before (left) and after (right) spraying, the lower ones were the recipient maize population and the non-cultivated outside plot that were treated with the herbicide.

The average pollen spread from the central square expressed in the percentage of out-crossed (normal endosperm) kernels in the recipient sugary population was 0.14 and 0.23 in 1993 and 1994, respectively (Table. 1, 2). There was a decreasing tendency in the spreading percentage with increasing distance from the center. In some directions, presumably owing to the prevailing winds, higher values were observed. On the other hand, there were variations deviating from the main trends as well, probably due to the activity of insects, birds, wind-storms etc.

Table 1. Rate (%) of normal (su/susu) endosperm kernels in the sample squares in eight major direction of the compass at increasing distances from the middle square in 1993.

Direction	2m	4m	8m	16m	32m	Average
			from the middle square			
N	0,2	0,0	0,1	0,1	0,2	**0,12**
NW	0,2	0,0	0,1	0,0	0,1	**0,08**
W	0,6	0,0	0,1	0,0	0,0	**0,14**
SW	0,2	0,0	0,0	0,0	0,2	**0,08**
S	0.1	0,0	0,1	0,2	0,0	**0,08**
SE	0,5	0,1	0,0	0,0	0,0	**0,12**
E	0,9	0,1	0,9	0,3	0,0	**0,44**
NE	0,1	0,1	0,0	0,16	0,1	**0,13**
Average	**0,35**	**0,04**	**0,16**	**0,10**	**0,08**	*0,14*

Table 2. Rate (%) of normal (su/susu) endosperm kernels in the sample squares in different directions and at increasing distances in 1994.

Direction	2m	4m	8m	16m	32m	Average
			from the middle square			
N	0,30	0,29	0,35	0,03	0,11	**0,22**
NW		0,29	0,11	0,20	0,22	**0,21**
W	0,33	0,20	0,03	0,04	0,04	**0,13**
SW		0,10	0,15	0,09	0,01	**0,09**
S	0,4	0,00	1,88	0,02	0,00	**0,58**
SE		0,75	0,00	0,06	0,17	**0,25**
E	1.33	0,16	0,04	0,13	0,13	**0,36**
NE		0,06	0,12	0,03	0,02	**0,06**
Average	**0,59**	**0,26**	**0,34**	**0,09**	**0,09**	*0,23*

4. Conclusions

The PAT gene was properly expressed under field conditions. The heterozygotic transgenic plants produced normal ears, while the non-transformed (sensitive to the herbicide) plants stopped growth and withered after spraying. Both monocot and dicot weeds present at the time of spraying were eliminated by the herbicide application.

Despite the pollen shedding by the recipient plants, the pollen derived from the central four square meter area could penetrate to a small degree to the edge of the experimental field. Therefore, it is foreseeable that the pollen from the transgenic fields will spread to the neighbouring maize fields.

Acknowledgements

This work was supported by a grant from the Committee for the National Technical Development, 1052 Budapest, Szervita tér 8., and this paper is based on the final report made in 1994 in Hungarian. The authors wish to thank for the skilful assistance to Mariann Ábrahám,

References

[1]   S. Mórocz *et al.*, An improved system to obtain fertile regenerants via maize protoplasts isolated from a highly embryogenic suspension culture. *Theor. Appl. Genet.* **80** (1990) 721-726.

[2]   EP/23.06.90/ EP 90111945: Improved *Zea mays* (L.) genotypes with capability of long term, highly efficient plant regeneration.

[3]   G. Donn *et al.*, Stable transformation of maize with a chimaeric, modified Phosphinithricin-acetyltransferase gene from *Streptomyces viridochromogenes*.   Abstracts VII. International Congress on Plant Tissue and Cell Culture, Amsterdam, June 24-29, 1990. p.53 .

[4]   EP/23.06.90/ EP 90111946: Fertile transgene Maispflanzen mit artfremden Gen sowie Verfahren zu ihrer Herstellung.

[5]   S. Omirulleh *et al.*, Improved maize transformation system based on morphogenic protoplasts. Abstracts of 8th Int. Protoplast Symp. June 20-26 1991 Uppsala, Sweden, *Physiologia Plantarum* **82** (1991) (1):A31.

[6]   S. Omirulleh *et al.*, Activity of a chimeric promoter with the doubled CaMV 35S enhancer element in protoplast-derived cells and transgenic plants in maize. *Plant Mol. Biol.* **21** (1993) 415-428

[7]   M.V. Golovkin et.al., Production of transgenic maize plants by direct DNA uptake into embryogenic protoplasts. *Plant Science*, **90** (1993) 41-52.

[8]   M. Belgin *et al.*, Production of transgenic maize plants by direct DNA uptake into protoplasts. Proceedings. Molecular-Genetic analysis of plant metabolism and development, 1993.

[9]   D. Dudits *et al.*, Transgenic maize plants from protoplasts: new products for plant breeding. *Hungarian Agricultural Research* **2** (1993) 4-8.

[10]   S. Omirulleh *et al.*, Regeneration of transgenic maize plants from embryogenic protoplasts after polyethylene glycol-mediated DNA uptake In:Potrykus,I., Spangenberg, G. (Eds.) Gene Transfer to Plants. Springer, 1995, pp:99-105.

[11]   B. Szarka *et al.*, Field experiments with maize genotypes carrying foreign genes. Scientific Days of Plant Breeding „93" Budapest, .January 11-12. 1994 (Hu).

[12]   S. Mórocz S., and D. Dudits, Progress report on „ Experimental risk assessment on the production and environmental effects of the transgenic plants carrying foreign genes" grant project of the National Committee for Technical Development (Hu), 1994.

Ten Generations of Transgenic *Triticale*

J. Zimny, S. Sowa, I. Menke-Milczarek, A. Czplicki, S. Oleszczuk
*Plant Breeding and Acclimatization Institute - Radzików, PL-00950 Warszawa P.O.Box
1019, Poland*

Abstract: Transgenic *Triticale* plants have been regenerated *in vitro* after particle
bombardment of scutellar tissue of hexaploid *Triticale*. Here we report the analysis of
their next generations. All transgenic plants were fertile. They were grown to maturity
and set seeds. Enzymatic and molecular progeny analyses provided evidence for
inheritance of the introduced genes to the next generations. Transgene expression was
found often to be unstable and declined in some lines. In contrary, other lines showed
stable expression in subsequent generations. Enzyme activity was tested histochemically
and for progeny by Basta treatment. The inheritance of the *uid*A gene in different plants
ranged from 2.0:1.0 to 15.6:1.0. The inheritance of the *bar* gene in different plants,
measured by herbicide application test was not always the expected 3:1. Southern blot
analysis on several lines showed loss of the *bar* gene in some plants and gene silencing
of *uid*A in other cases. The position of T-DNA insertion appears to be random within
the nuclear genome. In our study four independently transformed lines were analyzed
and T-DNA insertion sites were visualised directly by *in situ* hybridisation and mapped
to the four different chromosomes. We have used androgenesis to obtain homozygous
non-segregating transgenic lines. From 5000 anthers 295 green plants were regenerated.
Ninety-nine of them were fertile. They have been selfed and multiplied over 8
generations. A total of 1500 TDH plants tested showed transgene expression in 10'th
generation.

1. Introduction

The idea of creating *Triticale* as a crossing between wheat and rye is already 125 years
old. The actual breeding of *Triticale* started later on in 1948 by O'Mara and it was twenty
years later when CIMMYT, Mexico produced the first acceptable *Triticale* cultivar, which
was called "Armadillo". This cultivar showed very good characteristics like high fertility,
good yield, early maturity, improved test weight and a gene for dwarfness. Over the years,
other traits like nutritional quality, disease resistance and pre-harvest sprouting were also
improved and today *Triticale* varieties are cultivated in over one million hectares in Russia,
Poland, USA, Canada, West Germany, Argentina, Mexico, France, Portugal and Spain. The
most widely grown *Triticale* in the world nowadays is the polish cultivar "Lasko". It is
adapted to and has excellent yield potential not only in all the areas where wheat is grown, but
also in marginal production environments, such as in acid soils, in high elevations in the
tropics, under semi-arid conditions, and in sandy soils [1]. Bread making quality of *Triticale*
could be improved by introduction of the *Glu-D1d* gene encoding high molecular weight
glutenin. This gene is present in most high quality wheats. Recently a segment of
chromosome 1D with *Glu-D1d* was transferred to 1R and 1A of *Triticale* through
translocation [2,3] but it would be necessary to be able to introduce by transformation
specifically this gene to any desired line or cultivar.

2. Regeneration Systems

The development of regeneration systems in cereals was an area of research strongly pursued at the beginning of the 80's. The reason for that was that the possibility of regenerating whole plants from cells cultured *in vitro* was seen as a way to generate genetic variability, also called somaclonal variation, to be use for practical breeding. Also, regeneration was seen as a crucial prerequisite for the generation of transformed plants.

In *Triticale*, plants have been regenerated *via* organogenesis or *via* somatic embryogenesis and that has been achieved from explants such as young inflorescences, leaf bases, microspores, immature scutella, and cell suspension cultures. From all these explants, immature scutella proved to be the best starting material for optimal *in vitro* regeneration of *Triticale*.

For the development of an immature scutella regeneration system parameters like the culture medium composition, developmental stage of the explant and genotype were analysed [4] and as an outcome of this work a general procedure for regeneration from immature scutella was developed and also 3000 regenerated plants were produced. These plants were self-pollinated and part of their progeny was used for analysing the occurrence of somaclonal variation. At that time there was very little known on the mechanism by which these somaclonal variants were created and on how useful this variation was for practical purposes. As a general thought, the occurrence of somaclonal variation was linked with disorganised growth of calluses or suspension cells by which morphological, cytological and molecular alterations were induced. In the case of *Triticale*, among the 3000 regenerants, we found aluminium tolerant plants as well as deoxynivalenol tolerant plants (author's data, not published). Also, 26 somaclonal variants were screened for *Septoria* susceptibility and they appeared significantly more resistant to *Septoria nodorum* blotch than the controls but more susceptible to *Septoria tritici* blotch. Other laboratories found somaclonal variants for plant height, time of flowering, morphology of whole plants and spike, leaf waxiness, hairy neck and fertility.

3. Transformation

Immature scutella were used for transformation with the construct pDB1 [5], it contains the *uid*A gene under the control of the rice actin-1D promoter and the *bar* gene under the control of the cauliflower mosaic virus 35S promoter.

For bombardment, plasmid DNA was precipitated onto gold particles of an average size between 0.4 and 1.2 μm (Heraeus/Germany). The precipitation mixture contained 2 mg gold particles, 5 μl of plasmid DNA (μg/μl), 20 μl spermidine-free base ($0.1M$) and 50 μl $CaCl_2$ ($2.5M$). The particles were prepared according to Becker et al. [5]. For each shot 3.5 μl of the suspension was spread onto the surface of the macrocarrier. In the experiments a PDS 1000/He particle gun (BioRad) was used. The delivery conditions were as follows: 2.5 cm distance between rupture disc and macrocarrier, 5.5 cm distance between stopping screen and target cells, 1300 or 1500 psi helium pressure and 27 inch Hg partial vacuum. [6]

4. Analysis

From 465 bombarded scutella about 4000 plantlets were regenerated. Three hundred plants survived the selection. These regenerants were screened for enzyme activity by the histological GUS assay, and by spraying the plants with a herbicide (Basta) solution.

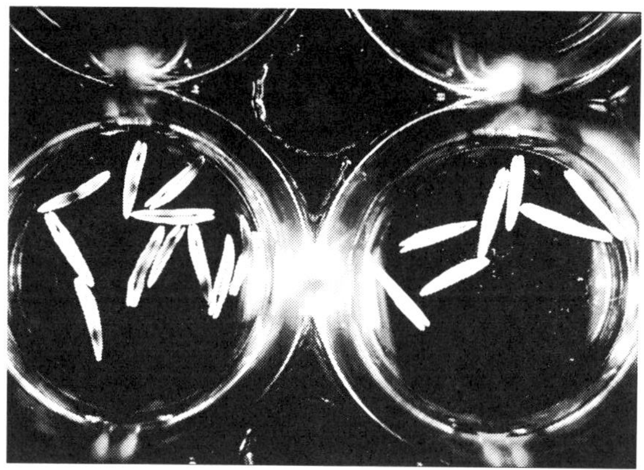

Figure 1: Anthers of GUS positive (left) and GUS negative (right) *Triticale* plants after x-gluck treatment.

Twenty five regenerants showed GUS-activity and survived repeated Basta spraying (Figure 1). Segregation of *uid*A gene ranged in different plants from 1.0:2.0 to 1.8:1.0 (GUS$^+$:GUS$^-$), and was specific for individual plants (Figure 2). Southern blot analysis showed the presence of both marker genes introduced into the genome of analysed plants.

All transgenic plants were fertile. They were grown to maturity and set seeds. Pollen and progeny analyses provided evidence for inheritance of the introduced genes to the next generation.

The application of transformation systems to various purposes is frequently hampered by instability of transgenes and their expression. It is important to better understand the factors effecting the stability and expression of inserted genes. One significant factor seems to be the position of transgene in the genome, known as the "position effect". This position effect can be studied by *in situ* hybridisation which show the physical position of the transgene within the genome. The routine isotopic *in situ* hybridisation protocols were established in many laboratories during the 1970s. The original ISH technique using isotopic probes is sensitive and very useful for detecting single-copy DNA sequences. Non isotopic *in situ* hybridization was introduced in plants in 1985. This technique can be used to identify chromosomes, detect chromosomal abnormalities or to determine the specific sequences on chromosomes. DNA or RNA sequences are first labelled with reporter molecules. The probe and the target chromosomes are denatured. Complementary sequences in the probe and target are allowed to reanneal. After washing and incubation in labelled reagents the signal is visible at the site of probe hybridisation.

Figure 2: Instability of the *uid*A gene under the control of consitutive Act-1 promoter in different trangsgenic plants. In one plant no expression in the leaves (right).

For signal detection the enzymatic reaction or fluorochromes can be used (Fluorescent In Situ Hybridisation -FISH). The integrated transgenes were detected in all four *Triticale* lines by this technique. Some lines showed single integration sites. However, two lines had several integration sites on the same chromosome arm. Rye chromosomes could easily be identified by the stronger staining of the telomeric heterochromatin with propidium iodide and DAPI.

The segregation of introduced genes in the T_2 generation was highly deviant. For some lines the expression can disappear and for others remain stable. To obtain non-segregating transgenic lines we have used androgenesis. From 5000 anthers 295 green plants were regenerated, of which 99 of them were fertile Androgenic, doublehaploid transgenic lines. These have been selfed and multiplicated over three generations. One thousand six hundred TDH_3 (T_3 generation of transgenic double haploid) plants showed transgene expression. It means that transgenic homozygous lines did not show any segregation of trait as determined by the introduced *uid*A gene, and that the method of androgenesis can be used to stabilise the transgenic lines. These lines were multiplicated over the next generations and they were found to be stabile in ten generations.

5. Summary

All transgenic plants were fertile. They were grown to maturity and set seeds. Enzymatic and molecular progeny analyses provided evidence for the inheritance of the introduced genes to the next generations. Transgene expression was found often to be unstable and declined in some lines. Other lines showed stable expression of the transgene in subsequent generations.

The position of T-DNA insertion appears to be random within the nuclear genome. In our study four independently transformed lines were analyzed and T-DNA insertion sites were visualised directly by *in situ* hybridisation and mapped to four different chromosomes. We have used androgenesis to obtain homozygous non-segregating transgenic lines. From 5000 anthers 295 green plants were regenerated. Ninety-nine of them were fertile. They have been selfed and multiplied over 8 generations. A total of 1600 TDH plants tested showed transgene expression in 10'th generation.

Acknowledgement

The presented work has been done in cooperation with Prof. H.Lörz and Dr. Dirk Becker from the Institute of General Botany, AMP II, University of Hamburg, Germany, and Dr. Carsten Pedersen from Riso National Laboratory, Roskilde, Denmark.

References

[1]   R.L. Villareal *et al.*, Advances in Spring *Triticale* Breeding. In: Wheat Program, CIMMYT, El Batan, Mexico, 1990, 43-87.
[2]   A.J. Lukaszewski and Ch. Curtis, Transfer of the *Glu-D1d* gene from chromosome 1D of bread wheat to chromosome 1R in hexaploid *Triticale*. *Plant Breeding* **109** (1992) 203-210.
[3]   A.J. Lukaszewski and Ch. Curtis, Transfer of the *Glu-D1* gene from chromosome 1D to chromosome 1A in hexaploid *Triticale*. *Plant Breeding* **112** (1994) 177-182.
[4]   J. Zimny and H. Lörz, High frequency of somatic embryogenesis and plant regeneration of rye (*Secale cereale* L.). *Z. Pflanzenzüchtng* **102** (1989) 89-100
[5]   J. Zimny *et.al.*, Fertile transgenic *Triticale (x Triticosecale* Wittmack). *Mol. Breeding* **1** (1995) 155-164
[6]   D. Becker *et al.*, Fertile transgenic wheat from microprojectile bombardment of scutellar tissue. *Plant Journal* **5** (1994) 299-307.

Potato Transformations

Z. Bánfalvi, A. Molnár, A. Lovas, R. Dóczi, L. Lakatos, G. Hutvágner
Agricultural Biotechnology Center, H2101 Gödöllö, P. 0. Box 411, Hungary

Abstract. Potato *(Solanum tuberosum) is* the most important non-cereal world food crop, although it is prone to over a hundred diseases caused by viruses, bacteria, fungi, mycoplasms, insects and nematodes. Thus, a considereable effort has been made to improve its disease resistance by selective breeding techniques and, more recently, by genetic engineering methods. In our experiments, we used leaves and slices of *in vitro*-grown microtubers for *Agrobacterium mediated* gene transfer. As targets of transformation we chose two cultivars: Désirée and Keszthelyi 855, a new Hungarian potato cultivar. We have modified a microtuber disc transformation method in a way that is potentially suitable now for production of transgenic potato plants without introduction of an antibiotic resistance gene used as a selective marker in other gene transfer technologies.

1. Introduction

Transformation systems for potato have been developed which allow for reasonably efficient genetic modification. Published systems rely on the recovery of transgenic plants from leaf protoplasts [1], callus cultures [2], leaf discs [3,4,5,6,7,8,9, 10,11], shoot cultures [8,12,13,14,15] or tuber discs [15,16,17,18]. In the application of these technologies for obtaining transgenic plants, a selectable marker, an antibiotic resistance gene, is cointroduced with the gene of interest. In the case of potato, kanamycin resistance is the most widely used selection marker that is conferred by transgenic expression of neomycin phosphotransferase, the product of the *nptII* gene from the bacterial transposon Tn5. Although, several reports including [19] analyzed the desirability of the antibiotic resistance genes in transgenic plants and concluded that legislative clearance of this transgenic trait is acceptable, there is still a general public concern with respect to the relative safety of the large scale application of antibiotic resistant transgenic plants.

Here we report leaf- and microtuber disc transformation experiments on two potato cultivars: Désirée, that we chose because of its general use and economic importance, and Keszthelyi 855, a new Hungarian potato cultivar. Regenerated plants obtained from these experiments were analyzed for the presence and copy number of the transgene by Southern-, and for the expression of the transgene by Northern hybridizations. It was shown that the modified microtuber disc transformation method is potentially suitable for production of transgenic potato plants without introduction of an antibiotic resistance marker.

2. Materials and Methods

2.1. Potato transformation and regeneration

The modified binary vectors were transferred into *A. tumefaciens* strain C58C I containing pGV2260 [20] or pGV/MP90 (originated from the MPI für Züchtungsforschung, Cologne, Germany) as described previously [21,22].

Leaf transformation was performed according to Dietze et al. [11]. Leaves of 4 weeks old *in vitro* plants grown in MS medium were cut from the stem. Cuts of 1-2 mm were made across the midrib in two places and the base of the leaves were cut off. The leaves were co-cultivated with the appropriate *Agrobacterium* strain for 2 days in MS liquid medium. Callus induction was achieved in CIM-, shoot induction in SIM medium containing 50 mg/l kanamycin. Regenerated shoots were transferred to the root induction medium, RIM.

For microtuber transformation tuber slices from *in vitro* cultures of S. *tuberosum* were co-cultivated with *A. tumefaciens* on RM medium [23] solidified with 0.5% of agar. After 2 days co-cultivation the tuber slices were washed with 500 mg/l cefotaxime and placed to regeneration medium [patent application No. P 98 00280]. Transgenic plants were selected after regeneration by rooting in RM medium containing 100 mg/l kanamycin.

2.2 DNA, RNA isolation and analysis

Genomic DNA from leaves was prepared as described by [24]. Southern hybridization was performed in 1M NaCl, 1% SDS, 10% dextran sulfate, 50 mM TrisHCl pH 7.5. After overnight incubation at 65°C, the filter was washed in distilled water for 2 min at room temperature and in 2 X SSC, 1% SDS for 15 min at 65°C.

Total RNA was extracted according to [25] and separated on formaldehyde-agarose gels according to [26]. Blotting and hybridization were performed as described by Amasino [27].

3. Results and Discussion

3.1. Leaf disc transformation

Two and seven independent leaf disc transformation experiments were performed with the potato cvs. Keszthelyi 855 and Désirée, respectively (Table 1). In sum, 320 shoots emerged from 280 leaf explants 5-6 weeks after co-cultivation with *A.tumefaciens.* These shoots were cut off and transferred onto root induction medium after 7-8 weeks when they were 1-1.5 cm long. The *Agrobacterium* strains used contained 10 different constructs with different potato genes and promoters in sense- or antisense orientation or fused to a reporter gene. However, all the constructs were the derivative of the vector pGA482 (Pharmacia), pCP60 (kindly provided by P. Ratet), or pBinl9 [28], all of them carrying the *nptII* gene as marker gene for transformation. Based on this, callus induction and shoot regeneration were performed in the presence of kanamycin in the medium.

From 280 kanamycin resistant regenerants derived from 5 independent experiments 120 were analyzed by Southern or Northern hybridizations. Beside the *nptII* gene, majority of the lines carried the "target" gene as well, only 23 were not transgenic in this respect. Fig. 1 shows the result of a typical hybridization experiment when 13 lines were tested: 5 possessed the insertion in single-copy, 6 in multiple copies, 2 lines were not transgenic. Rearrangements occurred in 4 out 79 transgenic lines tested. Based on Northern hybridization, however, the level of transgene expression varied in the individual lines even within one experiment (Fig. 2), probably due to position effect.

Table 1. Efficiency of potato leaf transformation

Cultivar	Const.	Number of explants	Number of KmR reg.
Keszthelyi 855	1.	20	13
	2.	20	5
Désirée	3.	30	40
	4.	30	37
	5.	30	48
	6.	30	40
	7.	50	83
	8.	10	16
	9.	30	15
	10.	30	23

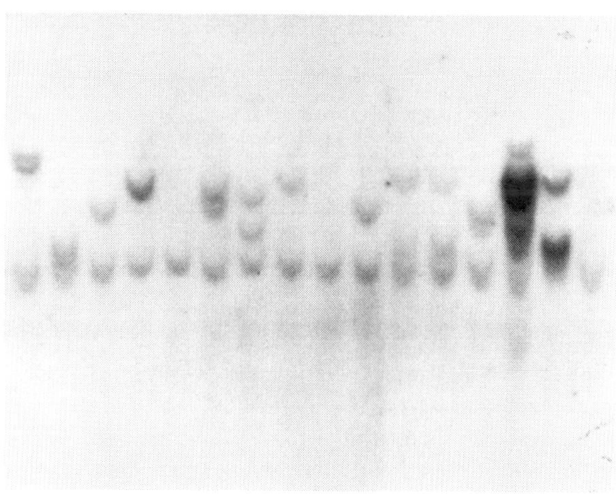

Fig.1 Southern hybridization for transgene detection. L, independent lines derived from leaf disc transformation; T, independent lines derived from microtuber disc transformation; K, non-transformed control. Lines 5 and 9 are not transgenic for the tested gene.

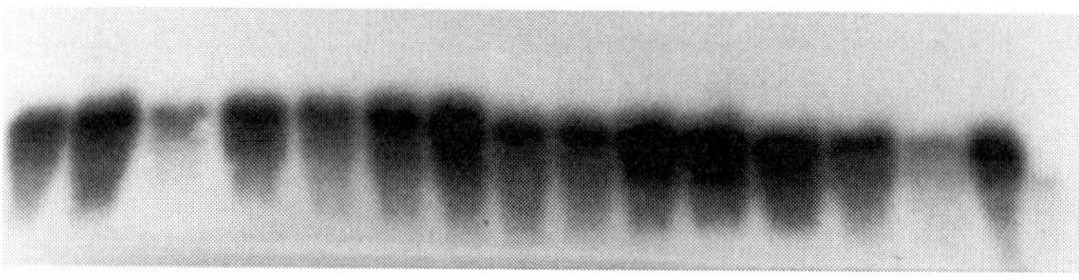

Fig. 2 Northern hybridization for detection of transgene transcript level. 1-15, independent transgenic lines; K, non-transformed control.

3.2. Mictuber disc transformation

Ten independent microtuber disc transformation experiments were performed with the potato cv. Keszthelyi 855, and one with the cv. Désirée (Table 2). Nine different pGA482 constructs were introduced, including the vector itself. Approximately 580 regenerants were obtained on 240 tuber slices without antibiotic selection. Regenerated shoots 1-1.5 cm in size were transferred onto rooting medium supplemented with kanamycin. Out of 575 regenerants 86 proved to be kanamycin resistant. Efficiency of transformation varied between 10 to 30%. No significant difference was detected between the cvs. Keszthelyi 855 and Désirée. Several kanamycin resistant transgenic lines were tested by Southern- (Fig. 1., lanes T1 and 2), or Northern hybridizations and data similar to those characteristic for leaf disc transformation were obtained.

Table 2. Efficiency of potato microtuber transformation

Cultivar	Const.	Number of tuber slices	Number of regenerants	Number of KmR reg.	% of KmR reg.
Keszthelyi 855	11.	20	64	9	14
	12.	20	42	5	12
	13.	20	43	8	17
		30	69	22	32
	14.	15	47	5	11
	15.	40	156	15	10
	16.	20	22	5	23
	17.	15	34	5	15
	18.	20	25	3	12
	19.	20	34	5	15
Désirée		20	39	4	10

The modified microtuber disc transformation method has two advantages over previously published protocols. Since transgenic plants can be recognized by standard molecular techniques including PCR or by phenotypical changes depending on the gene introduced, this method is potentially suitable for production of transgenic potato plants without introduction of an antibiotic resistance gene used as a selective marker in other gene transfer technologies. Since there is a general public concern with respect to the relative safety of the transgenic plants and especially to the large scale application of the antibiotic resistant plants, we hope that our transformation protocol can facilitate the public acceptance of transgenic plants and molecular breeding. The other advantage is in the possibility of isolation of multi-transformed potato lines that can be useful in basic-, as well as in applied potato research.

References

[1]   Fehér *et al.*, PEG-mediated transformation of leaf protoplasts of *Solanum tuberosum L.* cultivars. *Plant Cell, Tissue and Organ Culture* **27** (1991) 105-114.

[2]   T. Komari, Transformation of callus cultures of nine plant species mediated by *Agrobacterium. Plant Science* **60** (1989) 223-229.

[3]   G. An *et al.*, Transformation of tobacco, potato, and *Arabidopsis thaliana* using a binary Ti vector system. *Plant Physiology* **81** (1986) 301-305.

[4]   E.A. Shahin and R.B. Simpson, Gene transfer system for potato. *HortScience* **21** (1986) 1199-1201.

[5]   R. Tavazza *et al.*, Genetic transformation of potato (*Solanum tuberosum*): an efficient method to obtain transgenic plants. *Plant Science* **59** (1988) 175-181.

[6]   M. De Block, Genotype-independent leaf disc transformation of potato (*Solanum tuberosum*) using *Agrobacterium tumefaciens. Theoretical and Applied Genetics* **76** (1988) 767-774.

[7]   R.G.F. Visser *et al.*, Efficient transformation of potato *(Solanum tuberosum)* using a binary vector in *Agrobacterium rhizogenes. Theoretical and Applied Genetics* **78** (1989) 594-600.

[8]   R.G.F. Visser, *et al.*, Transformation of homozygous diploid potato with an *Agrobacterium tumefaciens* binary vector system by adventitious shoot regeneration on leaf and stem segments. *Plant Molecular Biology* **12** (1989) 329-337.

[9]   H. Wenzler *et al.*, A rapid and efficient transformation method for the production of large numbers of transgenic potato plants. *Plant Science* **63** (1989) 79-85.

[10]   E.S. Higgins *et al.*, Early events in transformation of potato by *Agrobacterium tumqfaciens. Plant Science* **82** (1992) 109-118.

[11]   J. Dietze et.al., *Agrobacterium*-mediated transformation of potato (Solanum *tuberosum). In:* 1. Potrykus, G. Spangenberg (eds.), Gene Transfer to Plants. Berlin: SpringerVerlag, 1995, pp. 24-29.

[12]   G. Ooms et.al., Genetic modifications of potato development using Ri T-DNA. *Theoretical and Applied Genetics* **70** (1985) 440-446.

[13]   G. Ooms et.al., Genetic transformation in two potato cultivars with T-DNA from disarmed *Agrobacterium. Theoretical and Applied Genetics* **73** (1987) 744-750.

[14]   C.A. Newell et.al., *Agrobacterium*-mediated transformation of *Solanum tuberosum L.* cv. 'Russet Burbank'. Plant *Cell Reports* **10** (1991) 3 0-34.

[15]   J.P. Spychalla and M.W. Bevan, Agrobacterium-mediated transformation of potato stem and tuber tissue, regeneration and PCR screening for transformation. In: Plant Tissue Culture Manual. Kluwer Academic Publishers, pp. B 11: 1- 18, 1993.

[16]   S. Sheerman and M.W. Bevan, A rapid transformation method for *Solanum tuberosum* using binary *Agrobacterium tume/aciens* vectors. *Plant Cell Reports* **7** (1988) 13-16.

[17]   W.J. Stiekema *et al.,* Introduction of foreign genes into potato cultivers Bintje and Désirée using an *Agrobacterium tumefaciens* binary vector. *Plant Cell Reports* **7** (1988) 47-50.

[18]   B.K. Ishida *et al.*, The use of in vitro-grown microtuber discs in *Agrobacterium*-mediated transformation of Russet Burbank and Lemhi Russet potatoes. *Plant Cell Reports,* **8** (1989) 325-328.

[19]   J.-P. Nap *et al.*, Biosafety of kanamycin-resistant transgenic plants. *Transgenic Research* **1** (1992) 239-249.

[20]   R. Deblaere *et al.*, Efficient octopine Ti plasmid-derived vectors of *Agrobacterium* mediated gene transfer to plants. *Nucleic Acid Research* **13** (1985) 4777-4788.

[21]   R. Hugen and L. Willmitzer, Storage of competent cells for *Agrobacterium* transformation. *Nucleic Acids Research* **16** (1988) 9877.

[22]   G.S. Ditta *et al.*, Broad host-range DNA cloning system for gram-negative bacteria: construction of a gene bank of *Rhizobium meliloti. Proceedings of the National Academy of Sciences USA* **77** (1980) 7347-7351.

[23]   T. Murashige and F. Skoog, A revised medium for rapid growth and bioassays with tobacco tissue cultures. *Physiology Plantarum* **15** (1962) 473-479.

[24]   M. Shure *et al.*, Molecular identification and isolation of the Waxy locus of maize. *Cell* **35** (1983) 225-233.

[25]   W.J. Stiekema *et al.,* Molecular cloning and analysis of four potato tuber mRNAs. *Plant Molecular Biology* **11** (1988) 255-269.

[26]   J. Logemann *et al.*, Improved method for the isolation of RNA frorn plant tissues. *Analitical Biochemistry* **163** (1987) 16-20.

[27]   R.M. Amasino, Acceleration of nucleic acid hybridization rate by polyethylene glycol. *Analitical Biochemistry* **152** (1986) 305-307.

[28]   M. Bevan, Binary *Agrobacterium* vectors for plant transformation. *Nucleic Acids Research* **12** (1984) 8711-8721.

*Use of Agriculturally Important
Genes in Biotechnology
G. Hrazdina (Ed.)
IOS Press, 2000*

Transformation of Foreign Gene and Sexual Transfer of Transgene in Wheat

J. Pauk[1], R. Mihály[1,2], A. Mészáros[1,3], G. Schwarz[4] and R. Hansch[4]

[1]Cereal Research Non-profit Co., Szeged, Hungary
[2] Agricultural University, Gödöllö, Hungary
[3]Agricultural College of Debrecen Universitv, Hódmezóvásárhely, Hungary
[4]Technical University of Braunschweig, Germany

Abstract. The first genetically engineered food crops have been introduced into the market in some countries, but several questions have been under discussion. The fate of transgene is an up-to-date question in plant genetic and breeding research. This lecture gives a report on the sexual transfer, selection and use of the *bar* gene in non-transformed genotypes for the development of homozygosity in the offspring generations. For a breeder, the ability to transfer foreign gene and its Mendelian inheritance is a very important aspect. In crossing experiments the homogenous T_2 transgenic line 124 was crossed with different nontransgenic females. The isolated hybrid embryos from the crosses were tested in medium containing bialaphos. This experiment showed the dominant character of bar gene. All the hybrid embryos germinated onto 5mg/1 bialaphos medium expressed resistance, while the control variety (Góbé) showed sensitivity to the bialaphos as expected. The fertile T_1, generation expressed 3:1 segregation ratio for *bar⁺ / bar⁻* plants. The transformed foreign gene *(bar)* was stable in the wheat genome after two self pollinations and PAT assays. It is therefore, possible to transform transgene successfully via sexual way into non-transgenic genotypes. Cell-level selection of *bar* gene was possible in anther and microspore culture. The regenerated haploids and spontaneous diploids are herbicide resistant. Finally, by using *in vitro* haploid technique (anther and microspore culture), it's possible to produce homogenous *bar* resistant wheat lines by cell-level selection.

1. Introduction

Fertile transgenic wheat (*Triticum aestivum* L.) plants have been produced during the past few years [1,2,6,7]. At the end of 20" century the fundamental discoveries made in the field of molecular biology and cell and tissue culture have initiated a biological revolution in plant breeding and agricultural production. The first genetically engineered food crops have been introduced into the market in some countries [3], but several questions have been under discussion. The fate of transgene is an up-to-date question in plant genetic and breeding research.

Earlier, six independent fertile transgenic lines of spring wheat bearing selectable marker genes *bar* have been obtained. This paper gives a report on the sexual transfer, selection and use of the *bar* gene in non-transformed genotypes for the development of homozygosity in the offspring generations.

2. Material and Methods

2.1. Plant Material

Donor plants of CY-45 spring wheat (*Triticum aestivum* L) line that was previously selected for its high response to somatic tissue culturing was used. The wheat line was grown in a phytotron chamber and in greenhouse and immature embryos were excised, cultured, bombarded, bialaphos selected and regenerated as described earlier by Vasil et al. [6]. Crossing (emasculation, pollination) and embryo cultures were made according to protocols in cell and tissue culture handbooks. The microspore and anther cultures were made using our previous published methods [4,5].

2.2. Transgenic experiments

To optimize the transformation system, transient expression experiments were carried out using pAHC25 plasmid containing the UidA reporter gene, and the *bar* marker gene. In the stable transformation experiments, pAHC 20 plasmid containing the selectable bar gene for resistance to the herbicide (Finale 14 SL 1-50g/l ammonium glufosinate) was used. The selectable *bar* gene encodes the enzyme phosphinothricin acetyltransferase (PAT) that inactivates phosphinothricin (PPT), the active ingredient of the herbicide by acetylation. Plasmid DNAs were precipitated, absorbed to gold particles and delivered to target tissue using a DuPont PDS- I OOO/He device.

3. Results

3.1. Development of homogenous transgenic wheat lines

Six transgenic spring wheat lines (106-3a, 116, 117, 124, 128, 129) were obtained from transformation experiments of bombarded immature embryo-derived calli using pAHC20 plasmid molecule, followed by bialaphos selection. All of the R, regenerants were identified as transgenic plants using PAT assay, RNA-RNA hybridization and herbicide spray.

Five of the six R_0, lines produced complete fertile heads while one line (106-3a) was partially fertile. The offsprings of the complete fertile transgenic (T) lines (116, 117, 124, 128, 129) were used in the sexual gene transfer experiments. The T_1, generation of 4 transgenic lines (116, 124, 128, 129) showed a 3:1 segregation in PAT assay. Line 117 showed close to a 3:1 segregation ratio. The PAT positive lines were self-pollinated and the subsequent generation was checked by direct herbicide spray. In the T_2, generation, the four complete fertile lines (116, 124, 128, 129) expressed a homogenous herbicide resistance. The line 117 was homogenous. The T_2, homogenous herbicide resistant plants were used as male parent in the crossing experiments.

3.2. Sexual transfer of the bar transgene

For the breeder, the ability to transfer a foreign gene and its Mendelian inheritance is a very important aspect. In crossing experiments the homogenous T_2, transgenic line 124 was crossed with different non-transgenic females. The isolated hybrid embryos from the crosses were tested in a medium containing bialaphos (Table 1). The result of this experiment showed the dominant character of *bar* gene. All the hybrid embryos germinated on 5mg/l bialaphos medium expressed resistance, while the control variety (Góbé) showed sensitivity to the bialaphos as expected. In the next experiment different transgenic lines (male) were tested on a wild type (Gar. non tansgenic) variety (Table 2). The sexual transfer of the transgene in case of line 116, 124 and 128 was good, all the hybrid embryos expressed resistance on the bialaphos embryo culture. However, two-thirds of the hybrid embryos of transgenic line 117 were sensitive to the bialaphos, indicating that line 117 was not homogenous.

3.3. Selection of the transgene in anther and microspore cultures

To develop the cell-level selection system of the bar gene, anther and microspore culture experiments were made using transgenic lines and bialaphos selection. T_2 PAT resistant transgenic lines were used as donors in anther culture (Table 3.). Three treatments were applied. In the first treatment bialaphos was not used in the anther culture, in the second treatments the medium was supplemented with bialaphos added into liquid medium at the time of isolation, while in the third treatment, bialaphos supplement was added 2 weeks after isolation. In all the treatments there was no significant difference in embryoid production. Similar results were obtained in microspore cultures, but the 3 mg/l bialaphos gave the best selection result.

Table 1. Sexual transfer of transgene *(bar)* in wheat by non-transgenic x transgenic crossing. T Hybrid embryos were tested in bialaphos included embryo culture.

Genotype or combination	Tested embryos	Germinated embryos on 5mg/l bialaphos	Fertile plants
Control variety:			
Góbé	12	0	0
Crossing comb.:			
RK-1	9	8	8
RK-2	20	20	20
RK-3	5	5	5
RK-4	2	2	2
RK-5	12	11	11
Total:	48	46	46

Table 2. Sexual transfer of transgene *(bar)* in wheat by non-transgenic x trarisgenic crossing. Test of different transgenic lines (male) with Gar. non-transgenic female

Genotype or combination	Tested embryos	Germinated on 5mg/l bialaphos	Transplanted in soil	Bialaphos sensitive
Gar. Non-trans.	12	0	0	12
Gar. x 116	9	9	9	0
Gar. x 117	6	6	2	4
Gar. x 124	12	12	12	0
Gar. x 128	9	9	9	0
Total:	48	36	32	16

Table 3. Bialaphos selection for *bar* gene from anther culture of different *bar* transformed wheat lines. Number of embryolds on the different bialaphos included medium.

Transgenic wheat lines	Plated anthers onto P-4mf medium	Bialaphos content in medium		
		0 mg/l	5 mg/l	After 2 weeks − 5 mg/l
116	3 x 50	175	140	147
117	3 x 150	340	331	147
124	3 x 100	356	157	298
128	3 x 50	138	162	226
129	3 x 50	98	137	112
Total:		1107	927	1175

3.4. Production of transgenic haploids and double haploids in anther culture

The haploids and spontaneous diploids regenerated from selective anther and microspore culture experiments expressed total herbicide resistance to Finale 14SL. Via chromosome doubling of transgenic haploids, homogenous diploid (DH) wheat lines were obtained. The stability of the *bar* transgene in doubled haploid wheat genome is under study.

4. Conclusions

The fertile T_1 generation expressed 3:1 segregation ratio for *har⁺ / bar⁻* plants. The transformed foreign gene *(bar)* was stable in the wheat genome after two self pollinations and PAT assays. It is therefore, possible to transform transgene successfully *via* sexual reproduction into non-transgenic genotypes. Cell-level selection of the *bar* gene was possible in anther and microspore culture. The regenerated haploids and spontaneous diploids are herbicide resistant. Finally, by using *in vitro* haploid technique (anther and microspore culture), it's possible to produce homogenous *bar* resistant wheat lines by cell-level selection.

Acknowledgements

The authors wish to thank Dr. G. Schwarz, Dr. K. Schledzewski, M. Csosz, K. Talpas, É. Kóta for their excellent assistance. The funding by BMF/OMFB and OTKA through research grants is highly appreciated.

References

[1]   D. Becker *et al.*, An in-vitro test for endocytotic activation of murine epidermal langerhans cells under the influence of contact allergens. *Plant J.* **5** (1994) 299-307.
[2]   N. S. Nehra *et al.*, Self-fertile transgenic wheat plants regenerated from isolated scutillar tissues following microprojectile bombardment with 2 distinct gene constructs. *Plant J.* **5** (1994) 285-297.
[3]   N. S. Nehra *et al*, *Plant Breeding Abstracts* **65** (1995) 803-808.
[4]   J. Pauk *et al.*, Androgenesis in hexaploid spring wheat F2 populations and their parents using a multiple-step regeneration system. *Plant Breeding* **107** (1991) 18-27.
[5]   M. Puolimatka *et al.*, Effect pf ovary co-cultivation and culture medium on embryogenesis of directly isolated microspores of wheat. *Cer. Res. Comm.* **24** (1997 393-400.

[6]   V. Vasil *et al.*, Rapid production of transgenic wheat plants by direct bombardment of cultured
       immature embryos. *Bio/technol* **11** (1993) 1553-1558.
[7]   J.T. Weeks *et al.*, Rapid production of transgenic wheat plants by direct bombardment of cultured
       immature embryos. *Plant Physiol* **102** (1993) 1077-1084.

Development of a Highly Regenerable Germplasm and Genetic Transformation of Alfalfa

J. Farago, J. Kraic, P. Hauptvogel
*Research Institute of Plant Production, Bratislavská cesta 122, 921 68 Piestany,
Slovak Republic*

Abstract. It is well documented that regeneration potential through somatic embryogenesis in alfalfa is genetically determined. The supposed genetic background of the regeneration potential in alfalfa enabled us to use a recurrent selection technique to produce a highly regenerative population of genotypes from the currently released Slovak cultivar Lucia. We used petiole explants from young *in vitro* seedlings and a four step *in vitro* culture system to screen the frequency of regenerative genotypes in the cultivar. In the first cycle of screening 13 regenerants of alfalfa were obtained among the approximately 4% regenerating genotypes. The flowers of the original regenerants were self-pollinated to produce seeds of S1 generation, or open-pollinated to produce seeds of OP1 generation. In the second cycle of screening for somatic embryogenesis in the S1 and OP1 progeny of primary regenerants, we obtained comparatively higher proportion of regenerative genotypes (up to 83,3 %). We selected two highly regenerative genotypes, Rg9/I-14-22 and Rg11/I-10-68 for further work aiming in genetic transformation of alfalfa. *Agrobacterium tumefaciens* AGL1 was used to deliver the plasmid pPDE1001Ov, containing the *npt*II expression cassette and a cDNA of a gene *ov* coding for Japanese quail ovalbumin, into cells of petiole or leaf tissues. Different inoculation solutions and cocultivation techniques were used to compare the transformation efficiency. Embryogenic cultures were produced from 1,4-10,3 % explants plated on a medium containing 50 mg.l^{-1} kanamycin. In total, 116 putatively transgenic lines of alfalfa were obtained.

1. Introduction

The techniques of genetic engineering allow the transfer of heritable traits between completely unrelated species. However, a number of requirements must be fulfilled in order to produce stably transformed plants. One of them is the appropriate tissue culture system to enable transgenic plants to be regenerated from transformed tissues.

Somatic embryogenesis offers such a system because of the proposed single cell origin of developing somatic embryos [1]. This regeneration system is well documented [2, 3, 4 and many others] in alfalfa (*Medicago sativa L.*). Moreover, the ability of alfalfa to regenerate plants from callus tissue has been shown to be under genetic control [5], which enables to select genotypes with high regeneration frequency [6]. Unfortunately, most of the currently available highly regenerative germplasms possess poor agricultural performance. Therefore, the use of highly regenerative genotypes selected from high-value commercial varieties can reduce the costly and time-consuming backcross programmes to transfer the genes for regenerative ablity from one variety to another.

The objectives of our work were i) to screen the currently cultivated Slovak and Czech alfalfa cultivars for *in vitro* regeneration through indirect somatic embryogenesis, ii) select from a currently released commercial cultivar *Lucia* [7] a range of highly regenerating genotypes, iii) to develop by several cycles of somatic embryogenesis a germplasm with high regenerative capacity and iv) to use the best regenerating genotypes for preliminary transformation experiments.

2. Material and Methods

2.1 Tissue culture and plant regeneration

Seeds of alfalfa (*M. sativa L.*) cultivars Palava, Lucia, Vanda, Regia and Zuzana were obtained from the Breeding Station Spacince, SK and from The Gene Bank of Slovakia, Piestany, SK.

Twenty to forty seeds of each cultivar were surface sterilized in 70% (v/v) ethanol for 1 min and 0.1% (w/v) mercuric chloride and germinated aseptically in test tubes. Somatic embryogenesis was induced from petiole-derived calli on modified B5 based media [8] using a four step protocol. At the first step, the petioles derived from seedlings were planted onto GMK medium [GM + 8.0 mg.l^{-1} 2,4-dichlorophenoxyacetic acid (2,4-D) + 8.0 mg.l^{-1} kinetin + 0.5 mg.l^{-1} α-naphthaleneacetic acid (NAA); 4] for callus induction. Calli were subcultured after 30 days onto B5m medium [B5 + 0.2 mg/L 6-benzylaminopurine (BAP)] for induction of somatic embryos. At the third step, the calli with induced globular somatic embryos were subcultured for 28 days onto GMR medium (GM without growth regulators) for maturation of somatic embryos. Well developed cotyledonary stage somatic embryos were finally transferred onto GMZ medium (half-strength GM) for conversion into plantlets. After 14-21 days rooted regenerants were potted in a soil-perlite (3:1) mixture and maintained in the greenhouse until maturity.

2.2 Recurrent selection of embryogenic genotypes

In the first cycle of somatic embryogenesis, 13 mature regenerants (R$_0$ generation) of cv. Lucia were obtained. At flowering, part of the flowers of each regenerant was self-pollinated to obtain the S$_1$ seeds and another part was left to cross pollinate to get the OP$_1$ seed (R$_1$ generation). In the second cycle of screening, representative samples of S$_1$ and OP$_1$ seeds were used to check the regenerative ability of the progenies of the first cycle-derived regenerants. The protocol for somatic embryogenesis induction was essentially the same as that of in the first cycle of screening.

2.3 Genetic transformation

From the second screening of regenerative genotypes, two highly regenerative lines, BR9/I-14-22 and BR11/I-10-68, were selected for preliminary genetic transformation experiments. *Agrobacterium tumefaciens* strain AGL1 was used to deliver the plasmid pPDE1001Ov [9] into petiole- and/or leaf segment-derived explants of alfalfa. The plasmid contains a selectable marker gene *nptII* encoding neomycin phosphotransferase (NPTII) enzyme and a cDNA of Japanese quail ovalbumin [10], both under the control of CaMV 35S promoter and *nos* terminator. Wounded petioles and leaf segments of alfalfa were immersed in an overnight culture of *A. tumefaciens* suspension containing the pPDE1001Ov plasmid for 1 minute. Different media (H$_2$O, 0.85% (w/v) NaCl, liquid

GMK, liquid GMK+10% (w/v) sucrose+100 μM acetosyringone) and different procedures (with or without vacuum infiltration) were used for inoculation of explants with bacteria. After blotting on sterile filter paper, tissues were co-cultured with *A. tumefaciens* cells for 4 days. Selection of transformed calli was performed on GMK medium supplemented with 50 mg.l^{-1} kanamycin, 500 mg.l^{-1} cefotaxime and 500 mg.l^{-1} carbenicillin. Kanamycin resistant (Kan^r) calli were subcultured in two-week intervals. After 28 and another 10 days they were transferred onto selective B5m and GMR media, respectively. The B5m medium contained full-, and GMR medium half-strength concentrations of antibiotica. Regenerated plants were maintained as shoot cultures on MS0.25 [MS [11] + 0.25 mg.l^{-1} γ-indolylbutyric acid (IBA)].

Selected putative transgenic plants were assayed for the presence of NPTII protein through leaf bleach assay [12] and/or using the NPTII ELISA Assay Kit (5 Prime-3 Prime, Inc., Boulder, USA). For leaf bleach assay, leaf samples of putatively transformed T_0 plants were placed in a 24 well culture plate (Iwaki Glass, Japan). Each well was filled with 0.5 ml of an aqueous solution of 300 mg.l^{-1} paromomycin. Leaf samples from the nontransformed BR9/I-14-22 and BR11/I-10-68 plants were also included. The NPTII ELISA assay was performed according to the recommendations of the manufacturer.

3. Results

3.1. Screening of cultivars for somatic embryogenesis in vitro

The five most commonly cultivated commercial cultivars of alfalfa in Slovakia (See „Material and Methods") were screened *in vitro* for the ability to regenerate plants through indirect somatic embryogenesis. Of the cultivars tested, only two, *cvs*. Lucia and Palava showed a very low level of embryogenic capacity, up to 4% of the responding genotypes (results not shown). Also the number of somatic embryos per responding callus was very low, generally 2 to 3 cotyledonary stage somatic embryos per callus.

3.2. Recurrent selection for highly regenerative genotypes

Thirteen regenerated plants of *cv*. Lucia were obtained after conversion of somatic embryos in the first cycle of screening. All the regenerants showed normal morphology and were fertile. A representative sample of S_1 and OP_1 seed was used for the second cycle of screening for regenerative genotypes in the R_1 generation. The results of the second cycle of screening are presented in Table 1. The percentage of embryogenic genotypes in two controls (*cvs*. Lucia and Palava) were not very different from that recorded in the first cycle of screening, i.e. 6.4 and 2.5%, respectively. However, the percentage of embryogenic genotypes was dramatically increased in both the OP_1 (14.3 - 50.0%) and S_1 progenies (27.3 - 83.3%) of the regenerants. Also, the average frequency of embryogenic calli was significantly improved in the self- and open-pollinated progenies of regenerants (4.4 - 61.1%) comparing to two controls (0.2 and 1.7%, resp.).

Table 1. Screening for *in vitro* regeneration ability of S_1 and OP_1 progeny of the 13 regenerants of alfalfa cv. Lucia obtained through induction of somatic embryogenesis from petiole derived explants.

Progeny	Number of genotypes tested	Number of explants cultured	Average frequency of callus forming explants	Average frequency of embryogenic calli	Frequency of regenerating genotypes
Palava	68	568	71.1	0.2	2.5
Lucia	75	710	68.4	1.7	6.4
Rg1/S1	14	84	42.9	30.0	44.4
Rg2/S1	33	198	12.1[a]	61.1	66.7
Rg3/S1	6	84	98.8	36.1	66.7
Rg4/S1	11	143	91.6	33.6	72.7
Rg5/S1	10	137	83.2	32.5	50.0
Rg6/S1	10	117	82.9	45.4	60.0
Rg7/S1	14	162	97.5	8.9	45.4
Rg8/S1	31	434	93.3	26.2	68.0
Rg9/S1	14	185	87.6	31.5	63.6
Rg10/S1	19	203	94.1	18.3	46.1
Rg11/S1	14	179	89.9	4.4	27.3
Rg12/S1	8	123	88.6	5.5	33,3
Rg13/S1	6	77	97.0	35.8	83.3
Rg1/OP1	30	180	15.0	9.7	16.7
Rg2/OP1	30	180	0[a]	0	0
Rg3/OP1	10	151	86.1	17.2	50.0
Rg4/OP1	12	187	92.5	17.3	50.0
Rg5/OP1	9	150	99.3	16.1	33.3
Rg6/OP1	16	231	98.3	24.8	37.5
Rg7/OP1	11	106	98.1	23.1	28.6
Rg8/OP1	16	191	89.0	17.6	50.0
Rg9/OP1	15	166	100.0	18.1	38.5
Rg10/OP1	17	199	82.4	11.0	46.1
Rg11/OP1	15	167	86.2	6.2	50.0
Rg12/OP1	14	173	75.7	6.1	14.3
Rg13/OP1	10	88	93.2	9.8	16.7

[a]The low frequency of callogenesis is due to high frequency contamination of primary explants.

3.3. *Genetic transformation (preliminary results)*

Two highly regenerative lines, BR9/I-14-22 and BR11/I-10-68, have been chosen for genetic transformation experiments with alfalfa.

In two separate experiments, 754 explants of both selected lines were inoculated and consecutively co-cultured with *A. tumefaciens* AGL1. Non-inoculated petiole and leaf segment explants were used as controls. Results of the experiments are presented in Table 2. Petiole segments proved to be more suitable explant sources for use in genetic transformation experiments (9% Kan^r embryogenic callus formation on selection medium) in accordance with the higher regeneration ability of petiole-derived explants (73%) compared with leaf segments (22%). Also, with line BR9/I-14-22 we were able to obtain slightly more Kan^r embryogenic calli (9.0%), than with the line BR11/I-10-68 (6.7%). That means, probability of obtaining transformed alfalfa plants seems to be determined by the regeneration capacity of plant tissues *in vitro*.

Table 2. Summary of *A. tumefaciens* mediated genetic transformation experiments with two regenerative lines (BR9/I/14-22 and BR11/I-10-68) of alfalfa *cv.* Lucia. Two types of explants (leaves and petiole segments) were used to co-cultivation with the AGL1 strain of *A. tumefaciens* carrying the binary plasmid pPDE10010v. Different inoculation solutions and modes of inoculation were applied before co-cultivation of explants with the bacteria.

Genotype	Explant	Co-cultivation	Inoculation solution	Vacuum infiltr.	No. explt.[e]	No. (%) calli[f]	No. (%) EC[g]	No. regs.[h]
BR9/I-14-22	Leaves	Posit. contr.[a]	GMK-L[c]	VI-[d]	36	16 (44)	8 (22)	ND[i]
		Negat. contr.[a]	GMK-L	VI-	18	5 (28)	0 (0)	ND
		Cocultured[b]	GMK-L	VI-	109	36 (33)	1 (1)	12
	Petioles	Posit. contr.	GMK-L	VI-	48	35 (73)	35 (73)	ND
		Negat. contr.	GMK-L	VI-	24	5 (21)	0 (0)	ND
		Co-cultured	GMK-L	VI-	144	35 (24)	13 (9)	73
		Co-cultured	GMK-L	VI+[d]	135	17 (13)	6 (4)	5
BR11/I-10-68		Co-cultured	H_2O[c]	VI-	60	13 (22)	1 (2)	0
		Co-cultured	NaCl[c]	VI-	60	14 (23)	1 (2)	15
		Co-cultured	GMK-L	VI-	60	16 (27)	4 (7)	0
		Co-cultured	Sa[c]	VI-	60	18 (30)	3 (5)	0

[a] The positive controls were cultured on non-selective media without antibiotics; the negative controls were cultured on selective media amended by 50 mg•l⁻¹ kanamycin, 500 mg•l⁻¹ cefotaxime and 500 mg•l⁻¹ carbenicillin; [b] Cocultivation of explants with *A. tumefaciens* was carried out on non-selective GMK medium; [c] Alfalfa leaf- and petiole-derived explants were inoculated in a liquid GMK medium (GMK-L); in redistilled water (H_2O); 0.85% (w/v) NaCl solution or GMK-L supplemented with 10% (w/v) sucrose and 100 µM acetosyringone (SA); [d] Inoculation of explants in bacterial suspension was (+) or was not (-) under application of vacuum infiltration; [e] Number of explants initially plated; [f] Number of explants forming calli; the number in parentheses is the percentage of callus forming explants; [g] Number of embryogenic calli; the number in parentheses is the percentage of embryogenic explants; [h] Number of kanamycin resistant shoots/lines derived from embryogenic calli by conversion of cotyledonary somatic embryos on MS0.25 medium containing half the concentrations of each of the antibiotika; [i] Not defined

The use of vacuum infiltration during the inoculation step showed no positive effect neither on Kan[r] callus formation nor on the frequency of embryogenic Kan[r] calli. Liquid GMK (GMK-L) proved to be the most suitable solution for inoculation of bacterial suspension cultures with alfalfa explants. However, we were not able to convert somatic embryos of BR11/I-10-68 forming on explants inoculated with *A. tumefaciens* in GMK-L.

In total, 116 putatively transformed Kan[r] alfalfa lines were obtained and maintained as shoot cultures on MS 0.25 medium supplemented with 50 mg•l⁻¹ kanamycin.

Selected Kan[r] plants producing profound root systems on kanamycin media were tested preliminarily for the expression of the *npt*II gene using a leaf bleach assay. After 4 days the leaves of control, non-cocultivated plants completely bleached, while that of putatively transformed ones remained green or bleached only at the marginal areas or around wound sites. NPTII ELISA assay was also performed with selected lines.

PCR and Southern blot analyses for integration of the transgenes in the plant genome are under way.

4. Discussion

To speed up cultivar development in alfalfa using tissue culture methods, Phillips [13] suggests that screening be performed on materials already adapted to the region. It is generally believed that regenerative genotypes can be found in any alfalfa germplasm, if enough genotypes are screened [3,14]. On the other hand, several authors [15, 16] have reported that only a few genotypes in certain cultivars have been found to possess regeneration capacity.

In our laboratory, screening of five alfalfa cultivars of Slovak and Czech provenience for *in vitro* response was performed. The results showed that the best responding variety was *cv.Lucia,* one of the latest released Slovak commercial variety [7]. We have obtained 13 regenerants of R_0 generation. They displayed normal and vigorous growth in the greenhouse, and morphologically resembled the plants. All the regenerants flowered and set seeds.

A substantial evidence, that *in vitro* responses in alfalfa are under simple genetic control [5] leads to the suggestion that convential breeding techniques can be used to obtain tissue culture-responsive plants. Firstly, Bingham et al. [17] showed, that the capacity to regenerate plants *in vitro* via somatic embryogenesis can be enhanced in a population of plants by recurrent selection. Our results confirm the genetic nature of control of somatic embryogenesis in alfalfa and a high heritability of this trait, since the frequency of regenerative genotypes in *cv.* Lucia increased after one repeated cycle of selection from about 2-6% in the original cultivar to 27.3 - 83.3 % in S_1 progeny and 16.67 - 50.0% in OP_1 progeny, respectively.

Successful genetic transformation of alfalfa, mediated by *A. tumefaciens,* is well documented [18, 19, 20]. However, most of the transformed alfalfa plants originate from model genotypes, such as A70-34 or RA3. In North America, clones from the cultivar *Regen-S* developed by Bingham et al. [17] have been used almost exclusively for tissue culture work with alfalfa. In Europe research has been reported on clones isolated by Arcioni et al. [21] and Novák & Konecná [22]. Only a few working groups report genetic transformation of plants derived from commercial cultivars [18].

Therefore, our aim is to obtain a highly efficient transformation system that would allow the addition of agronomically important traits to populations derived from commercial cultivars having more desirable characteristics for yield, persistence and disease resistance and thus to avoid a time consuming process with backcrossing the transferred trait into agronomically useful genetic background.

The process of genetic transgormation of alfalfa is affected by a wide variety of factors, among them one of the most serious seems to be the regenerative ability of plant tissues used for transformation [23, 24, our results]. Also, providing the appropriate conditions for *A.tumefaciens* cells to attack the plant cells, is one of the key factors affecting successful transfer of T-DNA region from agrobacteria to the plant cells. Therefore, the use of different inoculation conditions may lead to dramatic differences in response of plant cells to agrobacterial infection, as shown in our results.

The availability of an efficient transformation system for commercial cultivars will allow the introduction of important foreign genes into alfalfa and it represents an attractive perspective for genetic improvement of this important forage legume.

References

[1] B. Haccius, Question of unicellular origin of nonzygotic embryos in callus cultures. *Phytomorphology* **28** (1978) 74-81.
[2] J.W. Saunders and E.T. Bingham, Production of alfalfa plants from callus tissue. *Crop Sci.* **12** (1972) 804-808.
[3] D.C.W. Brown and A. Atanassov, Role of genetic background in somatic embryogenesis in *Medicago. Plant Cell Tissue Organ Cult.* **4** (1985) 111-122.
[4] J. Kraic and J. Farago, Secondary embryogenesis and germination of alfalfa somatic embryos. Proc.Int.Conf."Process of Plant Embryogenesis in situ and in vitro", Nitra (Slovakia), 1993, p.158.

[5]   B. Reisch and E.T. Bingham., The genetic control of bud formation from callus cultures of diploid alfalfa. *Plant Sci.Lett.* **20** (1980) 71-77.

[6]   E.T. Bingham, Registration of Regen-S alfalfa germplasm useful in tissue culture and transformation research. *Crop Sci.* **29** (1989) 1095-1096.

[7]   E. Polak, [Final report on the breeding of new alfalfa (*Medicago sativa* L.) cultivar Lucia.]. Záverecná správa, VÚRV Piestany, 1990, pp.13, (in Slovak).

[8]   O.L. Gamborg et.al., Nutrient requirements of suspension cultures of soybean root cells. Exp.Cell.Res. **50** (1968) 151-158.

[9]   J. Simúth et.al., Preparation of transgenic tobacco transformed with the Japanese quail ovalbumin cDNA mediated by *Agrobacterium tumefaciens*. Biológia (Bratislava) **48** (1993) 241-246.

[10]   J. Mucha et.al., The sequence of Japanese quail ovalbumin cDNA. Nucleic Acids Research **18** (1991) 5553.

[11]   T. Murashige and F. Skoog, A revised medium for rapid growth and bioassays of tobacco tissue cultures. Physiol. Plant. **15** (1962) 473-497.

[12]   M. Cheng et. al, Genetic transformation of wheat mediated by *Agrobacterium tumefaciens*. Plant Physiol. **115** (1997) 971-980.

[13]   G.C. Phillips, Screening alfalfas adapted to the south western United States for regenerator genotypes. In Vitro **19** (1983) 265 (Abstr.).

[14]   D.H. Mitten et.al., *In vitro* regenerative potential of alfalfa germplasm sources. Crop Sci. **24** (1984) 943-945.

[15]   T.H.H. Chen and J. Marowitch, Screening of *Medicago falcata* germplasm for *in vitro* regeneration. J.Plant Physiol. **128** (1987) 271-277.

[16]   E.G.M. Meier and D.C.W. Brown, Screening of diploid *Medicago sativa* germplasm for somatic embryogenesis. Plant Cell Rep. **4** (1985) 283-288.

[17]   E.T. Bingham et.al., Breeding alfalfa which regenerates from callus tissue in culture. Crop Sci. **15** (1975) 719-721.

[18]   E.A. Shanin et.al., Transformation of cultivated alfalfa using disarmed *Agrobacterium tumefaciens*. Crop Sci. **26** (1986) 1235-1239.

[19]   K. D'Halluin et.al., Engineering of herbicide-resistant alfalfa and evaluation under field conditions. Crop Sci. **30** (1990) 866-871.

[20]   B.D. McKersie et.al., Superoxide dismutase enhances tolerance of freezing stress in transgenic alfalfa. Plant Physiol. **103** (1993) 1155-1163

[21]   S. Arcioni et.al., *In vitro* selection of alfalfa plants resistant to *Fusarium oxysporum* f.sp. *medicaginis*. Theor.Appl.Genet. **74** (1987) 700-705.

[22]   F.J. Novak and D. Konecna, Somatic embryogenesis in callus and cell suspension cultures of alfalfa (*Medicago sativa* L.). Z.Pflanzenphysiol. **105** (1982) 279-284.

[23]   R. Desgagnes et.al., Genetic transformation of commercial breeding lines of alfalfa (*Medicago sativa*). Plant Cell Tissue Organ Cult. **42** (1995) 129-140.

[24]   S. Du et.al., Effect of plant genotype on the transformation of cultivated alfalfa (*Medicago sativa*) by *Agrobacterium tumefaciens*. Plant Cell Rep. **13** (1994) 330-334.

Use of Agriculturally Important Genes in Biotechnology
G. Hrazdina (Ed.)
IOS Press, 2000

New Aspects of Breeding Crops for Disease Resistance: The Role of Antioxidants

Z. Király
Plant Protection Institute, Hungarian Academy of Sciences, Budapest, Hungary

Abstract. One new way of breeding crops for stress and disease resistance is based on *in vitro* selection of callus tissues on agar media containing paraquat. This latter compound produces superoxide and other reactive oxygen species (ROS) upon illumination. Regenerated ROS-resistant plants exhibited relative stress and disease resistance to necrotrophic pathogens and several stresses that cause plant cell and tissue necrotization. ROS-resistant tobacco and potato strains had high antioxidant activities, as compared to the original paraquat sensitive strains. This characteristics is in correlation with their disease and stress resistance. Another way of creating crops resistant to necrotic symptoms is to produce transgenic plants that express the gene ferritin on a high level. Ferritin, the gene product, stores free iron hindering thereby the Fenton reaction: $Fe^{2+} + H_2O_2 \rightarrow Fe^{3+} + OH^- + OH^\bullet$, which means a primary antioxidant defense. Thus, if free iron is not available, less hydroxyl radical ($OH^\bullet$) will be produced, and consequently, less cell necrotization will take place upon stress or infection. Thus, transgenic plants will be resistant to necrotic symptoms. This was the case with transgenic tobacco strains expressing ferritin on a high level. Quantitative resistance against 3 pathogens and stresses were detected.

1. Introduction

Sustainable agriculture requires the production of crop plants in balance with the environment. This requirement has to be fulfilled in spite of the fact that world population expanded already to 6 billion and increasing continuously, however the arable land is decreasing because of the growing new cities and increasing road construction, industrial area etc. In addition, there are as yet no dramatic breakthrough research bases for achieving high yields in many staple crops. Many scientists believe today that plant breeding for stress and disease resistance will have key role in increasing crop production and will not contradict requirements of environmental and human health protection.

2. Biochemical aspects of cell necrotization caused by stress and disease

It was shown recently by several laboratories that necrotic disease symptoms formed upon infection or stresses are associated with accumulation of superoxide radicals ($O_2^{\bullet-}$), hydrogen peroxide (H_2O_2), hydroxyl radical ($OH^\bullet$) and other reactive oxygen species (ROS) [1-14]. It is tempting to suppose that ROS *per se* cause cell and tissue necroses (Table 1).

However, the signalling process from infection to ROS-evolution is poorly understood at present. According to one hypothesis the signal transduction cascade may involve protein kinases [12, 15-19]. This process would be analogous to the activation of tyrosine kinases of mammalian T-cells. It seems reasonable that activation of these enzymes could lead to the activation of an NADPH oxidase. This latter enzyme might be involved in the release of reactive oxygen species, such as $O_2^{\cdot-}$ as well as H_2O_2, $OH^{\cdot}$ etc. also in plants.

ROS that are activated by infection or stress may have a dual effect: causing death in the infected, toxin-exposed or stressed host plant, as well as in the infecting pathogens [12,20-22]. This dual effect seems to be analogous to phagocytosis by human neutrophils [23]. There may be parallels in the way both animals and plants recognize and control pathogens.

Table 1. Physiological changes in plants as a result of infection, abiotic stress or herbicide treatment.

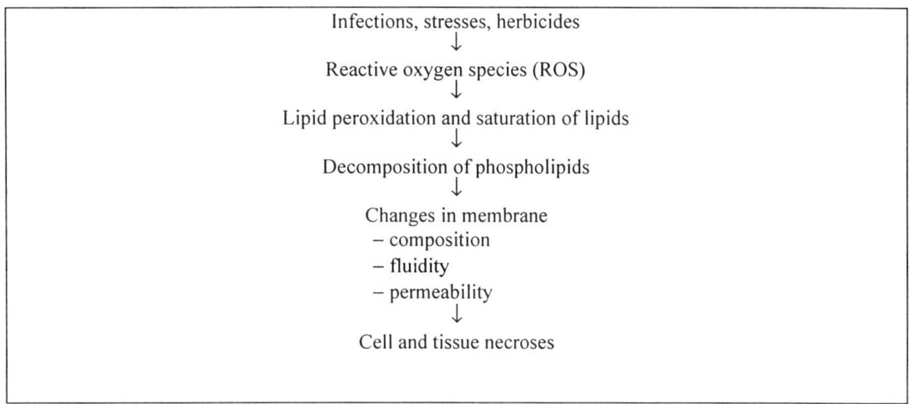

It is known that necrotic plant symptoms refer to host susceptibility against necrotrophic pathogens (bacteria and facultative fungal parasites). On the other hand, early necrotic disease symptoms, such as the hypersensitive response (HR), are associated, but not causally related, to plant resistance against both biotrophic and necrotrophic pathogens. One can suppose that suppression or inhibition of plant tissue necrotization by creating ROS-resistant crops or by treatment of plants with antioxidants may result in increased resistance to disease symptoms or stresses. However, suppression of hypersensitive necrosis does not mean that the host becomes susceptible, because the hypersensitive response (HR) is a consequence, not the cause, of disease resistance [24,25].

As regards the different ROS, hydrogen peroxide [12] and superoxide [9] might be involved in induction of host cell and tissue death. There is no direct evidence for the involvement of hydroxyl radical ($OH^{\cdot}$) in induction of tissue death caused by pathogens. However, the role of this radical in oxidative stress seems very probable. Tenhaken et al. [26] called the attention to the role of H_2O_2 in causing plant cell death as well as antioxidative defense. According to their hypothesis, the hydrogen peroxide in higher doses is a local trigger of cell death. However, at lower doses it is a diffusible signal for induction in adjacent cells of genes encoding antioxidant enzymes, such as glutathione peroxidase and glutathione S-transferase.

3. Antioxidants suppress ROS and cell death

The role of antioxidants in arresting necrotic processes in plant tissues seems to be important in plant breeding. It was shown that antioxidants are activated in concert with ROS and necrosis after herbicide treatment in tobacco [27,28] and as a consequence of virus and bacterial infection [29-32], or infection of barley with powdery mildew [33,34]. Recently, it has been demonstrated that transgenic plants with reduced capability to detoxify ROS because they expressed two antioxidant RNAs in antisense form, became hyperresponsive to pathogen attack. In other words, plants exhibited greater sensitivity to host cell necrotization [32]. Heiser and Elstner [35] point out that at advanced stages of stress or infection the balancing antioxidant activity (scavenging processes) may be gradually lost and chaotic oxygen free radical processes may dominate. On the other hand, active scavenging action (antioxidative process) may control necrotization after infection. Thus, the steady-state level of ROS within plant cells depends on an interplay between generation of ROS and the activity of antioxidant enzymes and/or the level of non-enzymatic antioxidants, such as ascorbic acid, glutathione, etc.

It was shown by Barna et al. [36] that there exists a correlation between tolerance of tobacco plants to ROS and relative resistance to necrotic disease symptoms caused by necrotrophic pathogens and to stresses. ROS-tolerant tobaccos were created first by Furusawa [37] who selected callus tissues against paraquat *in vitro*, and then regenerated whole tobacco plants. It was known that paraquat, a potent herbicide, produces superoxide anions upon illumination.

4. In vitro selection of callus tissues for ROS resistance

As mentioned above, ROS-tolerant tobaccos of Furusawa turned out to be resistant to diseases and stresses [36] (see Table 2).

Table 2. Effect of infections and stresses on paraquat (ROS) resistant and sensitive tobacco leaves.

	Paraquat $(O_2^{\bullet-})$ Ion Leakage(%)	Necrotized Leaf Area (%)			Fusaric acid Ion Leakage (%)	*Botrytis cinerea* Diseased Leaf Area	Tobacco Necrosis Virus (Number of Lesions)
		Freezing	Heat	HgCl$_2$			
ROS-resistant	37	5.0	7.4	16.6	64	143	3.2
ROS-sensitive	67	48.0	45.0	56.7	88	628	8.8

The superoxide anion $(O_2^{\bullet-})$ produced by paraquat probably was dismutated to H_2O_2 by subseqent spontaneous or superoxide dismutase (SOD)-catalyzed dismutation. Furthermore, one can suppose that in the presence of free iron, hydrogen peroxide produced hydroxyl radical $(OH^{\bullet})$ which is the most harmful radical damaging nucleic acids, lipids and proteins. It turned out that ROS-resistant tobaccos and weeds exhibited high antioxidant capacity in comparison to ROS-sensitive plants, diminishing thereby cell and tissue damages caused by infections and stresses [27-36]. These results lend further support for the hypothesis that ROS evolution and tissue necrosis in plant are in a cause-and-effect relationship.

On the basis of this idea we selected potato callus tissues on agar media containing paraquat (unpublished). The procedure is summarized in Table 3.

Several Desirée potato strains have been selected *in vitro* and regenerated, then checked in the laboratory for resistance to paraquat, *Alternaria alternata, A. solani* and *Botrytis cinerea*. The paraquat and disease resistant strains have been tested in the field for yield and virus resistance (Table 4).

Table 3. Crops resistant to ROS are resistant to disease and stress associated with necrosis.

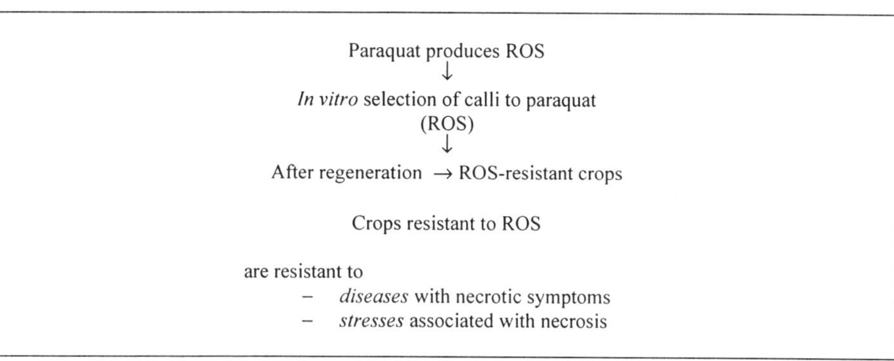

Table 4. ROS-resistant Desirée potato strains are resistant to virus diseases in field trials.

Potato strains	No. of samples	PLRV symptoms	PVY symptoms	DAS-ELISA extinction value (PVY)
Control	13	–	13	1,79
Res.13.	27	0	0	0,05
48.	7	0	0	0,04
49.	21	0	0	0,04
54.	7	0	0	0,05
57.	15	0	0	0,05

PLRV = potato leaf roll *polerovirus*
PVY = potato Y *potyvirus*

The mechanism of disease and stress resistance of tobacco and potato is based on increased antioxidant capacity of our ROS (paraquat)-resistant plants. Superoxide dismutase (SOD) showed 40-50 per cent higher activity in ROS-resistant tobacco leaf extracts than in the ROS-sensitive ones. As a result of stress caused by paraquat or acifluorfen, activities of some detoxifying and antioxidant enzymes, such as glutathione S-transferase, ascorbate peroxidase and glutathione reductase, were elevated in the ROS-resistant tobacco leaves. In addition, amounts of non-enzymatic antioxidants, such as ascorbic acid and glutathione, were on a higher level in the ROS-resistant leaves than in the sensitive ones. Furthermore, activities of three antioxidative enzymes, such as catalase, glutathione reductase and glutathione S-transferase, in ROS-resistant potato leaves taken from field experiments were 50-150 per cent higher, as compared to the ROS-sensitive leaves. Thus, up-regulated enzymatic and non-enzymatic antioxidants may participate in resistance to oxidative stress and cell death caused by ROS upon infection and adverse environmental actions (Table 5).

Table 5. Role of antioxidants in disease resistance of plants.

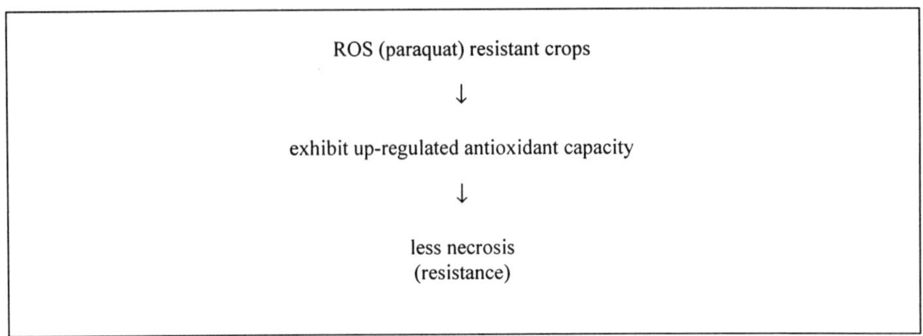

In summary, ROS (paraquat)-resistant tobacco and potato plants exhibited relative resistance to infection caused by viruses (TNV, PVY, PLRV), fungi (*Alternaria alternata, A. solani, Botrytis cinerea*) as well as to stresses, such as acifluorfen, paraquat, $HgCl_2$, fusaric acid, freezing and heat. It is noteworthy that paraquat resistant grape was recently created by Zs. Haydu, Viticulture and Enology Research Institute, Kecskemét, Hungary which exhibited resistance to powdery mildew infection. Similar work is in progress in the Agricultural Research Institute, Hungarian Academy of Sciences, Martonvásár with corn by B. Barnabás. It would seam that ROS-resistance and disease resistance of tobacco and potato is associated with up-regulated antioxidant capacity of those resistant plants.

5. Transgenic plants overexpressing the iron-binding protein, ferritin are resistant to stress and pathogens

Another way of creating crops resistant to necrotic symptoms is to produce transgenic plants that express the gene ferritin on a high level. The gene product of ferritin stores free iron hindering thereby the Fenton reaction: $Fe^{2+} + H_2O_2 \rightarrow Fe^{3+} + OH^- + OH^•$, which means a primary antioxidant defense. It seems keenly important to keep free iron out of the way, so that toxic free radicals such as $OH^•$ are not made. Thus, if free iron is not available, less damaging hydroxyl radical ($OH^•$) will be produced, and consequently less cell necrotization will take place upon infection or stress. One can suppose that transgenic plants will be resistant to symptoms caused by pathogens or stresses. This was the case with several transgenic tobacco strains expressing ferritin on a high level. We detected quantitative resistance against three pathogens as well as to iron and paraquat stresses [38]. The idea of the procedure is summarized in Table 6.

SR1 tobacco plants have been transformed with the gene ferritin isolated from alfalfa. This work has been done in the Institute of Plant Biology, Biology Center, Hungarian Academy of Sciences, Szeged, Hungary. Transformants exhibited normal photosynthetic function and chlorophyll content and retained photosynthetic function upon free radical toxicity generated by iron excess or paraquat treatment. Progeny of transgenic tobacco plants accumulating ferritin in their leaves exhibited resistance to necrotic damage caused by tobacco necrosis virus (TNV) and fungal (*Alternaria alternata* and *Botrytis cinerea*) infection.

Table 6. Ferritin and disease/stress resistance.

Transgenic tobacco overexpressing the gene ferritin:

- Stores free iron
- Inhibits Fenton reaction
- Inhibits OH$^•$ production
- Inhibits tissue necrotization caused by
    - iron or paraquat
    - tobacco necrosis virus (TNV)
    - *Alternaria alternata*
    - *Botrytis cinerea*
    - fusaric acid toxin
- Increases resistance to necrotic symptoms

In summary, it was shown that by sequestering intracellular iron involved in generation of the very reactive hydroxyl radicals (OH$^•$) through a Fenton reaction, ferritin protects plant cells from oxidative damage caused by abiotic and biotic stresses (infections), making thereby transgenic plants resistant to necrotic symptoms.

References

[1]   A.C. Allan and R. Fluhr, Two distinct sources of elicited reactive oxygen species in tobacco epidermal cells, *Plant Cell* **9** (1997) 1559-1572.

[2]   M.E. Alvarez *et al.*, Reactive oxygen intermediates mediate a systemic signal network in the establishment of plant immunity, *Cell* **92** (1998) 773-784.

[3]   C.J. Baker and E.W. Orlandi, Active oxygen in plant pathogenesis, *Annu. Rev. Phytopathol.* **33** (1995) 299-321.

[4]   C.S. Bestwick *et al.*, Localized changes in peroxidase activity accompany hydrogen peroxide generation during the development of a nonhost hypersensitive reaction in lettuce, *Plant Physiol.* **118** (1998) 1067-1078.

[5]   G.P. Bolwell, Role of active oxygen species and NO in plant defence responses, *Curr. Opin. Plant Biol.* **2** (1999) 287-294.

[6]   S.F. Chamnongpol, *et al.*, Defense activation and enhanced pathogen tolerance induced by H_2O_2 in transgenic tobacco, *Proc. Natl. Acad. Sci. USA* **95** (1998) 5818-5823.

[7]   N. Doke, Involvement of superoxide anion generation in the hypersensitive response of potato tuber tissues to infection with an incompatible race of *Phytophthora infestans* and to the hyphal wall components, *Physiol. Plant Pathol.* **23** (1983) 345-357.

[8]   E.F. Elstner and W. Osswald, Mechanism of oxygen activation during plant stress, *Proc. Royal Soc. Edinburgh* **102B** (1994) 131-154.

[9]   T. Jabs et.al., Initiation of runaway cell death in an *Arabidopsis* mutant by extracellular superoxide, *Science* **273** (1996) 1853-1856.

[10] Z. Király *et al.*, Effect of oxy free radicals on plant pathogenic bacteria and fungi and on some plant diseases. In: Gy. Mózsik *et al.* (ed.), *Oxygen Free Radicals and Scavengers in the Natural Sciences.* Akadémiai Kiadó, Budapest, 1993, pp. 9- 19.

[11] C. Lamb and R.A. Dixon, The oxidative burst in plant disease resistance, *Annu. Rev. Plant Physiol. Plant Mol. Biol.* **48** (1997) 251-275.

[12] A. Levine *et al.*, H_2O_2 from the oxidative burst orchestrates the plant hypersensitive disease resistance response, *Cell* **79** (1994) 583-593.

[13] N. Milosevic and A.J. Slusarenko, Active oxygen metabolism and lignification in the hypersensitive response in bean, *Physiol. Mol. Plant Pathol.* **49** (1996) 143-158.

[14] G. Wu *et al.*, Activation of host defense mechanisms by elevated production of H_2O_2 in transgenic plants, *Plant Physiol.* **115** (1997) 427-435.

[15] A.L. Ádám *et al.*, Rapid and transient activation of a myelin basic protein kinase in tobacco leaves treated with harpin from *Erwinia amylovora*, *Plant Physiol.* **115** (1997) 853-861.

[16] D.D. Dunigan and J.C. Madlener, Serine/threonine protein phosphatase is required for tobacco mosaic virus-mediated programmed cell death, *Virology* **207** (1995) 460-466.

[17] W. Ligterink *et al.*, Receptor-mediated activation of a MAP kinase in pathogen defense of plants, *Science* **267** (1997) 2054-2057.

[18] K. Suzuki and H. Shinshi, Transient activation and tyrosine phosphorylation of a protein kinase in tobacco cells treated with a fungal elicitor, *Plant Cell* **7** (1995) 639-647.

[19] T. Xing *et al.*, Regulation of plant defense response to fungal pathogens: Two types of protein kinases in the reversible phsphorylation of the host plasma membrane H^+ - ATPase, *Plant Cell* **8** (1966) 555-564.

[20] Z. Király *et al.*, Pathophysiological aspects of plant disease resistance, *Acta Phytopath. Entomol. Hung.* **26** (1991) 233-250.

[21] M.C. Mehdy, Active oxygen species in plant defense against pathogens, *Plant Physiol.* **104** (1994) 467- 472.

[22] D.D. Tzeng and J.E. DeVay, Role of oxygen radicals in plant disease development, *Adv. Pl. Pathol.* **10** (1993) 1-34.

[23] F. Morel *et al.*, The superoxide-generating oxidase of phagocytic cells: Physiological, molecular and pathological aspects, *Eur. J. Biochem.* **201** (1991) 523-546.

[24] Z. Király *et al.*, Hypersensitivity as a consequence, not the cause, of plant resistance to infection, *Nature* **239** (1972) 456-458.

[25] A. Bendahmane *et al.*, The *Rx* gene from potato controls separate virus resistance and cell death responses, *Plant Cell* **11** (1999) 781-791.

[26] R. Tenhaken *et al.*, Function of the oxidative burst in the hypersensitive disease resistance, *Proc. Natl. Acad. Sci. USA* **92** (1995) 4158-4163.

[27] Y. Shaalthiel *et al.*, Cross tolerance to herbicidal and environmental oxidants of plant biotypes tolerant to paraquat, sulfur dioxide and ozone, *Pest. Biochem. Physiol.* **31** (1988) 13-23.

[28] G. Gullner *et al.*, Enhanced inducibility of antioxidant systems in a *Nicotiana tabacum* L. biotype results in acifluorfen resistance, *Z. Naturforsch.* **46C** (1991) 875-881.

[29] G. Gullner *et al.*, Induction of glutathione S-transferase activity in tobacco by tobacco necrosis virus infection and by salicyclic acid, *Pestic. Sci.* **45** (1995) 290-291.

[30] A. Ádám *et al.*, Consequence of $O_2^{\bullet-}$ generation during a bacterially induced hypersensitive reaction in tobacco: deterioration of membrane lipids, *Physiol. Molec. Pl. Pathol.* **34** (1989) 13-26.

[31] J. Fodor *et al.*, Local and systemic responses of antioxidants to tobacco mosaic virus infection and to salicyclic acid in tobacco: Role in systemic acquired resistance, *Plant Physiol.* **114** (1997) 1443-1451.

[32] R. Mittler *et al.*, Transgenic tobacco plants with reduced capability to detoxify reactive oxygen intermediates are hyperresponsive to pathogen infection, *Proc. Natl. Acad. Sci. USA* **96** (1999) 14165-14170.

[33] H.M. El-Zahaby *et al.*, Effects of powdery mildew infection of barley on the ascorbate-glutathione cycle and other antioxidants in different host-pathogen interactions, *Phytopathology* **85** (1995) 1225-1230.

[34] H. Vanacker *et al.*, Pathogen-induced changes in the antioxidant status of the apoplast in barley leaves, *Plant Physiol.* **117** (1998) 1103-1114.

[35] I. Heiser and E.F. Elstner, The biochemistry of plant stress and disease. In: P. Csermely (ed), *Stress of Life. Ann. New York Acad. Sci.* **851** (1998) 224-232.

[36] B. Barna *et al.*, Juvenility and resistance of a superoxide-tolerant plant to diseases and other stresses, *Naturwissenschaften* **80** (1993) 420-422.

[37] I. Furusawa *et al.*, Paraquat-resistant tobacco calluses with enhanced superoxide dismutase activity, *Plant Cell Physiol.* **25** (1984) 1247-1254.

[38] M. Deák *et al.*, Plants ectopically expressing the iron-binding protein, ferritin, are tolerant to oxidative damage and pathogens, *Nature Biotech.* **17** (1999) 192-196.

Evaluation of Transgenic Cucumbers Expressing the Thaumatin Gene

S. Malepszy, M. Szwacka

*Department of Plant Genetics, Breeding and Biotechnology, Warsaw Agricultural
University, 02-787 Warsaw, Poland*

Abstract. We have analyzed the expression of the structural gene of the sweet-
tasting plant protein (preprothaumatin II) from *Thaumatococcus daniellii* in
transgenic cucumber (*Cucumis sativus* L.) plants, obtained from *Agrobacterium*-
mediated leaf microexplant transformation, using RNA and protein blot and fruit
flavour analysis. Thaumatin is initially translated in a precursor form
preprothaumatin II with amino and carboxyterminal peptides that are cleaved of
during the protein's maturation [1]. The naturally occuring thaumatin II represents a
processed form. Our data demonstrate that a DNA copy of preprothaumatin II-
encoding mRNA is expressed in cucumber leaf and fruit tissues under transcriptional
control of the CaMV35S promoter. Biologically active recombinant thaumatin (r-
thaumatin) is produced in fruits of transgenic cucumbers of levels that induce the
sweet-taste phenomenon. Leaf discs of transgenic cucumber plants exhibited
inhibited development of disease symptoms after inoculation with the spore
suspension of fungus *Pseudoperonospora cubensis* which is the cause of downy
mildew, a disease of cucumber [2].

·1. Introduction

Higher plants are able to biosynthesize intensely-sweet compounds of various
structural types, notably terpenoids, flavonoids and proteins [3]. There is a great deal of
interest in naturally occuring sweet substances for potential use in diabetic and dietetic
foods, beverages and medicines. Proteins: thaumatin and monellin are the two sweetest
known to man, about 100,000 times sweeter than sugar on the molar basis and 300 times
on a weight basis [4]. The food processing industry found several applications for the
sweet tasting fruit extracts of *T. daniellii* [5]. The two predominant thaumatins, thaumatin
I and II have been well characterized. They are composed of 207 amino acids in a β-barrel
form with eight disulphide bonds [6,7]. These eight disulphide bonds may explain why
under some conditions thaumatins may be boiled for several hours and regain sweetness
upon cooling [5]. Production of another intensely-sweet plant protein, monellin, in
transgenic tomatoes and lettuce was reported by Penarrubia et al. [8].

Edible fruits, seeds and vegetables of the transgenic plants modified to produce the
sweetening protein, thaumatin, monellin etc. are useful in preparing food compositions
with enhanced sweetness, thus reducing sweetening of food with sugar or other agents.

In plants from the *Cucurbitaceae* family improved fruit taste is an important target for
engineering. For this trait it is necessary to express introduced genes at high levels in
specific tissues. Preferential expression of these genes under the control of fruit-specific
promoter would provide novel taste patterns.

To demonstrate the feasibility of the targeted gene expression in a commercially
important crop, cucumber (*C. sativus* L.) plants were transformed a with chimaeric
preprothaumatin II gene regulated by the CaMV 35S promoter.

The biological function of thaumatin in plants remains unknown. However, proteins having remarkable homology to the thaumatin sequence (*Thaumatin-Like*, TL) are present in many plant species and belong to the fifth class of pathogenesis-related proteins, PR-5 [9,10].

2. Methods

pRUR528 (a gift of Dr. A. Ledeboer) was digested with EcoRI and HindIII. The fragment encoding preprothaumatin II was directionally cloned into the polylinker site of pROK2 (constructed by Dr. A. Palucha, Institute of Biochemistry and Biophysics PAS, Warsaw, Poland) to form pRUR528 encoding CaMV35S5'-preprothaumatin II - nos3' chimaera. pRUR528 was transferred to *Agobacterium tumefaciens* LBA4404 by electroporation. *C. sativus* L. cv. Borszczagowski [11] was inoculated with *A. tumefaciens*: pRUR528.

Isolation of total DNA was performed by the method described by Dellaporta et al. [12]. Total RNA was isolated using the method described by Linthorst et al. [13]. Protein gel blots and resistance assays were prepared as described earlier [14]. The fruits were harvested 5-8 weeks after self-pollination from plants cultivated more than one month in the absence of kanamycin and evaluated immediately organoleptically.

3. Results

Transgenic cucumbers were generated through *Agrobacterium*-mediated transformation. Neomycin phosphotransferase II (*npt* II) gene was used as a selectable marker in binary plasmid pRUR528 containing one unselected gene, preprothaumatin II cDNA in sense orientation. We confirmed the presence of thaumatin mRNA in 11 out of 12 of the examined primary transformed cucumber plants, but the levels of mRNA were different. Only four plants produced high levels of thaumatin mRNA and two of them expressed very high levels of thaumatin protein [14]. Hybridization analysis of the T_1 progeny showed that an increase in transgenic copy number did not always have a concomitant decrease in thaumatin mRNA expression levels [15].

Freshly harvested fruits of T_1 and T_2 cucumber lines were evaluated organoleptically. We confirmed the sweet flavour phenotype of *T. daniellii* in some of them. Western blots confirmed the presence of thaumatin antigenic protein that comigrated with authentic thaumatin in SDS-PAGE. Table 1 shows T_2 lines expressing high levels of thaumatin protein in fruit tissues. The data indicate on the existence of positive correlation between levels of r-thaumatin protein in cucumber fruit tissue and the intensity of sweetness of the fruit. We obtained a subsequent generation of these lines.

Table 1. Data of the molecular analysis of T_2 plants exibiting improved taste

Plant	Sweetness intensity (0-3) scale	Levels of thaumatin	
		mRNA	Protein
210 03	2	Very high	Very high
04	3	High	High
212 01	3	High	High
224 09	2	High	Low
nontransformed	0	Lack	Lack

0 (low) – 3 (high)
*content equivalent to 0.25 µg protein standard

Eight pRUR528 cucumber T_2 lines were submitted to the resistance tests using *in vitro* assays. At day 10 after inoculation we observed significant differences in the infected leaf disc area between controls (susceptible cucumber cultivar, Wisconsin SHR-18) and some transgenic lines [14]. A decrease of the downy mildew symptoms have been observed in the case of four tested lines (nos. 210, 212, 215 and 225).

Plants of T_3 generation were analysed by PCR to confirm the presence of thaumatin II cDNA and were organoleptically evaluated in order to verify that functional r-thaumatin was produced in fruits of the T_3 generation. Thaumatin T_3 line (no. 224), which set fruits of improved taste.

4. Discussion

From recent reports it is known that the expression of foreign genes has been linked to the methylation, environmental factors, gene interactions and gene loss [16-22]. Southern hybridization detected intact T-DNA present in all T_1 plants analysed, therefore different levels of thaumatin gene expression were not caused by the loss of the gene. The decreased expression of the thaumatin gene could be caused by methylation of the 35S promoter. The silencing of the AI gene in petunia plants by methylation of the 35S promoter was suggested by Matzke and Matzke [20]. The variability in the expression of the transgene has been recently shown by Hobbs et al. [17]. They reported that low levels of GUS expression seen in tobacco plants were due to multiple insertion that tended to have increased methylation. In this work transgenic cucumber line with 4 inserted copies of T-DNA (no. 104) exhibited a higher level of thaumatin gene expression that the lines with single copy (nos. 219, 225 and 235). These results suggest different mechanisms of thaumatin gene silencing than the methylation.

Witty used the thaumatin gene construct to cause a change in the phenotype of *Solanum tuberosum* plants [23]. Biologically active thaumatin was produced in the transgenic potatoes inducing a sweet-taste phenotype [23]. The results of this work demonstrate that the thaumatin II gene is effective in conferring the sweet-taste phenotype in cucumber. The activity of r-thaumatin in *C. sativus* L. suggests that molecular folding and disulphide bonds form correctly because thaumatin is tasteless after disulphide bond reduction [5].

We conclude that the CaMV35S5'-preprothaumatin II-nos3' chimaeric gene may be thought of as a palatability gene of possible commercial importance.

References

[1]　L. Edens *et al.*, Synthesis and Processing of the Plant Protein Thaumatin in Yeast, *Cell* **37** (1984) 629-633.

[2]　J. A. Lucas *et al.*, The Downy Mildew: Host Specificity and Pathogenesis, In: K.S. Singh, K. Kohmoto, R. P. Singh (eds.), Pathogenesis and Host Specificity in Plant Diseases, Elsevier Science Ltd, 1995, pp. 217-238.

[3]　A. D. Kinghorn *et al.*, Sweet Constituents of Some Medical Plants, *Revista Latinoamericana de Quimica* **25** (1997) 49-61.

[4]　S. K. Adesina, *Thaumatoccocus daniellii* Source of Sweet Proteins, In: M. Witty and J. D. Higginbotham (eds.), Thaumatin, CRC Press, Inc., 1994, pp. 19-35.

[5]　K. Etheridge, The Sales and Marketing Talin, In: M. Witty and J. D. Higginbotham (eds.), Thaumatin, CRC Press, Inc., 1994, 47-59.

[6]　L. Edens *et al.*, Cloning of cDNA Encoding the Sweet-Tasting Plant Protein Thaumatin and Its Expression in *Escherichia coli*, *Gene* **18** (1982) 1-12.

[7]   A. M. De Vos and M. Hatada, Three-Dimensional Structure of Thaumatin I, an Intensely Sweet Protein. *Proc. Nat. Acad. Sci.* USA **82** (1985) 1406-1409.

[8]   L. Penarrubia *et al.*, Production of the Sweet Protein Monellin in Transgenic Plants, *Bio/Technology* **10** (1992) 561-564.

[9]   L. C. van Loon, Pathogenesis-Related Proteins, *Plant Molecular Biology* **4** (1985) 111-116.

[10]   B. J. C. Cornelissen *et al.*, A Tobacco Mosaic Virus-Induced Protein is Homologous to the Thaumatin Protein, *Nature* **321** (1986) 531-

[11]   S. Malepszy, Cucumber (*Cucumis sativus* L.), In: Y. P. S. (ed.), Biotechnology in Agriculture and Forestry 6, Springer Verlag. Berlin-Heidelberg, 1988, pp. 277-293.

[12]   S. L. Dellaporta *et al.*, A Plant DNA Minipreparation: Version II, *Plant Molecular Biology Reporter* **1** (1983) 19-23.

[13]   H. J. M. Linthorst *et al.*, Tobacco Proteinase Inhibitor I Genes are Locally, but Not Systemically Induced by Stress, *Plant Molecular Biology* **21** (1993) 985-992.13

[14]   M. Szwacka *et al.*, Thaumatin Expression in Transgenic Cucumber Plants, *Proceedings of the IXth International Congress on Plant Biotechnology and In Vitro Biology in the 21st Century*, Jerusalem, Israel, 14-19 June 1998, pp 609-612.

[15]   M. Szwacka *et al.*, Transgenic Cucumber Plants Expressing the Thaumatin Gene, *FOOD BIOTECHNOLOGY, An International Symposium*, Zakopane, Poland, 9-12 May 1999 (in press).

[16]   K. A. Feldman and M. D. Marks, Agrobacterium-Mediated Transformation of Germinating Seeds of *Arabidopsis thaliana*: a Non-Tissue Culture Approach, *Molecular General Genetics* **208** (1987) 1-9.

[17]   S. L. A. Hobbs *et al.*, The Effect of T-DNA Copy Number, Position and Methylation on Reporter Gene Expression in Tobacco Transformants, *Plant Molecular Biology* **15** (1990) 851-864).

[18]   N. J. Kilby *et al.*, Promoter Methylation and Progressive Transgene Inactivation in Arabidopsis, *Plant Molecular Biology* **20** (1992) 103-112.

[19]   F. Linn *et al.*, Epigenetic Changes in the Expression of the Maize A1 Gene in *Petunia hybrida*: Rule of Numbers of Integrated Gene Copies and State of Methylation, *Molecular General Genetics* **222** (1990) 32 9-336.

[20]   M. A. Matzke and A. J. M. Matzke, Differential Inactivation and Methylation of a Transgene in Plants by Two Suppressor Loci Containing Homologous Sequences, *Plant Molecular Biology* **16** (1991) 821-830.

[21]   P. Meyer *et al.*, Endogenous and Environmental Factors Influence 35S Promoter Methylation of a Maize A1 Gene Construct in Transgenic Petunia and Its Colour Phenotype, *Molecular General Genetics* (1992) 345-352.

[22]   M.P. Ottawiani *et al.*, Differential Methylation and Expression of the β-glucuronidase and Neomycin Phosphotransferase Genes in Transgenic Plants of Potato cv. "Bintje", *Plant Science* **88** (1983) 73-81.

[23]   M. Witty, Sensory Evaluation of Transgenic *Solanum tuberosum* Producing r-Thaumatin II, *New Zeland Journal of Crop and Horticultural Science* **18** (1990) 77-80.

This work was supported by grant no. 227 P06 96 11 from State Committee For Scientific Research.

Enhancement of Disease Resistance in Apples

H.S. Aldwinckle and J.L. Norelli
J.P. Bolar, G.E. Harman, E. Borejsza-Wysocka, and J.-P. Reynoird
*Dept. of Plant Pathology, Cornell University
Geneva, NY 14456 USA*

Abstract. Commercial apple production worldwide is dependent on high inputs of chemical pesticides to control diseases. Genetic engineering is being used to create disease-resistant strains of the currently favored cultivars to reduce pesticide use. Initial research was on the bacterial fire blight disease (*Erwinia amylovora*) using lytic protein genes (LP) to increase resistances, and has continued with research on the fungal apple scab disease (*Venturia inaequalis*) using genes for chitinolytic enzymes. The lytic protein genes, attacin E, and the cecropin analog SB-37 were transferred to the fire blight susceptible Royal Gala cultivar by *Agrobacterium*-mediated transformation. Field tests of transgenic lines containing each of the lytic protein genes have shown significantly increased resistance to fire blight following inoculation. In the case of cecropin SB-37-transgenics, several lines have been identified that are significantly more resistant than the Royal Gala parent. To date, the best fire blight resistance has been observed with attacin-transgenics. Growth chamber, greenhouse and field evaluation of attacin transgenic lines has shown significant increases in the fire blight resistance of some lines in specific tests. In 1998 field trials, one attacin-line (T138) had only 5% shoot blight compared with approximately 60% in non-transgenic Royal Gala controls and approximately 40% in the moderately resistant Liberty controls. Several lytic protein transgenic Gala lines fruited in 1999, and fruit appeared identical to non-transgenic Royal Gala. It is now being analyzed for quantitative fruit quality parameters. Genes for chitinolytic enzymes from *Trichoderma harzianum* were transferred to the scab susceptible Marshall McIntosh cultivar by *Agrobacterium*-mediated transformation. In greenhouse tests, an endochitinase gene (*Ech42*) gave a high level of resistance to scab in transgenic Marshall McIntosh plants, but caused significantly reduced plant vigor. An exochitinase gene (N-acetyl-β-D-glucosaminidase (*Nag1*) gave a lower level of scab resistance, but no vigor reduction. Lines transgenic for both genes were selected for low *Ech42* and high *Nag1* expression, and had a high level of scab resistance without significant growth reduction. Synergism between *Ech42* and *Nag1* for scab resistance was observed in lines containing both genes. Selected lines are now being evaluated in the field for resistance, growth, and fruit quality.

1. Introduction

Bacterial and fungal diseases are constant threats to apple production in all production areas worldwide. Some diseases, like fire blight, caused by the bacterium, *Erwinia amylovora,* are not adequately controlled, and result in considerable losses annually [19]. The best available control measure, used in certain countries, is application of the antibiotic streptomycin to the blossoms. In many areas this has resulted in development of resistance to streptomycin in *E. amylovora* [16]. Fungal diseases, especially the most prevalent and damaging disease worldwide, apple scab caused by *Venturia inaequalis* [11] are controlled with multiple applications of chemical fungicides, which are costly to producers, and have raised environmental and health concerns. A desirable solution to these problems is the creation of resistant cultivars. However, conventional breeding of apple is very long term and cannot reproduce the desirable qualities of our best commercial cultivars. Furthermore, resistance based on native resistance genes tends to be unstable due to presence of matching virulence genes in pathogen populations [17]. Genetic engineering offers an attractive alternative to conventional breeding for the creation of resistant cultivars since it is faster, it can use genes from many sources, and it will preserve the desirable qualities of the transformed cultivar.

The introduction of lytic protein genes into plants by genetic engineering has been reported to enhance resistance to phytopathogenic bacteria [6, 9,11,13,18]. However, there are few reports on the field performance of these lytic protein-transgenic plants. Here we report on the fire blight resistance of Royal Gala apple lines transgenic for the cecropin-like gene SB-37, and the attacin E gene under field conditions.

Chitinolytic enzymes from the biological control organism *Trichoderma harzianum* have *in vitro* activity against several plant pathogenic fungi [10], including *Venturia inaequalis,* the causal agent of apple scab. We have transformed the scab-susceptible cultivar, Marshall McIntosh, with the *T. harzianum* endochitinase and exochitinase (N-acetyl-β-D-glucosaminidase) genes. We report their effect on resistance to scab when expressed in transgenic apple. A preliminary report of increased scab resistance of endochitinase-transgenic Royal Gala has been published [20].

2. Materials and Methods

Genes encoding the lytic proteins were transferred to Royal Gala apple by *Agrobacterium*-mediated transformation [15] (Table 1).

Table 1. Plasmids Used to Transform Royal Gala Apple for Fire Blight Resistance

Plasmid	Promoter	Signal Peptide	Lytic Protein
pWIAtt	Pin	none	attacin E
pCa2Att	35S	none	attacin E
pWIC38	Pin	none	cecropin SB-37
pCa2C38	35S	none	cecropin SB-37
pBPRB37	35S	SP1	cecropin SB-37
pBCCB37	35S	SCC	cecropin SB-37
pBPRB37	35S	SP1	cecropin Shiva-1

Plasmid = *Agrobacterium tumefaciens* binary vector. Lytic protein genes were cloned into the *Hind*III site of pBI121 vector containing a neomycin phosphoryl transferase (*nptII*) marker gene for kanamycin resistance and a ß-glucuronidase (*uidA*) reporter gene.
Pin = wound inducible promoter of proteinase inhibitor II gene of potato.
35S = CaMV constitutive promoter with duplicated upstream sequence.
SP1 = signal peptide sequence of PR1b protein of tobacco.
SCC = signal peptide sequence of native cecropin B from *Hyalophora cecropia.*

Transgenic lines and control plants were grafted onto seedling rootstock, and planted on the Research Farm, NYSAES, Cornell University, Geneva, NY in 1996 (3-yr-old) and 1997 (2-yr-old). Three to ten plants of each line were planted in a completely randomized design. Plants were inoculated on 4 June 1998. Vigorously growing shoot-tips were inoculated by cutting the two youngest leaves of the shoot with scissors which had been dipped in 5 X 10^7 cfu of *E. amylovora* strain Ea273/ ml. Eight weeks after inoculation the length of the necrotic lesion was recorded and expressed as the % of the current season's shoot length as a measure of disease. Three to five shoots were inoculated per plant on 1 to 9 plants of each transgenic line, planted in a completely randomized design. Individual inoculated shoots were the unit of replication.

Both cDNA and genomic clones of *T. harzianum* endochitinase (*Ech42*) and exochitinase (N-acetyl-β-D-glucosaminidase [*Nag1*]) genes were cloned either singly, or in combination, into plasmid binary vectors under the control of the enhanced cauliflower mosaic virus 35S promoter [3]. Constructs were made of the chitinase genes with a translational enhancer sequence from alfalfa mosaic virus, their native signal peptide sequence to direct the enzymes to intercellular space, and *npt*II as a selectable marker. The chitinase constructs were transferred into the apple cultivar Marshall McIntosh by *Agrobacterium*-mediated transformation [2], and transgenics were identified by PCR, ELISA for NPT II protein, and Southern analysis. The level of chitinase expression was measured by methylumbelliferyl enzyme assays and by western analysis. Own-rooted transgenic plants were inoculated with a conidial suspension of *V. inaequalis,* incubated in a mist chamber (18± 1°C and 100% relative humidity) for 48 h and later moved to a growth chamber or greenhouse. Scab resistance was evaluated based upon number of sporulating lesions, the percentage of leaf area infected, and the number of conidia recovered by rinsing leaves.

3. Results

3.1. Fire blight resistance

Among the 2-yr-old cecropin SB-37 transgenic Royal Gala plants, seven pin/SB-37 (pWIC38)-transgenic lines (TG250, TG547, TG253, TG228, TG546, TG549, and TG272), two 35S/signal peptide/SB-37 (pBPRB37)-transgenic lines (TG201 and TG154), and one vector-control pBI121-transgenic line (TG159) were significantly more resistant than the Royal Gala parent in field trials (Table 2). There was no significant difference in the fire blight resistance of pBPRB37-transgenic lines containing cecropin SB-37 fused to a signal peptide sequence and pCa2C38-transgenic lines containing the same cecropin without a signal peptide sequence.

Among 3-yr-old cecropin SB-37 transgenic Royal Gala plants, three pin/SB-37 (pWIC38)-transgenic lines (TG245, TG548, and TG549) and two vector-control pBI121-transgenic line (TG267 and TG171) were significantly more resistant than the Royal Gala parent in this field trial (Table 3). TG549 was significantly more resistant than Royal Gala in trials of both 2-yr-old and 3-yr-old plants (Tables 2 and 3). Vector-control transgenic lines TG267 and TG171 were evaluated as significantly more resistant than Royal Gala in the 3-yr-old trial but were indistinguishable from Royal Gala in the 2-yr-old trial. The pin/SB-37-transgenic lines TG245 and TG548 were not included in the 2-yr-old trial.

Of 13 transgenic lines that were included in both the 2-yr-old and 3-yr-old trials, 9 lines performed similarly in both trials. The performance of one pBI121 and three cecropin SB-37 transgenic lines was inconsistent between trials [16]. The same plant material was inoculated again in June 1999 with broadly similar results.

TG138, an attacin E transgenic line, was significantly more resistant than other lytic protein transgenics and expressed the greatest amount of attacin E protein (data not shown). Nevertheless, when several attacin E transgenic lines were analyzed there was no statistically significant correlation between resistance and the level of attacin E expression.

Table 2. Disease evaluation of 2-yr-old cecropin-transgenic Royal Gala apple plants in the field

Cultivar	Plasmid	N	% shoot blighted	Waller Group[z]
TG149	pCa2C38	21	81	a
TG267	pBI121	3	80	a
TG243	pBPRB37	40	78	ab
TG204	pBPRB37	29	69	abc
TG242	pBPRB37	16	67	abcd
TG182	pBI121	30	65	bcde
TG192	pBPRB37	14	61	cdef
TG145	pCa2C38	20	60	cdefg
Royal Gala		12	56	cdefgh
TG160	pBPRS1	33	55	cdefghi
TG142	pCa2C38	28	54	cdefghi
TG254	pWIC38	19	54	defghi
TG248	pWIC38	29	53	defghi
TG262	pWIC38	22	52	defghij
TG468	pCa2C38	29	51	efghijk
TG181	pBPRB37	29	51	efghijk
TG125	pCa2C38	14	49	fghijk
TG251	pWIC38	12	48	fghijkl
TG126	pCa2C38	34	47	fghijkl
TG545	pWIC38	22	47	fghijkl
TG172	pBI121	25	45	ghijkl
TG179	pBPRB37	17	44	hijkl
TG208	pBPRS1	32	44	hijkl
TG247	pWIC38	25	42	hijklm
TG193	pBPRB37	39	42	hijklm
TG171	pBI121	44	42	hijklmn
TG221	pWIC38	20	41	hijklmno
TG225	pWIC38	39	41	hijklmno
TG272	pWIC38	24	40	ijklmno
TG549	pWIC38	20	38	jklmno
Liberty		22	37	jklmno
TG546	pWIC38	10	37	jklmno
TG154	pBPRB37	30	36	klmno
TG228	pWIC38	25	36	klmno
TG201	pBPRB37	4	33	lmno
TG159	pBI121	29	28	mno
TG253	pWIC38	5	27	mno
TG547	pWIC38	5	27	no
TG250	pWIC38	28	26	o

[z]Observations with the same letter are not significantly different (p=0.05), as determined by Waller-Duncan K-ratio T test.

Northern analysis indicated that one hour after wounding *in vitro* grown plants of T138, the level of attacin E m-RNA increased in comparison to that of elongation factor, which is constitutively expressed. In contrast to the northern results, western analysis of the same plant material indicated that attacin E protein was expressed in non-wounded plants at a high level and an increase in protein levels in response to wounding could not be detected.

3.2. Scab resistance

The expression of endochitinase in transgenic apple resulted in high levels of scab resistance (Table 4) but also reduced plant vigor (data not shown). There was also a significant correlation between the level of exochitinase expression and scab resistance.

Table 3. Disease evaluation of 3-yr-old cecropin-transgenic Royal Gala apple plants in the field

Cultivar	Plasmid	N	% shoot blighted	Waller Group[z]
M.9		17	65	a
TG547	pWIC38	20	45	b
TG155	pCa2C38	9	43	bc
Royal Gala		9	42	bc
TG139	pBPRS1	17	41	bcd
TG265	ppWIC38	42	40	bcd
TG468	pCa2C38	23	38	bcde
TG242	pBPRB37	14	38	bcde
TG546	pWIC38	35	37	bcde
TG467	pCa2C38	22	37	bcde
TG266	pWIC38	34	36	bcde
TG545	pWIC38	20	36	bcde
TG253	pWIC38	9	35	bcde
TG227	pWIC38	25	35	cde
TG252	pWIC38	14	34	cdef
TG143	pBCCB37	10	33	cdef
TG171	pBI121	29	31	def
TG267	pBI121	15	31	def
TG549	pWIC38	18	31	def
TG548	pWIC38	10	29	ef
TG245	pWIC38	20	24	f
Liberty		24	8	g
M.7		22	6	g

[z]Observations with the same letter are not significantly different (p=0.05), as determined by Waller-Duncan K-ratio T test.

Table 4. Evaluation of disease resistance of own-rooted transgenic Marshall McIntosh lines, 14 days after inoculation with conidial suspension of *Venturia inaequalis*.

Transgenic line	ThEn-42 activity[w]	Number of lesions[x]	Percent area of infection[x]	Conidia recovered[x]
M. McIntosh [y]	0.03	71.8 a	57.5 a	248,833 a
T286 (vector) [z]	0.02	79.6 a	58.4 a	275,182 a
T566	0.09	71.6 a	51.8 a	139,526 a
T565	0.04	61.6 ab	44.7 ab	266,400 a
T563	0.35	37.0 bc	33.9 bc	10,900 ab
T562	0.57	32.8 cd	25.7 cd	21,440 ab
T568	3.28	15.8 cde	12.5 de	9,400 bc
T561	112.50	17.3 de	10.0 de	nd
T564	2.08	4.3 de	4.0 e	1,500 c
T560	78.60	0.0 e	0.0 e	4,750 c
Liberty	0.01	0.0 e	0.0 e	0 d

[w] ThEn-42 activity in nM MU per min per μg protein: Activity was determined by a microtiter-activity assay using the substrate 4-methylumbelliferyl-ß-D-N,N',N"-triacetylchitotrioside.

[x] Data presented are from the mean of four leaves per plant (five to seventeen plants/line). The conidial count data were transformed to natural logarithm and subjected to analysis of variance (ANOVA). Observations with the same letter are not significantly different (p=0.05), as determined by Waller-Duncan K-ratio T test. The experiment was repeated twice. nd: not determined.

[y] Non-transformed Marshall McIntosh.

[z] Transgenic Marshall McIntosh line transformed with pBI121 that does not contain the endochitinase gene.

The levels of resistance observed in exochitinase-transgenic lines were less than in endochitinase-transgenic lines, but exochitinase had no effect on plant vigor. When both enzymes were expressed *in planta* they acted synergistically to reduce disease. Transgenic Marshall McIntosh lines were identified with a low level of endochitinase expression and a high level of exochitinase expression that were resistant to scab and had negligible reduction in growth in a greenhouse trial.

4. Discussion and Prospects for Future Research

The results of the transgenic experiments show the potential for using lytic protein genes in apple to increase resistance to fire blight, while retaining normal fruit characteristics. More information is needed on field resistance and tree performance of transgenic apples. The similar level of resistance found in certain lytic protein and vector transgenic lines and the lack of significant correlation between detectable attacin E and field resistance suggests that biological variation, either random or somaclonal, may contribute to the observed resistance. Now that transgenic lines are flowering, progeny analysis from crosses will allow conclusive determination of the role of these transgenes in resistance.

The alfalfa mosaic virus leader sequence is being evaluated for its effect on gene expression [5], and a signal peptide sequence is being evaluated for its ability to target antimicrobial proteins to the intercellular space [8]. Besides cecropin-like genes, and attacin E, lysozyme genes from hen egg white and T4 bacteriophage have shown promising results in transgenic apple [7,14]. The *hrpN* gene of *E. amylovora* is being evaluated for its effect on fire blight resistance in M.26 apple rootstock [1].

The results reported here indicate the potential for use of genes for chitinolytic enzymes to enhance resistance to apple scab. The chitinase-transgenic Marshall McIntosh plants are currently being evaluated under field conditions for scab resistance, tree performance, and fruit quality and yield. To avoid the negative effects of endochitinase expression on apple vigor we are exploring the utility of other cell wall degrading enzymes and the development of pathogen-specific expression of the endochitinase. The glucanase gene of *Trichoderma harzianum* is being cloned and will be evaluated for its effect on apple scab resistance, both singly and in combination with exochitinase. Various promoters are being evaluated in apple for induction by *V. inaequalis* to allow for expression of endochitinase only under conditions of pathogen invasion.

The transgenic lines reported in this paper are experimental. Transgenic lines designed for use in commercial apple growing will likely differ in genes, promoters, and regulatory sequences from those described here. Before being commercialized, transgenic apple cultivars will go through rigorous deregulation requirements to ensure safety for consumers, the environment, and agriculture.

References

[1]   A.M. Abdul-Kader *et al.*, Evaluation of the *hrpN* gene for resistance to fire blight in apple. *Acta Horticulturae* **489** (1999) 247-250. (Proc. 8th International Workshop on Fire Blight. Kusadasi, Turkey. October, 1998).

[2]   J.P. Bolar *et al.*, Factors affecting the transformation of 'Marshall McIntosh' apple by *Agrobacterium tumefaciens*. *Plant Cell, Tissue and Organ Culture* **55** (1999) 31-38.

[3]   J.P. Bolar *et al.*, Expression of endochitinase from *Trichoderma harzianum* in trangenic apple increases resistance to apple scab and stunts growth. *Phytopathology* **90** (2000) 72-77.

[4]   E. Borejsza-Wysocka *et al.*, Transformation of authentic M.26 apple rootstock for enhanced resistance to fire blight. *Acta Horticulturae* **489** (1999) 259. (Proc. 8th International Workshop on Fire Blight. Kusadasi, Turkey. October, 1998).

[5]   R.S.S. Dalta, *et al.*, Improved high-level constitutive foreign gene expression in plants using an AMV RNA4 untranslated leader sequence. *Plant Science* **94** (1993) 139-149.

[6]   K. Düring *et al.*, Transgenic potato plants resistant to the phytopathogenic bacterium *Erwinia carotovora*. *Plant J.* **3** (1993) 587-598.

[7]   V. Hanke *et al.*, Transformation of apple cultivars with T4-lysozyme-gene to increase fire blight resistance. *Acta Horticulturae* **489** (1999) 253-256. (Proc. 8th International Workshop on Fire Blight. Kusadasi, Turkey. October, 1998).

[8]   K. Ko *et al.*, Effects of multiple transgenes on resistance to fire blight of Galaxy apple. *Acta Horticulturae* **489** (1999) 257. (Proc 8th International Workshop on Fire Blight. Kusadasi, Turkey. October, 1998).

[9]   J.M. Jaynes *et al.*, Increasing bacterial disease resistance in plants utilizing antibacterial genes from insects. *Bioassays* **6** (1987) 263-270.

[10] M. Lorito *et al.*, Genes from mycoparasitic fungi as a source for improving plant resistance to fungal pathogens. Proc. Natl. Acad. Sci. **95** (1998) 7860-7865.

[11] W.E. MacHardy, Apple Scab: Biology, Epidemiology, and Management. The American Phytopathological Society, St. Paul, Mn, 1996.

[12] F. Mourgues *et al.*, Activity of different antibacterial peptides on *Erwinia amylovora* growth, and evaluation of the phytotoxicity and stability of cecropins. *Plant Sci.* **139** (1998) 83-91.

[13] J.L. Norelli *et al.*, Transgenic 'Malling 26' apple expressing the attacin E gene has increased resistance to *Erwinia amylovora*. *Euphytica* **77** (1994) 123-128.

[14] J.L. Norelli *et al.*, Genetic transformation for fire blight resistance in apple. Acta Horticulturae **489** (1999) 295-296. (Proc. 8th International Workshop on Fire Blight. Kusadasi, Turkey. October, 1998)

[15] J.L. Norelli *et al.*, Leaf wounding increases efficiency of *Agrobacterium*-mediated transformation of apple. *HortScience* **31** (1996) 1026-1027.

[16] J.L. Norelli *et al.*, Effect of cecropin-like transgenes on fire blight resistance. *Acta Horticulturae* **489** (1999) 273-278. (Proc. 8th International Workshop on Fire Blight. Kusadasi, Turkey. October, 1998).

[17] L. Parisi *et al.*, A new race of *Venturia inaequalis* virulent to apples with resistance due to the *Vf* gene. *Phytopathology* **83** (1993) 533-537.

[18] J.P. Reynoird *et al.*, First evidence for improved resistance to fire blight among transgenic pears expressing the *attacin* E gene from *Hyalophora cecropia. Plant Sci.* **149** (1999) 23-31.

[19] T. Van der Zwet and S.V. Beer, Fire blight - Its nature, prevention and control. USDA Agricultural Information Bulletin. No. 631. (2nd revision) 1999, pp.91.

[20] K-W. Wong *et al.*, Chitinase- transgenic lines of Royal Gala apple showing enhanced resistance to apple scab. *Acta Horticulturae* **484** (1999) 595-599. (Eucarpia Sym. Fruit Breeding and Genetics, Oxford, UK, 1996).

*Use of Agriculturally Important
Genes in Biotechnology
G. Hrazdina (Ed.)
IOS Press, 2000*

Performance of Virus Resistant Transgenic Potatoes

B. Flis[1], A. Chachulska[2], A. Palucha [2]

[1]*Plant Breeding and Acclimatization Institute, Research Center Mlochów, 05–832 Rozalin,
Poland*
[2]*Institute of Biochemistry and Biophysics, Polish Academy of Sciences,
ul. Pawinskiego 5A, 02–106 Warsaw, Poland*

Abstract. The aim of the presented study was an evaluation of performance of transgenic potato clones with introduced genes, which might improve level of resistance to potato leafroll virus (PLRV) in *cv.* Bzura and to necrotic isolate of virus Y (PVY^N) in *cv.* Irga. The *cv.* Irga was transformed with the truncated PVY^N polymerase gene in the sense and antisense orientation. The *cv.* Bzura was transformed with 4 variants of the gene CP PLRV in both orientations. Agronomical and morphological traits of vine and tubers were evaluated during propagation in a greenhouse and in field experiments. Performance of the transgenic clones was compared with non–transgenic plants of both cultivars, which were established from seed tubers and tissue culture. Most of the evaluated traits of transgenic clones expressed variability (the most variable was tuber yield and the less – the duration of vegetation period). This variability seems to depend on the genotype of the original cultivar and the variant of the introduced transgene. The significance of deviations from the true types of transformed cultivars was not related to the phenotypic expression of the transgenes. Generally, the selection within a set of transgenic clones is a necessary step in the breeding of transgenic cultivars. The chance to select desirable individual seems to be higher within transgenic clones than in conventional breeding, since eight resistant clones were selected out of 33 originating from *cv.* Irga and two resistant clones out of 51 from *cv.* Bzura. Some resistant clones were close to type of the original cultivar.

1. Introduction

Genetic transformation is believed to be a method which enables selective improvement of a single trait of valuable potato cultivars while maintaining key characteristics. It is known that this might not be easy to achieve, since genetic transformation usually evokes some side effects. Within a population of regenerated plants with the same transgene variability was observed that results in altered performance of these clones in comparison with the non–transgenic plants. The main sources of variability are: (a) insertional mutagenesis, (b) pleiotropic effects and (c) somaclonal variation [1].

The aim of the presented study was to evaluate the performance of transgenic clones under greenhouse and field conditions, and to compare these clones with non–transgenic plants. The transgenic clones had introduced viral genes in order to improve resistance to potato leafroll virus (PLRV) or to necrotic strains of potato virus Y (PVY^N).

2. Material and Methods

Two Polish potato cultivars, Irga and Bzura, were transformed with viral genome sequences in order to improve their resistance to PVY^N or PLRV. All transgenic clones were produced at the Institute of Biochemistry and Biophysics. The constructs introduced into the genome of *cv.* Irga contained a truncated gene coding for PVY^N replicase in sense and antisense orientations [2]. The constructs introduced into the genome of *cv.* Bzura contained four variants (in both orientations) of the gene coding for the coat protein of the leafroll virus (PLRV) –[3]. Each transgenic clone originated from a transformation event – 33 clones were from *cv.* Irga and 51 – from *cv.* Bzura.

Two types of non–transgenic control plants were established: plants of the original cultivars regenerated from leaf cuttings propagated like transgenic plants and plants growing from seed tubers.

Transgenic and non–transgenic plantlets propagated *in vitro* were planted in pots to obtain minitubers. Next year minitubers were planted in large pots in a greenhouse in order to produce tubers for planting in field trials.

During the propagation in the greenhouse data were collected about tuber yield, mean tuber weight and general appearance of the plants.

Each transgenic clone in the field trial grew in a three–hill plot with four replicates in randomized complete block designs. Non–transgenic control plants were represented by doubled number of plots.

Plant performance was measured in terms of: (1) tuber yield (dag/hill), (2) mean tuber weight (g), (3) starch content (%), (4) appearance of plants (1–5, 1 – plants deformed, 5 – plants true to type), (5) plant vigour (1–5, 1 – weak, 5 – vigorous), (6) abundance of inflorescence (1–5, 1 – poor, 5 – abundant), (7) tuber shape (1–7, 1 – shortened, 7 – long), (8) vegetation period (days), (9) regularity of tuber shape (1–9, 1 – irregular, 9 – the most regular), (10) depth of eyes (1–9, 1 – deep, 9 – very shallow) and (11) secondary growth of tubers (% of total number of tubers/hill).

In general, the higher values of ranked traits had positive meaning. In case of tuber shape higher value means "more elongated."

3. Results

The mean value from transgenic clones from *cv.* Irga were not different from the mean value of non–transgenic control plants. However, some tendencies were noticed. The mean tuber yield of transgenic plants was higher than yield of non–transgenic plants. Transgenic plants flowered more abundantly and expressed very distinct tendency toward secondary growth of tubers. Only few transgenic clones produced phenotypically-changed plants and these changes were not clearly visible (smaller vine and leaflets or rolling apical leaflets) (Fig. 1).

Nevertheless, values obtained from transgenic plants varied significantly, with exception of abundance of inflorescence and secondary growth of tubers. Deviating transgenic clones were not numerous, since mean values did not differ from mean values of non–transgenic plants.

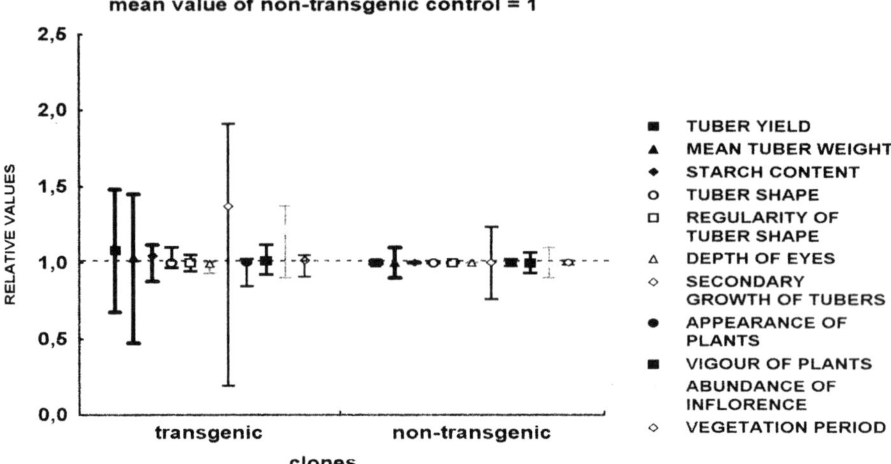

Fig. 1. Mean, minimum and maximum values of traits measured in the field experiment with transgenic clones from *cv.* Irga

There was some variability within non–transgenic plants (differences between seed tuber and tissue culture derived plants) and it was observed for the vigour of plants (significant difference), secondary growth of tubers, mean tuber weight and abundance of inflorescence.

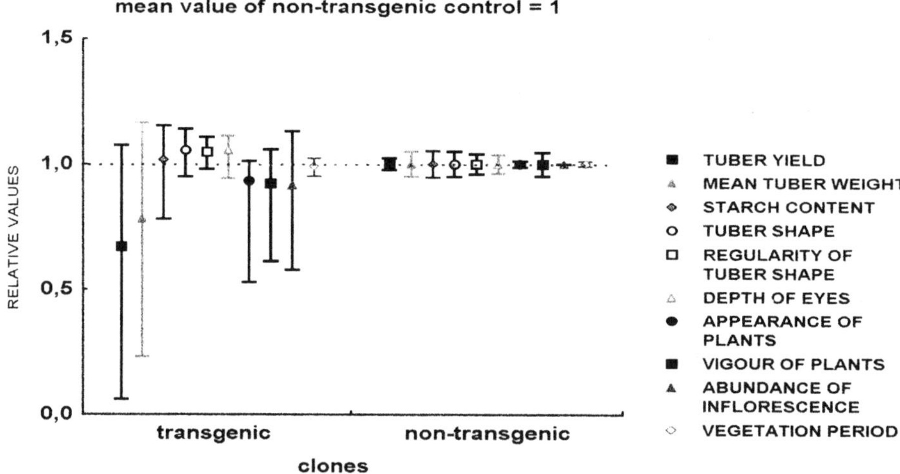

Fig. 2. Mean, minimum and maximum values of traits measured in the field experiment with transgenic clones from *cv.* Bzura

The group of eight highly resistant clones, which were selected from all tested clones [4], expressed tendency toward lower tuber yield and smaller tubers in comparison to other transgenic clones and higher starch content, more abundant flowering, more pronounced secondary growth of tubers in comparison to non–transgenic control plants. The values of other traits kept the same tendencies as observed for all transgenic clones. Most of the highly resistant clones were close to the type of original cultivar or deviated from the type only in respect of traits which are known to be sensitive to environmental factors.

Transgenic clones from *cv.* Bzura (Fig. 2) expressed significantly lowered mean values for tuber yield, mean tuber weight and vigour of plants in comparison with mean values of non–transgenic plants.

The mean values describing tubers of transgenic clones were higher than values of non–transgenic plants, what means that tubers were significantly more elongated and with tendency toward more regular shape and shallower eyes. A large group of transgenic clones, about 15%, produced abnormal plants with malformed or rugose leaflets and with short stem internodes. There was also a group of transgenic clones with less severe changes, i.e. more compact vine with smaller, but more abundantly growing leaflets.

Within transgenic clones significant variability for all tested traits was observed (the vegetation period was an exception).

It can be noticed that there was significant variability within non–transgenic plants. The non–transgenic regenerants had significantly higher starch content and values describing regularity of tuber shape and depth of eyes in comparison to non–transgenic plants from seed tubers (Fig. 2).

Table 1. Relation between measurments obtained in the greenhouse and in next year's field experiments

Trait	Correlation coefficients for	
	transgenic clones from cv. Irga	transgenic clones from cv. Bzura
Tuber yield	0.38 *	0.76 **
Mean tuber weight	0.45 **	0.41 **
Tuber shape	0.14	0.65 **
Appearance of plants	off-types observed only in the field	off-types observed in the greenhouse and in the field

* $p<0.0$; ** $p<0.01$

Two transgenic clones resistant to PLRV were selected. One of these clones differed from the original cultivar only in the respect of tuber morphology (elongated tubers with more regular shape and shallower eyes). The other one had very low tuber yield and its plants differed from the phenotype (compact vine with smaller leaflets). It is interesting that this clone at the first stage of propagation (*in vitro* plants grown in pots) did not differ from non–transgenic plants and produced sufficient number of minitubers. This was an indication that at first stages of propagation of transgenic plants their field performance is not predictable. This was confirmed by comparing measurements of some traits obtained in the greenhouse and in the field experiment. Generally, these measurements showed poor correlation (Table 1). For transgenic clones from *cv.* Irga measurements of tuber yield and mean tuber weight obtained in the greenhouse and in the field showed weak correlation. Surprisingly, tuber shape showed no correlation at all. Under greenhouse conditions abnormal plants were not identified. For transgenic clones of *cv.* Bzura tuber yield and tuber shape measured in both conditions showed good correlation, but again, rather weakly. Clones with atypical and abnormal plants were observed during propagation in the greenhouse.

Table 2. Mean values of traits measured for transgenic clones with different constructs introduced into *cv*. Irga

Introduced construct[a]	No of clones	Traits for which variability between groups of clones was observed			
		Tuber yield (dag/hill)	Mean tuber weight (g)	Plants vigour (1–5)	Regularity of tuber shape (1–9)
R1	15	430	69	4.3	6.4
R2	18	368	60	3.8	6.7

[a] R1 = sense orientation and R2= antisense

Some variability between groups of clones containing different constructs was observed. In the case of the transgenic clones from *cv*. Irga significant differences were observed for tuber yield, mean tuber weight and regularity of tuber shape. For other traits no differences were noticed. Clones with the construct R1 had higher tuber yield and bigger tubers with less regular shape and more vigorously growing plants than clones with the R2 construct (Table 2). It is important that plants of clones R1 produced higher mean tuber yield than non–transgenic plants.

Table 3. Mean values of traits measured for transgenic clones with different constructs introduced into *cv*. Bzura.

Introduced construct[a]	No of clones	Traits for which variability between groups of clones was observed			
		Tuber yield (dag/hill)	Tuber shape (1–7)	Regularity of tuber shape (1–9)	Plants vigour (1–5)
pRPC1s	2	454	4.8	6.9	3.9
pRA4a	13	448	5.2	6.0	4.0
pRA7s	12	414	5.3	6.2	3.7
pRA7a	7	377	5.2	6.5	3.8
pBCP1s	8	271	4.8	6.3	3.0
pRA4s	2	254	4.9	6.3	2.8
pBCP1a	2	243	4.0	6.4	3.0
pRCP1a	3	218	5.3	5.7	2.5

[a] s = sense and a= antisense orientation

Transgenic clones from cv. Bzura can be grouped into eight classes according to the introduced construct type. Variability between such groups was observed for 4 out of 10 measured traits (Table 3). Generally, variation was due to differences between mean tuber yield, that correlated with plant vigour. Plants of the groups pRA4a, pRA7s, pRA7a and pRCP1a produced tubers with significantly changed tuber shape (long tubers). Plants of the group pRCP1s produced tubers with the most regular shape.

4. Conclusions

We conclude that there is a substantial variability between transgenic clones. Mean values of some traits of transgenic clones differed significantly from values of non–transgenic plants. However, these mean values were not always shifted toward lower values. Especially, some clones produced higher tuber yield than non–transgenic plants.

It seems that the amount of variability among transgenic clones depends on genotype of the original cultivar, since transgenic clones from *cv* Bzura showed higher variability than clones from *cv.* Irga. The phenotypical expression of the transgene (i.e. high level of resistance) was combined with poorer performance (especially lower tuber yield), but resistant clones identical or close to the original cultivar in all respects have also been identified.

It seems that specific constructs evoke similar changes in all plants of all clones that contain it.

Generally, transformation is an effective method of introducing resistance, since 8 clones resistant to PVY^N were selected out of 33 and 2 clones resistant to leafroll virus were selected out of 51. However, the transformation process and/or transformation effects produce undesired variability, that requires selection within populations of transgenic clones. The key problem is that testing in the field is essential for proper evaluation of transgenic plants.

References

[1]   W.R. Belknap *et al.*, Field performance of transgenic potato. In: W.R. Belknap, M.E. Vayda and W.D. Park (eds), The Molecular and Cellular Biology of the Potato, 1994, pp. 233–243.
[2]   A.M. Chachulska *et al.*, Potato and tobacco cultivars transformation towards potato virus Y resistance. Biotechnologia **4**(39) (1997) 48–54.
[3]   A. Palucha *et al.*, Otrzymanie transgenicznego ziemniaka odpornego na infekcje wirusem lieciozwoju ziemniaka. (in Polish with English summary) Biotechnologia, **4**(39) (1997) 38-47.
[4]   B. Flis *et al.*, Transgenic potato clones resistant to potato virus Y. Abstracts 10th EAPR Virology Section Meeting, Baden, Austria, 1998, pp. 37–38.

*Use of Agriculturally Important
Genes in Biotechnology
G. Hrazdina (Ed.)
IOS Press, 2000*

Molecular Phylogeny as a Tool for Controlling Potato Virus Y Spread

W. Zagórski[1], A. Chachulska[1], A. Lipska-Dwuznik[1], B. Flis[2], A. Palucha[1]

[1]*Institute of Biochemistry and Biophysics, Polish Academy of Sciences, ul. Pawinskiego
5A, 02130 Warszawa, Poland*

[2]*Plant Breeding and Acclimatization Institute, Radzik6w, 05-870, Blonie, Poland*

Abstract. Selected regions of three Polish PVY isolates were sequenced. The analysis of the sequences allowed us to pin-point the near identity of the common potato virus strain isolate PVY^0-LW and the necrotic one PVY^N-Wi, and their divergence from the second necrotic isolate -PVY^N-Ny. PVY isolates (studied and those whose sequences were retrieved from databases) were clustered into three groups on the basis of the 5' and 3' terminal region sequence comparisons. The molecular taxonomy of PVY differs from the classical one. Group I is rather homogenous, grouping only necrotic isolates, group II is highly heterogeneous, encompassing mostly necrotic and common strain potato isolates. Group III is the smallest, also heterogeneous, composed mostly of non potato isolates of various geographical origins. The sequence analysis led to the proposition of localisation of group I-specific epitopes and to the exclusion of the terminal genome regions as responsible for determination of symptom severity and tobacco necrosis induction. Full-length cDNA genome copies of the three isolates studied were synthesised and amplified. The RFLP analysis of full genomes confirmed the polymorphism deduced from partial sequencing and allowed the localisation of a polymorphic region differentiating PVY^0-LW and PVY^N-Wi isolates, probably involved in the induction of necrosis in tobacco. Binary vectors carrying the PVY polymerase and NTR cDNAs were constructed in order to engineer plants resistant to PVY. Tobacco cv. Xanthi and SR1 and potato cv. Irga were transformed by an Agrobacterium mediated transformation system and transgenic lines were regenerated. The transgene-carrying lines were submitted to preliminary PVY^N-Wi resistance tests. Several tobacco and potato lines highly or partially resistant to PVY were selected and subjected to further analysis.

1. Introduction

In Poland PVY infects agriculturally important species such as tobacco, potato, pepper, and tomato. It is transmitted mechanically or by aphids in a non-persistent, stylet-borne manner [1].

The conventional phytopathological classification of PVY isolates is based on primary hosts and on symptoms induced in indicator plants. Potato isolates have been separated into three main strains according to the symptoms induced on potato and tobacco [1]. The term "PVY strain" defines a collection of isolates sharing symptomatic properties [2]. The PVY^N strain induces severe systemic "tobacco veinal necrosis" in *Nicotiana tabacum* and mild mosaic or mottling symptoms in most potato cultivars [1, 3]. Some PVY^N isolates were shown, however, to induce a quite severe potato tuber necrotic disease [4], and share biological properties that distinguish them from classical PVY^N representatives [4]. They are referred to as PVY^{NTN} isolates (N-for the necrotic group, TN-for tuber necrosis). The PVY^0 strain (common strain) induces a non-necrotic mosaic in tobacco and more severe symptoms on potato (such as crinkling, leaf dropping or severe necrotic mosaic) [1, 3]. The PVY^c strain comprises PVY isolates inducing "stipple streak" symptoms on potato cultivars bearing the Nc resistance gene. In *N. tabacum,* these isolates induce symptoms similar to those of the PVY^0 strain [1].

The PVY^0-LW isolate was identified in the early seventies in the potato cultivar Lipinski Wczesny, as a member of the common strain of PVY. It mainly causes local necroses, leaf dropping and distinct mosaic symptoms on primarily infected plants, inducing severe mosaic and leaf malformation on secondary infected plants.

The PVY^N -Ny isolate was identified in 1974 in the cultivar Nysa, as a representative of the N strain on the basis of its ability to induce tobacco veinal necrosis. After infection with PVY^N-Ny, most potato cultivars express distinct vein necrosis or severe mosaic and leaf malformation.

The PVY^N-Wi isolate was identified in 1984 in the cultivar Wilga, as a representative of numerous necrotic isolates found recently in seed potatoes grown in northern Poland. This new isolate is highly infective, it infects cultivars resistant to PVY^N -Ny, and accumulates to high concentrations in a shorter time than the older necrotic isolates, but it induces weaker symptoms. This group of isolates is spreading in Poland and has been responsible for 80-96% of PVY infections in potato fields during the last three years [5].

2. Results and Discussion

The new PVY isolate spreading fast in Polish fields (PVY^N -Wi) has unusual biological properties. Phytopathologically, it is classified as a necrotic strain, but immunologically - as a representative of the common strain. According to the simplest hypothesis, this new PVY version could represent a recombined genome of two halves coming from distant parents. A proper evolutionary positioning of this isolate needs therefore a molecular analysis of both termini of the genome. We decided to sequence both non-coding terminal regions of the atypical isolate and, for comparison, the same regions from two typical isolates from Polish collections, representing the necrotic and common PVY strains. The analysis was reinforced by sequencing of the 5' terminal regions coding for the P1 protein and 3' terminal regions, coding for the coat protein (CP). The non-coding genome regions are under selective pressure exerted on the sequence and secondary RNA structure, whereas the coding regions are additionally under pressure exerted on protein products. Therefore the genetic stability, mutation frequencies, and capacity to undergo RNA recombination of coding and non-coding regions may differ and a proper molecular comparison of the isolates analyzed required a significant sequencing effort. Analysis of molecular data shows that PVY^N-Wi is not a recombinant, and it shares its taxonomical position with common strain PVY^0-LW (Figure 1).

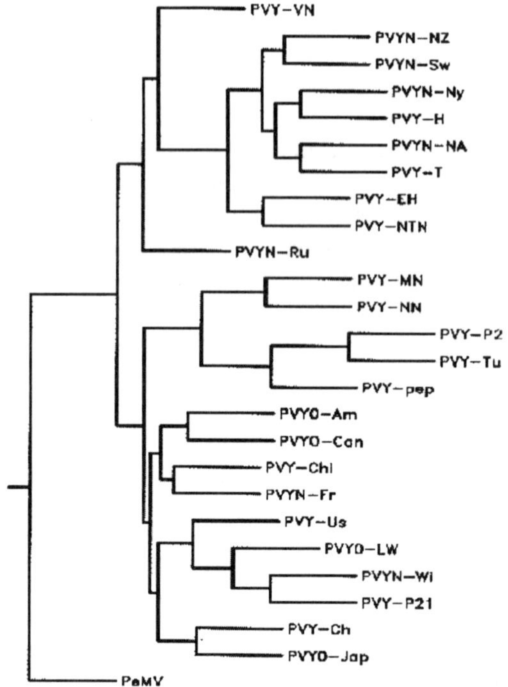

Figure 1. Phylogenetic parsimonious tree obtained with the SEQBOOT, PROTPARS and CONSENSE programs from the PHYLIP 3.5c package and drawn using the DRAWGRAM program. The tree is unrooted, the PeMV sequence is included as an outgroup to simulate the root of the tree.

The sequence comparison showed that the non-necrotic PVY^0-LW and necrotic PVY^N-Wi isolates display a near identity in the sequences analyzed (only two nucleotide and amino acid changes in the CP gene, some clones showing single nucleotide changes in the P1 gene). The second necrotic isolate PVY^N-Ny, falls into a separate taxonomic group. This group assignment of the isolates analyzed is unambiguous, being based on a large set of comparative data on the PVY sequences in two terminal genome regions. This reveals that the ability to induce tobacco veinal necrosis does not follow the molecular classification based on the analysis either of the 5'NTR and P1 gene or of the CP and 3'NTR sequences. The CP cistrons of the common PVY^O-LW and necrotic (PVY^N-Wi) isolates are almost the same (99% amino acid identity), devoid of putative necrotic type epitopes and the capsid antigenic properties of both isolates are identical. This confirms the suggestion by Van der Vlugt *et* al. [6] that the CP gene does not carry the tobacco necrotisation trait.

The identity of the 3'NTR sequences of PVY^0-LW (non-necrotic isolate) and PVY^N-Wi (necrotic isolate) is contradictory to the hypothesis that the tobacco necrotisation trait might be encoded by the PVY 3'NTR, as proposed by Van der Vlugt *et* al. [6]. This conclusion is supported by the analysis of the PVY^N -Ru isolate, which is assigned to

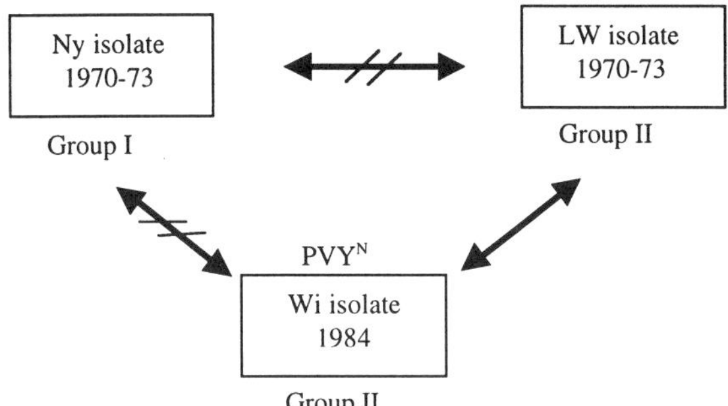

Figure 2. Putative origin of a new highly infectious PVY^N isolate (PVY^N-Wi), deduced on the basis of the genome structure analysis.

group I (Figure 1) according to the sequence data (97% homology with the 3'NTR of the common isolate PVY^0 - LW) and according to the 3'NTR secondary structures [6], and is phenotypically necrotic.

This strongly suggests that the capacity to induce tobacco veinal necrosis is coded by a specific, so far unidentified genome fragment, outside the CP and P1 genes and 5' and 3' NTRs. The natural history of PVY^N-Wi, which appeared more recently than PVY^0-LW, also indicates that the acquisition of this trait may result from a subtle mutational event or limited sequence rearrangements (Figure 2).

Sequence analysis results are in line with RFLP data (not shown) indicating near identity of PVY^0-LW and PVY^N-Wi . These two strains differ in the position of the *Ava*II restriction site in the RNA polymerase gene. Such subtle difference may arise by a quite limited number of point mutations. Taking into account that mutations overcoming natural resistance to necrotic isolates seem to be localized in the viral RNA polymerase gene, we constructed several lines of transgenic tobacco and potato plants (cultivar Irga) carrying a truncated or full length polymerase gene in either (+) or (-) orientation.

Thirty-seven tobacco transformants of R$_0$ generation were transferred to the soil and subjected to the resistance tests. Following the inoculation with PVY^N-Wi, the plants were monitored for several weeks for disease symptoms. In the majority of transformants, mild symptoms of PVY^N infection were noticed, consisting of vein clearing and banding or chlorotic mosaic, but veinal necrosis was observed very rarely. The immunological tests, performed about two weeks after inoculation, allowed identification of seven transformants that were partially or totally resistant to PVY, showing nil or very low ELISA values. The results of immunological tests, performed about two weeks after infection, were in line with the phytopathological evaluation. Plants negative in the ELISA tests were symptom-free and plants with a low ELISA titer showed attenuated disease symptoms. In certain transformants with initial high ELISA values a "recovery" phenotype could also be observed, featuring a decrease of ELISA values and symptom disappearance in newly developing leaves.

Thirty-four potato clones were tested for resistance to PVY. The disease symptoms started to develop in the fourth week post inoculation. They were limited to a systemic mosaic. Six clones were symptom-free with very low ELISA values, similar to those for the non-infected control plants and for inoculated resistant standard plants. Two of the resistant clones carry the sense RNA-expressing transgene, the other four are antisense RNA expressing clones. This shows the potential of antisense RNA expression in engineering potato resistance against novel versions of viral genomes.

References

[1]   J.A. De Bokx and H. Huttinga, Potato virus Y. CMI/AAB Descriptions of Plant Viruses, n° 242. 1981.

[2]   O.W. Barnett, A summary of potyvirus taxonomy and definitions. *Arch. Virol.,* Supplementurn **5** (1992) 435-444.

[3]   M. Chrzanowska, New isolates of the necrotic strain of potato virus y (PVY^N) found recently in Poland. *Potato Res.* **34** (1991) 179-182.

[4]   M. Le Romancer and C. Kerlan, La maladie des necroses annulaires superficielles des tubercules: une affection de la pomme de terre due au virus Y. *Agronomie* **11** (1991) 889-900.

[5]   M. Chrzanowska and T. Doroszewska, Comparison between PVY isolates obtained from potato and tobacco plants grown in Poland. *PhytopathoL Polon.* **34** (1997) 179-182.

[6]   R.A.A. Van Der Vlugt *et al.*, Taxonomic relationships between distinct potato virus Y isolates based on detailed comparisons of the viral coat proteins and Y-nontranslated regions. *Arch. Virol.* **131** (1993) 361-375.

Use of Agriculturally Important
Genes in Biotechnology
G. Hrazdina (Ed.)
IOS Press, 2000

Production of Herbicide Tolerant Rice by *in vitro* Selection

O. Toldi[1], S. Tóth[1 2], A.S. Oreifig[1 2], E. Kiss[2] and B. Jenes[1]
[1] *Agricultural Biotechnology Center, H-2101 Gödöllö, P.O. Box 411, Hungary*
[2] *University of Agricultural Sciences, H-2100 Gödöllö, Páter K. u. 1., Hungary*

Abstract. A novel procedure has been developed to produce tolerant rice (*Oryza sativa* L.) to the herbicide phosphinothricin (hereinafter: PPT) by *in vitro* selection. First, the sublethal and lethal concentrations of the PPT on 7 days-old seedlings were determined and morphogenetic events during PPT treatment were evaluated. Differentiation of 6-30 microshoots on 5-40% of the treated plant material was observed on a hormone-free culture medium supplemented with PPT at a sublethal concentration. We proved that the PPT is morphogenetically active, similarly to many other herbicides showing cytokinin-like effects in rice tissue culture. Subsequently, fertile plants were grown from these microshoots having PPT resistance under greenhouse conditions. According to our knowledge, this is the first report on production of tolerant rice plants to this herbicide without genetic transformation. Since PPT is an inhibitor of glutamine synthetase (GS), which plays a central role in the assimilation of inorganic nitrogen and in the regulation of the nitrogen metabolism, GS activity in PPT-resistant and PPT-sensitive plants was examined comprehensively. Differences in nitrogen assimilation, intracellular ammonia cycling and endogenous hormone balance between these plant groups are discussed.

1. Introduction

The use of herbicides to control weeds allows growers to employ more efficient crop management, and increases yields. Several classes of herbicides are quite effective for broad spectrum weed control, but they are either non-selective and kill crop plants or significantly injure some crops at the application rates required. The development of herbicide tolerant cultivars is a way that crops can be protected from herbicide damage. Three general methods have been used to generate herbicide tolerant crops: germplasm screening (mutant isolation by conventional breeding), *in vitro* selection (mutant isolation in cell cultures) and direct insertion of herbicide-tolerant genes using the genetic engineering approach (creating mutants) [1].

Since the culturing of transgenic crops in EU countries is still prohibited, we were interested in the development of a method for producing PPT-tolerant rice without genetic tranformation. Therefore, experiments were planned and carried out to monitor the main morphogenetic events during *in vitro* PPT-treatments of rice seedlings, because there had been reports about hormone-like activity of several herbicides applied in sublethal concentrations. This work was inspired by the published results of Hoshino and Mii [2], who presented evidence that PPT is a morphogenetically active herbicide on snapdragon (*Antirrihium majus* L.), where it showed cytokinin-like effects stimulating shoot differentiation and subsequent plant regeneration. However, these PPT-induced *Antirrhium* plants had not been tested for PPT resistance. The fact that herbicide resistant wheat calli had been isolated after selection on sublethal doses of PPT [3, 4] drove us to go further and screen plants regenerated on PPT for PPT resistance. That is, we hypothesized that plantlets responding positively to PPT-treatment will possess some degree of resistance to this herbicide.

2. Materials and methods

2.1. Plant material, surface sterilization of seeds and in vitro germination

Rice (*Oryza sativa* L.) cultivar Taipei 309 was chosen for the experimental work. Mature seeds were dehusked and then surface sterilized in 70% (v/w) ethanol for 1 min. and in 30-50% (v/w) Domestos (Unilever Hungary Kft.) solution supplemented with a few drops of Tween 20, for 30 min. The surface sterilized seeds were germinated on sugar-free GM medium [5] containing half-strenght MS elements, 10 mg/l thiamine, filter sterilized GA_3 (0.1 mg/l) and 7 g/l agar in plastic petri dishes. The pH of germination medium was adjusted to 5.8 with 1M KOH. The seeds were positioned embryo side facing upwards and half embedded in the medium. Germination occurred at 27°C under a 16 hr light/ 8 hr dark regime.

2.2. Microshoot induction and plant regeneration

Rice plants with coleoptiles reaching 5-8 mm in length were excised from the scutellum after 3-7 days of germination and placed onto a hormone-free MS medium [6] supplemented with MS macro and microelements, MS vitamins, 30 g/l sucrose, 8 g/l agar and filter sterilized PPT in different concentrations (1.0, 2.0, 3.0, 4.0, 5.0, 10.0 mg/l). The pH of the induction medium was adjusted to 5.8 with 1M KOH. Microshoot induction was carried out in plastic petri dishes (10 plantlets / petri dish) at 27 °C, in a 16 hr light / 8 hr dark regime. Differentiating microshoots were kept on the induction medium until they reached 30-40 mm height (this usually took 3-4 weeks).

2.3. In vitro and in vivo test to determine PPT tolerance

The R0 generation of microshoot originated plants cultured on PPT and the *in vitro* cultured but PPT non-treated control plants were placed onto hormone-free N6 medium [7] supplemented with N6 elements and vitamins, 30 g/l sucrose, 2.5 g/l Gelrite and 2.0 mg/l PPT (pH 5.8) in order to test their PPT tolerance under *in vitro* conditions. After four weeks of incubation at 27 °C, in a 16 hr light / 8 hr dark regime, when the PPT tolerant plants reached 60-80 mm height in VegBox plastic containers, they were divided into two groups. Some growing characteristics (the fresh and dry weight[1] of both the shoot and root system, plant height, degree of tillering) of plants in the first group and of the control plants were determined. At the same time, the second group of the plants was transferred into pots in greenhouse. After a short acclimatization period, several hundred of these plants were sprayed twice with a lethal dose (5 ml herbicide / 500 ml water / 1 m^2 soil surface) of FINALE® (a herbicide with PPT as active ingredient) to test their PPT tolerance. Other plants grown in the greenhouse were used to compare some agrobiological characteristics of microshoot-derived plants to those of *in vitro* culture-originated but PPT non-treated ones, and to the greenhouse-grown control plants. In this comparison, the fresh and dry weight (separately the shoot and root system), plant height, 1000 kernel weight and the number of grains per panicle were determined in mature plants after seed set.

---

[1] dry weight determinations were carried out by drying samples in a laminar air flow until constant weight was achieved.

2.4. PPT tolerance in progenies (R1) of microshoot derived plants

Seeds of R0 generation of PPT-treated microshoot derived plants and control plants were collected. These seeds were dried, then all the *in vitro* treatments (germination, microshoot induction, plant regeneration, PPT-tolerance tests) and measurements in the greenhouse were repeated on these R1 plants to decide whether PPT tolerance could be still detected in the next generation. The only difference between the treatments of the R0 and R1 generations was that the seeds of microshoot derived R1 plants were sown in a GM medium containing lethal dose of PPT (5.0 mg/l), instead of PPT-free medium, to test the PPT tolerance of R1 seedlings.

2.5. Determination of GS activity

Frozen leaf segments of R0 plants and R1 seedlings were homogenized, then the supernatant fraction after the centrifugation of the homogenate at 27000 g for 20 min was used to assay GS activity as described by Kamachi et al. [8]. One unit of enzyme activity was defined as that amount that synthetized 1 micromol of product / min at 30^0C.

2.6. Statistical analysis

The investigations were done in several repetitions, in most cases 3 or 5 times. This information is shown in each case in the appropriate section of 'Results' or in the legend of the corresponding tables. The data was evaluated by mean analysis, and standard deviations were calculated from mean differences [9].

3. Results

3.1. PPT-mediated induction of microshoot differentiation

After 3-7 days of *in vitro* germination, when the coleoptiles reached 5-8 mm in length, the plantlets were carefully separated from the seeds leaving the scutellum behind. One hundred plantlets of each treatment in 3 repetitions were placed in a vertical position onto PPT-free or microshoot induction media containing PPT in different concentrations. Statistical analysis was done after 1 month of incubation. The results demonstrated that PPT was effective in a narrow range of concentrations in generating cytokinin-like morphogenetic effects. This phenomenon was even more marked when the microshoot differentiation was evaluated by the number of plants regenerated on each explant. The significantly effective microshoot development on induction medium supplemented with 2.0 mg/l PPT was mostly due to the fact that the process of differentiation was prolonged and for this reason, several microshoots at different stages of development were present on the same explant. The plants incubated on the medium containing 1.0 mg/l PPT showed retarded growth compared with *in vitro* cultured but PPT non-treated control plants and, after a month of stagnation, all these explants died. The plantlet explants also died after about a month of stagnation if higher than 3 mg/l PPT was applied.

3.2. PPT tolerance of R0 generation of microshoot derived plants under in vitro conditions

The microshoot-derived plants were placed into VegBox plastic containers containing solid N6 medium supplemented by 2.0 mg/l PPT (the 2xPPT sample group represents these plant material). At the same time, some of the *in vitro* cultured, but PPT non-treated plants were also put on the same medium as control plant material (the 1xPPT sample group represents these plant material). As a second control, *in vitro* not cultured, greenhouse grown plant material was used (the 0xPPT sample group represents these plant material). The growth parameters of these plants were recorded and evaluated after four weeks of incubation by statistical analysis of 100 plants of each treatment in five repetitions.

Our data showed that the growing parameters of 1xPPT plants fell short of the results of both the microshoot-derived (2xPPT) and the PPT non-treated (0xPPT) control plants on selective medium containing PPT. At the same time, the dry weight of microshoot-derived plants grown in the presence of PPT (2xPPT) was basically the same as that of the control plants grown in the absence of PPT (0xPPT). This data demonstrate that the sublethal concentration of PPT had no inhibitory effect on the growth of microshoot-originated plants (2xPPT), while it strikingly inhibited the growth of the 1xPPT control group's plants. Based on this observation, we can state that the majority of the microshoot-originated plants are PPT-tolerant, while the control plants are all PPT-sensitive.

3.3. PPT tolerance of R0 generation of microshoot derived plants under greenhouse conditions

The microshoot derived R0 plants and the *in vitro* cultured but PPT non-treated plants were transferred into pots. Subsequently, after a short period of acclimatization, some of them were used for testing PPT tolerance under greenhouse conditions. A PPT-based herbicide FINALE® was applied (5 ml FINALE / 500 ml water / 1 m^2 soil surface) twice on microshoot derived and control plants. On average, 78.1% of sprayed plants developed from microshoots survived the treatment, while all of tissue culture or seed-derived control plants died.

There were significant differences between the microshoot-derived plants and the control plants in characteristics connected to vegetative growth as average plant height and fresh weight. However, the microshoot-derived plants showed similarities to plants grown *in vitro* without PPT in some characteristics, such as average fresh weight of panicles and thousand-kernel weight. These parameters obviously correspond to the reproductive fitness of the plants or in agrobiological means to the average yield. Since PPT has not decreased the yield of the microshoot originated plants itself, we can state that the majority of the microshoot-originated plants are PPT-tolerant, while the control plants are all PPT-sensitive.

3.4. PPT tolerance of R1 generation of microshoot derived plants in vitro and in vivo

Seeds harvested from microshoot derived plants and from greenhouse-grown control plants were surface sterilized, then germinated on GM medium supplemented by 5.0 mg/l PPT. None of the seeds of the control plants germinated when PPT was present in the germination medium. The embryos became swollen when the "pigeon heart" stage of development was reached, but then turned brown and died. On average, 85% of the control seeds germinated on PPT-free medium and a similar rate of germination (78.4%) was exhibited by the seeds of microshoot-derived plants on GM medium supplemented by 5.0 mg/l PPT. The PPT-tolerant plantlets underwent microshoot induction as described earlier, but this was not successful in the range of PPT concentration (1.0, 2.0, 3.0, 5.0, 10.0 mg/l PPT) which had been used earlier. Then, all the data of development were measured on the R1 generation of microshoot derived plants that were measured earlier in the case of R0 generation. In terms of PPT tolerance, the results were basically the same as reported

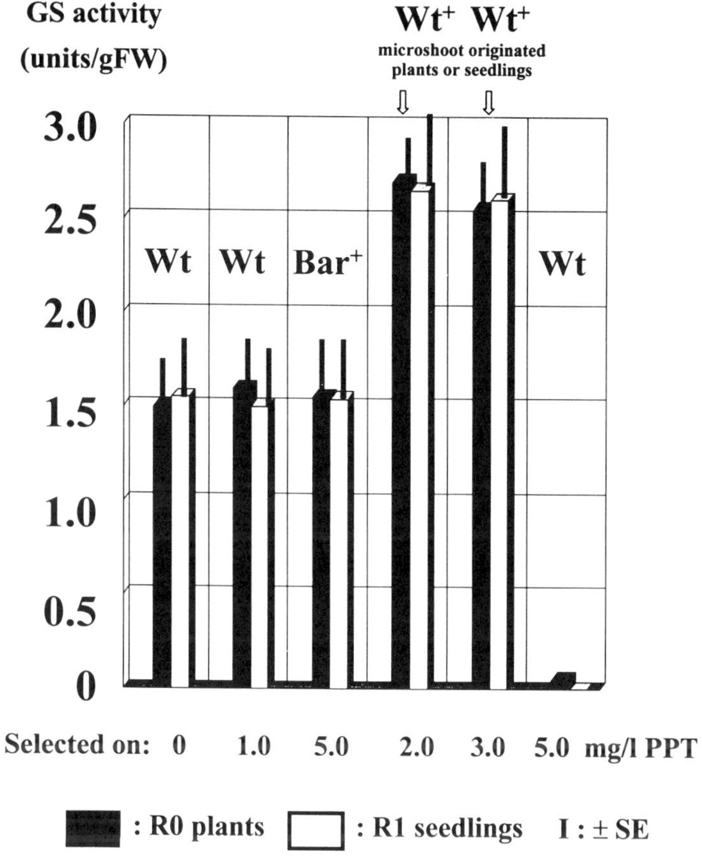

Figure 1. Differences in the activities of total glutamine synthetase (GS) in the R0 and R1 generations of PPT-sensitive and PPT-resistant rice plants

previously during the first 3 weeks of the plant development. However, the degree of PPT-tolerance gradually declined after that period until the original PPT-sensitivity has been reached by the 6th week of the plant development (data not shown).

3.5. GS activity measurements

Total glutamin synthase (GS) activities were measured from R0 and R1 generations of PPT-sensitive and PPT-tolerant rice plants in order to decide whether GS has any role in the PPT-tolerance or not (Figure 1).

The first column pair corresponds to the total GS activities of the PPT-sensitive control plants (the black columns represent the values of R0 plants and the white ones represent the values of the seedlings of their next generation). The PPT-resistant sample pair, which was produced through genetic transformation (Bar+) shows nearly the same values than that of the PPT-sensitive control plants. But, in contrast with the previous data, the total GS activity of the R0 generation of PPT-tolerant plantlets and their R1-generation seedlings were double than that of the previous samples.

4. Discussion

Running the preliminary experiments to test the morphogenetic effects of PPT, *de novo* differentiation of microshoots was observed on developing rice seedlings in the presence of PPT on hormone-free induction media. These events predicted the possibility of separating PPT-tolerant plants by *in vitro* selection, according to our hypothesis that plantlets responded positively to PPT-treatment will possess some degree of resistance to this herbicide. To prove this, first the sublethal (≤ 3.0 mg/l) and lethal (≥ 3.0 mg/l) concentrations of PPT were determined in 3-7 day-old rice seedlings. At the same time, morphogenetic events during PPT treatment were investigated. We observed that on 5-40% of the seedlings, 6-30 shoot primordias (microshoot) per explantum were differentiated on the hormone-free induction media containing PPT in sublethal concentrations (1.0-3.0 mg/l). These results supported those reported by Hoshino and Mii [2], in which they induced adventive shoots on hairy root explants of snapdragon, exploiting the morphogenetic effect of PPT. Our results show that PPT is morphogenetically active *in vitro* in a narrow interval of concentration similarly to many other herbicides. While most morphogenetically active herbicides have auxin-like activity, we first proved that PPT has a cytokinin-like effect in rice tissue cultures. Fertile rice plants were grown from the induced microshoots which showed resistance to PPT under greenhouse conditions. Isolation of spontaneous PPT-resistant cell-lines using *in vitro* tissue culture is not unusual among cereal species. For example, wheat embriogenic calli survived even 10-40 mg/l concentrations of bialaphos (whose active compound is PPT) in reported experiments [3, 4]. However, according to our knowledge, this is the first report on production of resistant rice plants to the herbicide PPT by *in vitro* selection. In our experiments, PPT induction acted as a sort of *in vitro* selection, because the relatively high proportion of resistant explants (5-40%) indicated that possibly not mutant plants, but those individuals of the population were isolated which had a "natural" tolerance to PPT [10]. More precisely, this relative tolerance may exists within the frame of the natural genetic diversity of the rice species. But how can we explain the nearly uniform inheritance of PPT-resistance by the next generation (R1) of microshoot-originated plants by this hypothesis? Existence of natural resistance can be explained by the fact that PPT is a competitive inhibitor of glutamin synthetase (GS), which plays a central role in the assimilation of inorganic nitrogen. Phosphinothricin impedes the binding of glutamic acid (a natural substrate of GS) to the active centrum of the enzyme, and in this way, it results in toxic accumulation of ammonia in the cytoplasm [11]. In the present case, it means that by supplying exogenous glutamic acid or by overproduction of GS, the inhibitory effects of PPT can be repressed [11, 12]. Our GS-activity data indicated that in our PPT resistant plants, either the GS was synthesized in greater amounts, or the isoform of the enzyme with higher activity took part in the reaction inhibited by PPT. Microshoot-originated plants accumulated more GS during long-term PPT-selection, that is, during their life-history, including seed filling. Plant glutamin synthetases are coded by a small multigene family and these have, among others, kernel specific isoforms [13]. Since our data confirm that the regulation of their expression is synchronised rather than independent, PPT-resistant parental generation can provide short-term protection for their progeny through accumulation of kernel-specific GS isoforms during seed filling, as was demonstrated in Figure 1. The theoretical basis of the above speculation was the visible analogy between our intergeneration PPT-resistance and the well-studied maternal effects on the development of *Drosophila* embryos, in which the origin of the phenomenon is the existence of the so-called epigenetic somatic memory (for review see [14]).

References

[1]   M.A.W. Hinchee, *et al.*, Herbicide-tolerant crops. In: S Kung, R Wu (eds.) Transgenic Plants, Academic Press Inc., Vol. 1., (1993) 243-263.

[2]   Y. Hoshino and M. Mii, Bialaphos. Stimulates shoot regeneration from hairy roots of snapdragon (*Antirrhinum majus* L.) transformed by *Agrobacterium rhizogenes*, *Plant Cell Reports* **17** (1998) 256-261.

[3]   V. Vasil, *et al.*, Herbicide resistant fertile transgenic wheat plants obtained by microprojectile bombardment of regenerable embryogenic callus. *Bio/Technology* **10** (1992) 667-674.

[4]   H. Zhou *et al.*, Stably transformed callus of wheat by electroporation-induced direct gene transfer. *Plant Cell Reports* **12** (1993) 612-616.

[5]   O. Toldi, *et al.*, Minibeet initiation from derooted sugarbeet (*Beta vulgaris* L) seedlings *in vitro*, *Plant Science* 97/2 (1994) 217-224.

[6]   T. Murashige and F Skoog, A revised medium for rapid growth and bioassays with tobacco tissue cultures, *Physiologia Plantarum* **15** (1962) 473-497.

[7]   C.C. Chu, The N6 medium and its application to anther culture of cereal crops. In: Proceedings of Symposium on Plant Tissue Culture, Science Press, Peking, (1978) 43-50.

[8]   K. Kamachi, *et al.*, A role for glutamine synthase in the remobilization of leaf nitrogen during natural senescence in rice leaves, *Plant Physiology* **96** (1991) 411-417.

[9]   J. Sváb, Statistical test of the mean. In: Biometrical Methods in Research Work, Agricultural Press, Budapest, (1981) 47-69.

[10]   O. Toldi, *et al.*, Phosphinothricin: a herbicide or a synthetic cytokinin? *Journal of Experimental Botany* (1999) **50**:56 (supplement).

[11]   C.J. Thompson and H. Seto, Bialaphos. In: Vining LC, Stuttard C (eds.) Genetics and Biochemistry of Antibiotic Production, Butterworth-Heinemann, (1995) 197-222.

[12]   K. D.'Halluin *et al.*, The *bar* gene as selectable and screenable marker in plant engineering. In: Methods in Enzymology, Academic Press Inc., Vol. 216., (1992) 415-426.

[13]   M.J. Muhitch, *et al.*, Immunolocalization of a unique form of maize kernel glutamine synthetase using a monoclonal antibody, *Plant Physiology* **107** (1995) 757-763.

[14]   R. Rivera-Pomar, H. Jackle, From gradients to stripes in *Drosophila* embryogenesis: filling in the gaps, Trends in Genetics **12** (1996) 478-483

*Use of Agriculturally Important
Genes in Biotechnology
G. Hrazdina (Ed.)
IOS Press, 2000*

Current Procedures for Applying Risk Assessment in Genetically Modified Crops

A. Aniol

Plant Breeding & Acclimatization Institute, Radzików, 05-870 Blonie, Poland

Abstract. The rapidly increasing utilization of transgenic cultivars in agriculture and emergence of genetically modified (GM) plants and plant products on the market calls for effective risk assessment procedures to be developed and applied in handling applications for GM plant release and marketing. The main producers of transgenic crops are at the same time also major exporters of agricultural products. Therefore there is an urgent need to develop internationally standardized and consistent schemes, test procedures and concepts of risk assessment. In this paper the procedures of risk assessment of transgenic are reviewed with special reference to procedures adopted by the European Union as exemplified in EU Directive 90/220, and procedures enforced in the USA. Special attention is given to the principle of substantial equivalence of the transgenic plant and the receiver plant as it was developed in the USA in the context of the dispute over labeling & consumer choice and precautionary principles as stressed in Europe. An example of general procedure is presented as proposed by the Danish group in the form of a three-step hierarchical frame for risk assessment. The main idea of the hierarchical procedure is that experimental tests and model simulations of varying complexity are applied at each level: individual plant, populations and the ecosystem. The perspectives of the adoption of generally accepted concepts and procedures of risk assessment are discussed.

1. Introduction

The products of modern biotechnology are increasingly available on the market, causing a lot of dispute on its impact on human health and on other organisms in the environment. The procedures and methods of risk assessment of introduction of genetically modified organisms (GMO) are constantly reviewed and improved. Safety requirements for GMO release in the European Union are becoming rigorous. The main difference between genetic modifications utilized trough natural processes of mating, segregation and selection and modern techniques of biotechnology is that in the later case the small fragment of "naked" DNA, isolated from an organism or synthesized, and is responsible for expression of a desirable character, e.g. resistance to herbicide or pest or disease, is directly incorporated into a native DNA of the donor organism.

Despite of all controversies and discussions on GMO's, all parts agree that release of GMOs into the environment and introduction of a GMO and its products in the market should be regulated by independent agencies in order to eliminate and/or minimalize the risk associated with this technology.

The hottest debate is associated with GMO application in agriculture and food production. At the same time this is the fastest growing branch of GMO application. Transgenic crops planted in the field increased in the past 3 years from 1.7 million ha. in 1996 to almost 30 million ha. in 1998 [3].

The biosafety issue in GMO utilization in agriculture is very important and difficult problem since it is inherently connected with the release of genetically modified crops or transgenic plants into the environment. This fact has ecological, agricultural, economic and political implications and this makes the discussions on risk associated with the introduction of transgenic cultivars very complex. Additionally, almost all transgenic cultivars are used as food or feed, and thus makes the dispute even more complicated.

The fundament question in this context is: what is risk and why we are talking about risk assessment of transgenic crops? There is an assumption that two parameters could be separated in the risk concept: the hazard associated with the genetic modification and the likelihood of this hazard to occur. Therefore risk is likelihood of the hazard being realized [4]. Considering this general assumptions in the discussion on transgenic crops, we first must answer whether this particular genetic modification poses any inherent hazards and if so, what kind of hazard it poses and what would be the likelihood of this hazard to take effect. Another important aspect is the social acceptance of the risk. The prevalent but not generally accepted opinion is, that genetic modifications obtained through modern biotechnology are of different nature then genetic modifications used by plant breeders in conventional breeding procedures. Therefore, despite that historically crop plants have not been subjected to formalized risk/safety analysis such procedure is regarded as needed in the case of transgenic crops. Improvement of crops by traditional plant breeding methods was and is possible when genes controlling the characters of interest are found within the crop species itself or within species that are sexually compatible. Different techniques applied in traditional breeding were developed (embryo culture, ovary culture) to overcome this difficulties in gene transfer, but still gene transfer was possible only between related species. Modern techniques of molecular biology based on recombinant DNA allow theoretically unlimited gene transfer between all living organisms, and even the introduction of "synthetic" genes. The biological consequences of this kind of manipulations are very difficult to anticipate and sometimes impossible to predict. The concerns about GMO's are the main reason for the establishment of risk assessment procedures applied for the release of transgenic crops. This concerns center on six issues [7]:

a. will transgenic plants be toxic or allergenic to wildlife, humans or domestic animals?
b. will the transgenic crop plants become weeds of agriculture?
c. will transgenic crops become invasive of natural habitats?
d. will engineered genes be transferred to related weeds or wild relatives?
e. will transgenes affect non-target organisms?
f. will transgenes affect ecosystems, biodiversity, or a regionally adapted genetic variation?

Most transgenic cultivars released up to date represent minor modifications of well-known varieties but the potential exists to develop cultivars with markedly changed qualities and characters. This potential of creating novelty in plant kingdom is the main reason of application of precautionary principles in releasing transgenic cultivars. In taking precautionary approaches we assume that it is better to be safe than sorry regarding the exposure of the environment or people to the risk.

2. Risk assessment in biotechnology

Historically, risk assessment procedures were developed well before the emergence of modern biotechnology. Many governments on central or local level are involved in regulation of activities relating to safety and well being of its citizens, for example rules for drinking water, road traffic or sanitary requirements. Large scale risk assessment associated with the hazards to environment and human health was estalished when chemicals were introduced into agriculture production and pest control. The principles applied and procedures established at that time are used as a template in regulating GMO use and release [1].

The first documented and successful transfer of genes using recombinant DNA took place in 1973 and this is symbolically regarded as a birthday of modern biotechnology [5]. Since the beginning of modern biotechnology the scientists involved in development of this techniques were aware of potential hazards to human health and the environmen. Because of these concerns, in 1976, only three years after the birth of modern biotechnology, in 1976 the first conference was held in Asilomar, California devoted to elaboration of safety rules for work and handling of recombinant DNA. At that time elaborated rules were dealing with contained use and laboratory handling of recombinant DNA, and genetic modifications mostly in microorganisms. In the mean time a rapid development of molecular biology and plant biotechnology took place, and the development and release of transgenic crops created new and important questions related to release of GMO to the environment. However, the so called points to consider that were set up by the Recombinant Advisory Committee (RAC) after the Asilomar conference were and are still a important reference for risk assessment procedures.

Four main categories of questions are included in the points to consider in risk assessment [5]:
- a. related to risk to humans – toxicity, allergenicity;
- b. related to risk to other plants - potential for sexual transfer of nucleic acids between:
1. transformed variety and unmodified varieties;
2. transformed variety and other compatible species;
- c. related to risk to environment - colonization ability, seed dormancy, resistance to pest control measures like pesticides, increased toxicity to other species;
- d. related to nature of the gene - possibility of horizontal transfer between unrelated species by a virus, bacterium or other vector.

There are some aspects of gene technology which are considered as having some potential for creating risk. These are the main reasons for risk assessment procedures to be required by all regulatory systems, whatever the differences might be between the countries and their legal systems.

To carry out a genetic modification, a recombinant genetic vector has to be constructed in order to carry the cloned gene into the target plant tissue in which the gene should be expressed e.g. the protein coded by the inserted gene is produced. Such vector, besides the gene in question, is composed of a number of other DNA elements. The vector also typically contains the promoter or enhancer (control element) and selection gene, often an antibiotic resistance gene that allows screening of the transformed vs. untransformed cells. There are different methods of inserting gene constructs in the target cells - via *Agrobacterium* infection, particle bombardment, membrane disruption by electric field (electroporation). In all methods integration of the inserted gene construct takes place at an unpredictable location in the chromosomes. In the outlined process there are some points which justify the procedure of risk assessment:
- a. the method allows the creation of unusual gene combinations between genes from species for which there is no possibility of natural progeny.
- b. unpredictable location of the inserted gene construct in target organisms genetic material might lead to development of new, unpredictable genetic combinations and/or reshuffling the recipient genetic material, and this might have adverse effect for the plant, the environment or human/animal health.

   c. the antibiotic resistance genes used as marker genes might create a hazard of spreading antibiotic resistance in the environment [9].

In the process of development of transgenic crops that are presently commercialized a rigorous selection process took place, in which all unwanted or adverse effects were eliminated. In new developments in crop biotechnology, new gene constructs will be used and new characters would be introduced into the crop plants, therefore this precautionary approach in risk assessment is justified.

Regulatory authorities in most countries require the evaluation of safety of transgenic plants and other GMO's.

As indicated above, the points to consider in risk assessment involve a wide range of issues, but there is a lack of defined models which could provide the ability to generalize. Therefore risk assessment is limited to a case by case procedure. There is often a presumption that a risk assessment must be a complicated document which provides all sorts of nightmare scenarios. Although there might be cases where complex assessment are necessary, often only few issues will require this and simplified version of the risk assessment procedures might be used. Usually, assessments of releases that are widespread and include organisms with unusual and/or unfamiliar characteristics with a high likelihood of survival in different habitats require a comprehensive, full evaluation. At the other extreme, for organisms that are familiar or are directly derived from familiar organisms that are known to have very limited hazard potential, as it is the case of transgenic crops, a simplified assessment can be considered. In such cases, the hazard component of the risk is constrained and one may concentrate on exposure. Therefore assessment for the release of organisms that could have global impact are generally comprehensive, while those limited to the site of production are often very simplified (small scale field trials are usually treated differently than commercial releases). The decision what kind of risk assessment procedure should be applied, comprehensive or simplified, is in each case a difficult one with many legal and political consequences. These are the main source of differences between the US and the European approaches to risk assessment in biotechnology and even more, they reflect differences in the social perception of novelty in those communities. Often governments provide guidance on what to consider for risk assessment, because there are legal consequences of misinterpreting or misapplying risk assessment, and this is influenced by the legal system which might be different in different countries. Sometimes, even trivial cases with high visibility may require greater than average scrutiny. This is a good public policy, even if it is not fully scientifically justified. There are different answers to the question: 'who is responsible for performing risk assessment and proposing risk management procedures'? Risks involving GMOs may be effected directly by a governmental agency, or the government may require that the applicant for the release should prepare it, or a committee of experts will do it on behalf of the government. Despite this variability in organizational matters, the legal systems and public perception of risk in biotechnology, procedures involved in risk assessment are very similar and are based on commonly accepted "points to consider".

3. Information required for risk assessment

As said before, the amount and kind of information required for risk assessment for GMO release is similar in many countries. EU requirements as expressed in an annex to the Directive 90/220 and modified in annex II of Directive 94/15 are shown as examples. The information required under this directive is divided in two parts: annex IIA refers to GMOs other than higher plants, and annex IIB refers to higher plants, e.g. organisms belonging to *Gymnospermae* and *Angiospermae*. Some differences in information required in this annexes basically reflect differences in the mode of utilization of GMO between contained use and environmental release. For plants, e.g. transgenic crops the body of information required consist of the following [2]:

A. General information- name and address of organization wishing to release a transgenic cultivar. Here it is important that the institutions carrying out the release have a sufficient level of expertise and experience to do this.

B. Information relating to the recipient or parental plant, namely:
-   scientific name and taxonomic details;
-   reproduction mode and generation time;
-   survivability (seed dormancy);
-   dissemination;
-   geographical distribution;
-   natural habitat, parasites, predators, competitors and symbionts;
-   potentially significant interactions with other components of ecosystem.
Information on the recipient species will establish the basis for comparison with transgenic plants. Information of donor is important for assessment, if for example the donor organism is a plant pathogen the question arises of the possibility of recombination between the integrated DNA from the pathogen and other pathogens that may infect the transgenic plant. In case of transgenic plants with genes coding for viral coat proteins there might be a question of the possibility of transcapsidation.

C. Information relating to the genetic modification.
-   description of the methods used for genetic modification;
-   nature and source of vector used;
-   size, source and intended function of each constituent fragment of the region intended for insertion.
Here, the information on antibiotic resistance genes used as selective markers is important.

D. Information relating to the genetically modified plant.
-   description of traits and characteristics introduced or modified;
-   information on DNA sequences actually inserted/deleted, size and structure of the insert, methods of its characterization, cellular location of insert (chromosome, mitochondria, chloroplasts), methods of its determination, number of copies of the insert;
-   information on the expression of the insert:
        parts of the plant where the insert is expressed and methods of its
        characterization;
-   information on how the genetically modified plant differs from the recipient plant in:
        mode and rate of reproduction, dissemination, and survivability;
-   genetic stability of the insert;
-   potential transfer of the genetic material from transgenic plant to other organisms;
-   information on any toxic or harmful effects on human health and the environment arising from this modification;
-   mechanism of interaction between the genetically modified plant and target
-   organisms (if applicable);
-   potentially significant interactions with non - target organisms.
-   description of detection and identification techniques for genetically modified plants;
-   information on previous releases of GMOs (if applicable).

Particularly interesting information would be here on the nature of pollen dissemination and distances over which pollen can give a successful pollination.

E. Information relating to the site of release.
- location and size of release;
- release site ecosystem (climat,flora,fauna);
- presence of sexually compatible wild relatives or cultivated species;
- proximity to officially recognized biotypes or protected areas which may be affected.

F.   Information relating to the release:
- purpose of the release;
- foreseen date(s) and duration of the release;
- method by which the genetically modified plant will be released;
- method of preparing and managing the release site, prior to, during and post-release, including cultivation practices and harvesting methods;
- approximate number of plants or plants per square meter.

G. Information on control, monitoring, post-release and waste treatment plans.
- any precautions taken: distances from sexually compatible plant species, any measures to minimize/prevent pollen or seed dispersal;
- description of post-release treatment methods for genetically modified plant material, including wastes;
- description of monitoring plans and techniques;
- description of any emergency plans.

H. Information on the potential environmental impact from the release of the genetically modified plants(GMHP).
- likelihood of the GMHP becoming more persistent than the recipient or parental plants in agricultural habitat or more invasive in natural habitats;
- any selective advantage or disadvantage conferred to other sexually compatible plant species, which may result from genetic transfer from the GMHP;
- potential environmental impact of the interaction between GMPH and target organisms (if applicable);
- possible environmental impact resulting from potential interactions with non-target organisms.

It is very important to determine the extent to which it is possible to monitor transgenes after release.

It must be stressed that risk assessment is not an exact science; it cannot be said with absolute certainty that the release of a transgenic plant will have no risk. The most important feature of risk assessment of transgenic plants is therefore to determine how the risk might be altered in transgenic plants as compared to unmodified crop plants. One of the major ecological concerns associated with the release of transgenic varieties is that the inserted information may be transferred to wild populations. New procedures allowing delivery and expression of an inserted sequence in the plastid genome, particularly in chloroplasts, eliminates this concern, introducing a so called biological containment of transgenes. The highly publicized and controversial "terminator" technology will have the same effect [8].

There are basically two approaches in making a risk assessment by integrating its hazard and exposure components. With both approaches the same conclusions should be reached when using the same facts. This two general assessment concepts can be classified as parallel and sequential analysis [6].

The parallel procedure involves the simultaneous and to some extent independent evaluation of different components of an assessment and these partial evaluations are finally integrated by a technical integrator. This procedure provides some advantages in certain situations: it separates risk characterization from risk management, allows independent evaluation of each component therefore ensuring the complete documentation of all aspects of risk characterization, and allows a development of expertise in a particular discipline.

In the sequential procedure, some elements of risk characterization are given priority, and if the results of such analysis indicate that a risk may be to high, measures for its reduction can immediately be considered, thus risk management is evaluated at the same time as risk assessment. The main advantage of the sequential review is that non-essential components can be defined and eliminated from review at the very beginning of the evaluation process. This may simplify the assessment and shorten the review time. On the other hand it will quickly identify key areas that require additional data.

4. Administrative and legal aspects of risk assessment.

The US, UK, and EC approaches to regulation of product of plant biotechnology are examples of different ways to obtain similar effects. This is not surprising, since the main objective, whatever the specificity of the country's legal and administrative systems, is the safety of people and environment. Also, as was said before, the basic questions which must be answered during the process of risk assessment are of the same nature everywhere and are listed in the "points to consider".

Therefore the main differences between countries are not related to the risk assessment process in terms of its objective and scope, but they relate to the following questions:
- who is performing the risk assessment?
- how is it regulated under the domestic law?
- who is responsible for decision making ?

The governments even of very rich countries can not probably afford to keep a permanent staff of best experts in all disciplines who can review the issues which might arise in every application for GMO release. Therefore the governments relay on experts from outside agencies to assist in more complex cases. The manner in which the experts are employed can vary significantly from country to country. The experts are usually employed in committees or boards, associated with the respective governmental bodies, that differ in status. The two main ways of handling such committees may be observed:
1. As an advisory body which could give advice on scientific matters but had no power to make decisions on behalf of the government. The weak point of this solution is that since these committees are purely advisory, the government is not compelled to act on their advice; the members of committees are usually selected from academic research institutions and are prone to raise questions based on intellectual curiosity, which are often not pertinent to the decisions which the government must make based on law and regulations. Regulatory decisions must be made partly on the basis of legitimate non-scientific considerations, therefore these expert advisory bodies remain advisory in function.
2. As a strong advisory committee which can provide advice that is close to regulatory function. It is usually in the form of a statutory, independent expert body appointed by a high ranking governmental agency under the rule of law (example is ACRE = Advisory Committee on Releases to the Environment in UK). Such committees are composed of experts from government and academic institutions.

The advantages of this second approach are that it utilizes existing authority to implement regulatory decisions and at the same time gathers research scientists as members, giving credibility to the decision making process. Among the disadvantages of this solution is a substantial operational cost, and in some sense the government abdicates some of its responsibility by placing the decision in the hands of such a committee. Usually, the members of a committee have expertise in the particular fields relevant to a given case but they may not always have the ability to extend their interpretations beyond the narrow focus of their field of expertise.

Usually it is the applicant whose has the responsibility of gathering and supplying the data needed for risk assessment. In some countries there is a firm set of data requirements; in others the preference is given to flexible data requirements- first time applicants are encouraged to contact the government agency in charge well in advance of an intended release in order to obtain specific guidance.

The situation in different countries can be summarized as follows:

USA

Shortly after the emergence of modern biotechnology (1986), a special governmental commission was established in order to review existing regulations to specify which acts and agencies would have jurisdiction over the GMO's.

The conclusion was that existing laws are adequate to provide oversight of products of modern biotechnology. The consequence of this finding was that the responsibility of GMO production and utilization was placed in the hands of four US agencies: the Environmental Protection Agency (USEPA), the Department of Agriculture (USDA), the National Institute of Health (NIH) and the Food and Drug Administration (FDA). The question of developing an omnibus legislation regulating GMO matters in the USA was suggested in the beginning and it still has its advocates, but a decade of field applications of plant GMOs without apparent adverse effects gives considerable confidence in the adopted procedures.

UK

The regulatory system in UK is of intermediate character. It partially operates under the existing law (Health and Safety at Work Act), but new statues affecting biotechnology were also enforced (Contained Use Act). An advisory committee (Advisory Committee on Genetic Modification =ACGM) established under the Health and Safety at Work Act has produced regulations for activities with GMOs for both laboratory and field work. Later on, when the number of transgenic crop varieties ready to be released increased, and public awareness on GMO matters increased also, a subcommittee was formed, with a statutory character (ACRE= Advisory Committee on Release to the Environment). Its activity is focused on environmental issues.

EC

EC regulations of biotechnology are also concerned with safe use of GMO as do member state regulations, but additionally they must seek to harmonize their procedures for export and import of GMOs and their products. This harmonization process not only must do away with member states conflicting requirements, but must also ensure that the products are sufficiently well characterized, and that the importing country can be assured that the product is safe to use. The UE directives 90/219 and 90/220 are an attempt to harmonize regulation of GMO matters in the Union.

5. The main topics of dispute on risk involving GMO release

The release and utilization of transgenic crops increased exponentially during the last few years and while transgenic cultivars of maize and soybean are sown on large portion of acreage in the USA, Canada, Argentina and China there is a very small production of transgenics in Europe. There even the import of grain from transgenic crops and its products is subject to restrictive policies and very hot public debates. Here I will not attempt to present all the different arguments which are exchanged between enthusiasts on one side and convinced opponents on the other. There is a mixture of arguments based more or less on science and some views are mainly of philosophical, political and economical nature. As an illustration I will outline here the main controversial views centered around the concepts of precautionary approach, substantial equivalence and transboundary movement of commodities.

All disputing parties agree that precautionary approaches are an important and valid principle and must be applied when risk assessment of transgenic crops is made. While industry representatives and a part of the scientific community perceive that this approach must be based on solid scientific evidence, the other side of public opinion composed of different ecological groups, some farmer organizations and the media argue that this principle should be applied in the sense that the absence of evidence of harmful effect of GMO should never be taken as the evidence of absence of hazard [9]. This approach would mean that no transgenic variety should be released at this moment.

The concept of "substantial equivalence" is defined as equivalence of a transgenic cultivar in terms of use and safety to a non-transgenic "conventional" variety of the same species that has been in use for long time and is generally regarded as safe. This concept goes together with the concept of "familiarity" defined as the knowledge of this plant species and the experience with its cultivation and use. The use of both concepts can be compared to use of a benchmark, or the controls in the risk assessment of chemical substances. These principles are used mostly for food and feed safety evaluations, but should also be applied as an ecological and agricultural substantial equivalence. In fact, these concepts are used for risk assessment evaluation of transgenic crops in North America. However, in Europe the concept of substantial equivalence is often regarded as premature.

Commodity

A commodity defined as grain for food, feed and processing (FFP) is another controversial issue, where risk assessment procedures are involved. This controversy is expressed during the negotiations of a Biosafety Protocol for an agreement on the international movement of GMOs (in the language of the Protocol the term LMO = Living Modified Organisms) and its products. This issue is linked to information content, which should be associated with commodity transfers. Exporter countries claim that a simplified form of information (notification) is sufficient, while some importing developing countries require the full set of information as required under the annex of AIA (Advanced Informed Agreement) . This information is needed for risk assessment, since in many developing countries commodities are often used as seed.

Another point of dispute on commodities is connected with the requirement of labeling in the EU. This requirement involves the politically hot issue of the consumer's right to choose. On the producer/exporter side, in order to meet this requirement, it would be necessary to separate transgenic cultivars from non-transenics, starting from the farm through elevator and shipment systems, which is not practiced for commodities up to now. Separation means additional, substantial and unnecessary costs (the industry claims) which will be eventually placed on the consumer. In some cases it is argued, that this additional cost might outweigh the benefits brought by a transgene.

6. The concept of hierarchical risk assessment of transgenic plants.

The great challenge confronting EU regulations, biotechnology regulatory system included, is the harmonization of member countries regulatory systems and the development of internationally standardized tests and procedures based on accepted common concepts. There is a proposal for such harmonized, hierarchical risk assessment procedure presented by B. Strandberg et al [7]. The main features of this approach are outlined below.

The proposed approach is similar to the hierarchical test systems established for risk assessment for toxic compounds. It is a stepwise procedure designed to produce the final body of information needed for decision making about marketing of transgenic crops. This approach is consistent with the idea of a step-by-step procedure, and is based in the body of information required by EU Directive 90/220. The fundamental idea of this proposal is that possible adverse effects of a transgenic plant and the risk which it poses is tested in stages on an increased scale and decreased safety from contained experiments in the laboratory, or closed systems in growth chambers and glasshouses, through small-scale field trials to commercial release. For each step a separate regulation is proposed. This stepwise procedure is designed to close the gap between members of its industry, who focus on scientific results perceived in an agricultural context and ecologists concentrated on long distance ecological effects on natural ecosystems. After completion the hierarchical risk assessment procedure will yield the final body of information needed for taking decisions on marketing. This procedure should fulfil the need for:
- a. a structured, coordinated approach for assessment of risk to humans and the environment;
- b. obligatory basic information involving experimental data;
- c. standardized test procedures specifying the conditions important to the inserted transgene;
- d. clarifying the acceptance criteria;
- e. experimental tests of transgenic plants at each step.

Step I

At this stage a range of background information on transgene, parental plant and effects on human health and the environment is gathered (see information requirements in E U Directive 90/220). This information must be supplemented by data from simple standardized and replicable tests on changed competitiveness, survival and reproduction, and toxicity. Such tests might be run in parallel during the efficacy tests performed by the companies. At step I transgenic plants which express toxins or pharmaceuticals must be tested for effects on non-target organisms.

The risk assessment of transgenic plants performed at this step should allow to separate them into three categories:
- those that do not show higher risk than their non-transgenic counterparts and are accepted for commercialization according to the principle of substantial equivalence;
- those that present high risk, exceeding acceptance criteria and are rejected. For this group of transgenics the analysis ends here;
- those for which the potential environmental risk is not sufficiently described. This group moves to step 2.

Step II

At this step the advance testing of the impact of transgenic crop on the environment is performed. The testing is done in different, competitive environments, the level of complexity of tests is higher and requires much more time. After such testing the risk assessment analysis is performed similarly as in step I.

Step III

Comprises risk assessment at a regional scale, where transgenic plants are used on large scale in adjoining areas. Some new situations might arise when different kinds of trasgenic cultivars would be cultivated together at different combinations over time. The risk assessment done on case by case basis might reveal that the risk is small, or that there is no risk at local, small scale cultivation of a transgenic crop, but this risk might rise substantially when large scale cultivation takes place or/and when the cultivation will move to regions with other climatic conditions. This might be specially important for transgenics with build-in tolerance to environmental stress.

In cases when marketing of a transgenic cultivar is accepted following risk assessment under Step II and Step III procedures, adequate monitoring programs must be set up and carried out.

References

[1]   F. Amifee, Assessment and development of policy on releases of genetically modified plants. Proceedings and Papers from 1996 Risk Assessment Research Symposium. http//:www.nbiap.vt.edu 1996.
[2]   Directive 94/15 EC, Annex IIB, http://biosafety.ihe.be.
[3]   C. James, Global Review of Commercialized Transgenic Crops: 1998. ISAAA Publ.No.8, 1998.
[4]   J. Kinderlerer, Tools of Regulation in Risk Assessment. Guide to Risk Assessment and Biosafety in Biotechnology. File A:/reg.html, 1997.
[5]   M. Levin, A Primer on Risk Assessment. File A:/raa.html, 1997.
[6]   M. Segal *et al.*, Current Procedures for Applying Risk Assessment. File:A:/pra.html, 1997.
[7]   B. Strandberg *et al.*, Hierarchical risk assessment of transgenic plants: proposal for an integrated system. BioSafety Journal. V.4, paper 2, 1998.
[8]   Technology Protection System (TPS). American Seed Trade Company, Delta & Pine Land Company/USDA, http://www.amseed.com, 1999
[9]   T. Traavik, Too early may be to late. The University of Tromso, Norway, 1999.

Questions Raised During the Implementation of the Hungarian Law on Biotechnological Activities

E. Balázs[1] and R. Lupócz[2]

[1]*Agricultural Biotechnology Center Gödöllö, Szent-Györgyi Albert Street 4, 2100, Hungary*
[2]*Ministry of Agriculture and Regional Policy, Budapest, Kossuth Lajos Square 11, 1055, Hungary*

Abstract. After years of preparation, a framework law on biotechnological activities was passed by the Hungarian Parliament in 1998. This law was implemented in January, 1999. Although the provisions of the law are originally based on two EC directives (90/219/EEC and 90/220/EEC), the subject of the Hungarian regulation is much broader than the EC's. In order to implement this law, Act No XXVII/1998, the competent ministry issued regulations in the fields of agriculture and food industry [ministerial decree 1/1999. (I.14) IVM and the amendment of the ministerial decree 1/1996. (I.9) FM-NM-IKM]. Due to the revision of the two EC directives and also to the initial experiences during the implementation of the Hungarian Law and its ministerial regulations, several questions were raised about different issues like labeling, threshold level, commodities and trade, LMO and products thereof, liability and compensations, etc. These issues will be discussed with special attention to risk assessment and management. Priority is given to the evaluation of their methodology. In addition to the introduction of the key elements of the law, several of the debated issues will be presented in detail.

1. Introduction

After years of preparation a framework law on biotechnological activities has been developed in Hungary. This law was passed by the Hungarian Parliament in 1998, and was implemented early in 1999. Although the provisions of the law are originally based on two EC directives (90/219/EEC and 90/220/EEC), the Hungarian regulation is much broader than the EC's. A ministerial regulation was issued to agriculture and the food industry in order to implement this law, Act No XXVII/1998. Due to the revision of the two EC directives and also to the initial experiences during the introduction of the Hungarian Biotechnology Law and its ministerial regulation, several questions were raised regarding issues like labeling, threshold level, commodities and trade, LMOs and products thereof, liability and compensations, etc. These issues are the key components of our risk assessment and management. Priority is given to the evaluation of their methodology. By implementing the Biotechnology Law, the safe introduction of modern biotechnology can be performed.

Hungary's Biotechnology Law arose from a variety of national agricultural conditions and technological advances. The present area of Hungary is 93,031 square kilometers or 9.3 million ha., with a population of 10 million inhabitants. Approximately 60% of this area is arable and under cultivation. Agriculture contributes an average of 6-8% per year to the Gross National Product of the country. Agricultural production in Hungary has declined over the past ten years; actually this figure is about one third of the figure from 1980. Hungarian agriculture went through an extensive reconstruction, which includes the re-privatization of agricultural land. Most of the former co-operatives declared bankruptcy and are currently being reorganized into a modern farming system. Agriculture has abandoned roughly 700,000 ha. of low quality soil, in spite of a strong economy which could result in agricultural subsidies for this land. This situation is evidence that a more intensive and environmentally friendly, sustainable agriculture must be developed.

Due to the rapid development of biotechnology, agriculture is facing the introduction of novel products. In 1990 an independent agricultural research institute was founded and began to evaluate the importance of the new technology. Hungarian agriculture traditionally depends on high quality breeding varieties. The continental climate of the country and the quality of the soil provide optimal conditions for quality seed production. Almost 60 percent of Hungary's arable land is used for agriculture, which means that a definitive surplus from the agricultural sector is possible.

Biotechnological research began with the production of somatic hybrids between the wild potato variety *Solanum brevidens* and potato. The use of recombinant DNA technology started at the Biological Research Center, Szeged. It was here that the isolation of plant plasmids started and the first maize intra-mithocondrial plasmid was created. Three years later, when the first transgenic tobacco was reported, alfalfa was transformed here. The research on different crops under laboratory conditions resulted in several crops which are resistant to viruses, herbicides and insects, and are considered to be ready for field releases.

In 1996, the Hungarian Parliament also adopted a new Nature Protection Law that contains a paragraph relating to GMOs. Because this paragraph states that an independent law on GMO related issues must be formulated and adopted, all experiments and field releases were reconsidered. The scientific community was asked to prepare such a law, which it did. The final version was discussed with the different non-governmental organizations that were asked to evaluate it. Hungary's Environmental Protection Law demands that any new regulations or laws regarding environmental issues that are considered for adoption must be evaluated by an official organization, the Hungarian Environmental Council. To comply with this order, the Hungarian Environmental Council had several debates and discussed individual opinions and views. Several workshops and round table discussions were organized under the Hungarian Biosafety Framework, which was supported by the UNEP-GEF program in 1998/1999. The Biotechnology Law that was formally adopted was based on the text of the two EC directives from 1990. After long preparation, a framework law was finalized between 1995 and 1998 in Hungary, was passed by the Hungarian Parliament in March, 1998, and took effect on January 1, 1999.

Although Hungary's Biotechnology Law was originally based on two European Community directives (90/219/EEC Council Directive on the contained use of genetically modified micro-organisms and 90/220/EEC Council Directive on the deliberate release of genetically modified organisms into the environment), the Hungarian regulations are much broader than the EC's. The Hungarian regulations cover all GMOs, with the exception of humans. Modification of protected and wild (game) organisms is prohibited. In addition, the law on biotechnology amends Act No. XC of 1995 on foodstuffs by adapting the new definitions relating to novel food and to the labeling of novel food from Regulation 258/97/EC of the European Parliament and the Council on novel food and novel food ingredients.

2. Main elements of the law

2.1 Authorizing system

A permit is required:
a. to establish a biotechnology laboratory;
b. to modify a natural living organism;
c. to use GMOs in contained systems;
d. to release GMOs into the environment;
e. to commercialize GMOs on the market;
f. to export and to import GMOs.

2.1.1 Institutions responsible for issuing permits for biotechnology activities are:
- an independent biotechnology committee consisting of 17 representatives from the competent ministries, the Hungarian Academy of Sciences, the National Committee on Technological Development and non-governmental organizations. This committee prepares the decisions and provides a verdict on permit applications;
- biotechnology authorities under the control of the competent ministries. Several authorities may be assigned depending on the industry in which the GMO application is proposed. Biotechnology authorities give the permits, control the applications, and in certain, defined cases, restrict or ban the GMO-activity, revoke the permit, fine, etc;
- three ministries: the Ministry of Agriculture and Regional Development, the Ministry of Health, and the Ministry of Economics, in accordance with the most common applications of biotechnology.

2.1.2 The Ministry of Environmental Protection also has an important role as a consultant with veto-authority that can give an expert-opinion on the activities seeking a permit. Biotechnology activity can be restricted or banned if there is new information relating to the risk of the activity (especially in case of danger to the health or to the environment). If banned, GMOs should be destroyed. Biotechnology-activity permits can be revoked and the holder can be fined if either the law or the provisions written in the permit are violated.

2.2 Categories for Biological Activity

Biotechnology activities are divided into three categories by the law. These are:
a) plant- and animal-breeding, food- and feed-production;
b) human health-care, medicine-production;
c) industrial use not included in the preceding two categories.

2.3 Deadlines for Permit Approval

Deadlines for deciding on permit applications where biotechnology activity relates to:
- genetic modification or contained use of GMOs: 90 days;
- the deliberate release of GMOs into the environment or commercialization of GMOs on the market: 180 days;
- the export or import of GMOs: 60 days;
- the establishment of GMO laboratories: 45 days;

Both draft permits and permits are published in the official journal of the authority. The public is invited to comment on the draft permits before they are approved. Deadlines for public comments on draft-permits (from the date of publication) are:

- genetic modification or contained use of GMOs: 30 days;
- deliberate release of GMOs into the environment or commercialization of GMOs on the market: 40 days.

3. Other elements of the law

3.1 Registration

A database has been set up for the registration of relevant information relating to permitted biotechnological activities such as:
- genetic modifications;
- contained use of GMOs;
- deliberate GMO release into the environment;
- commercialization;
- the establishment of laboratories.

The data will be provided by the competent authorities. The following five criteria, defined according to the EC directives, must be registered:
- description of the GMO;
- name and address of the applicant;
- purpose and location of the biotechnology activity;
- methods and plans for monitoring the GMO and for emergency response;
- evaluation of foreseeable effects (in particular any pathogenic and ecologically disruptive effects) of the biotechnology activity.

3.2 Labeling

Both feed and novel foods are excluded from this provision. (Special labeling requirements for GM novel food have been defined in the ministerial decree 1/1996. (I.9) FM-NM-IKM on foodstuffs.) Labeling must occur under the following conditions:
- The product consists of or contains a GMO where protein or DNA resulting from genetic modification is present.
- The product does not consist of or does not contain a GMO, but is produced from GMOs.

The labels for cases a and b are different. Labeling requirements took effect on July 1, 1999. Refer to section six for examples of the labels being used.

3.3 Transportation

Permits for transporting GMOs will be given as part of the biotechnology activity permit. Certain data are required from the applicant when defining the necessary conditions for transporting GMOs.

3.4 Biosafety officers

Biosafety officers should be employed by the companies and institutes that deal with biotechnology. It is the responsibility of the Biosafety officer to enforce the laws and the provisions written in the permit of a company or institute.

3.5 Waste management

Three categories of waste management are differentiated: hazardous waste from biotechnological activity; waste from GMOs used in food-production which is considered harmless; waste to be analyzed by the authorities which will determine the method of handling and neutralizing it.

3.6 Liability and criminal acts

Liability for the damages caused by biotechnology activities is defined in accordance with the Hungarian Civil Code's provisions regarding liability for hazardous operations (objective liability with the exception of vis major). Measures defined in the Criminal Code of the Hungarian Republic are used if the permitted biotechnology activity results in a crime.

3.7 Partial retroactive force

Within 60 days of the law on biotechnology activities taking effect, those contained uses, deliberate releases, commercialization and import of GMOs which were carried out before the new regulation became law, should have been reported to the competent authorities. The competent authority had the right to evaluate the reported activities in order to determine whether they had fulfilled the legal requirements. If the requirements were not met, the activity was banned and restoration of the original condition was ordered.

4. Decrees implementing the Biotechnology Law

4.1 Ministerial decrees following the law further regulate:
- labeling;
- genetic protection zones;
- conditions (technical, technological, environmental, natural protection, health) of biotechnology activities;
- organization, function and operation of the biotechnology committee;
- rules regarding the export and import of GMOs;
- conditions of employment and education of the biosafety officer;
- provisions for data-registration;
- fees for permits;
- biotechnology fine.

4.2 Main elements of the decree

The main elements of the decree 1/1999. (I.14.) FVM on the implementation of the rules of the Act No. XXVII of 1998 on biotechnology activities in agriculture and food-industry are:

4.2.1 Scope

The implementation decree covers the following activities in agriculture and food industry:
a) establishment of GMO laboratories;
b) genetic modifications;
c) contained use of GMOs;
d) deliberate release of GMOs into the environment;

e) commercialization of GMOs;
f) export or import of GMOs;
g) transportation of GMOs.

The following techniques of genetic modification are excluded from this law on condition that they do not involve the use of GMOs as recipient or parental organisms:
a) mutagenesis;
b) cell fusion (including protoplast fusion).

Novel foods are also excluded from this law; Act XC of 1995 on foodstuffs covers them.

4.2.2 Definitions

<u>Emergency</u>: an incident, in the course of or because of which, the GMO or product thereof could present an immediate or delayed hazard to human health or to the environment.
<u>Organism</u>: any biological entity that is capable of replication or of transferring genetic material.
<u>Product produced from genetically modified organisms (product thereof)</u>: a preparation consisting of or containing a GMO, or a combination of GMOs, which is placed on the market.

4.2.3 Genetic surgery/engineering

For the purpose of this decree, genetic surgery/engineering means *inter alia:*
* recombinant DNA-techniques using vector systems;
* techniques involving the direct introduction into an organism of heritable material prepared outside the organism, including micro-injection, macro-injection, micro-encapsulation, electroporation, and electrofusion.

Techniques which are not considered to result in genetic modification since they do not involve the use of recombinant DNA-molecules or genetically modified organisms are:
2   *in vitro* fertilization;
3   conjugation, transduction, transformation or any other natural process;
4   polyploid induction;
5   gynogenesis and androgenesis.

4.2.4 Rules from the decree additional to the Biotechnology Law

* The Secretariat of the Committee comes from the National Institute for Agricultural Quality Control (Országos Mezögazdasági Minösítö Intézet).
* Competent agricultural (biotechnology) authorities have representatives for agriculture, plant health, animal health and food-production in the Ministry of Agriculture and Regional Development.
* The environmental and public health authorities, as authorities with veto power, take part in the activities of the competent agricultural authorities. The former environmental and public health authorities are those belonging to the Ministry of Environmental Protection or to the Ministry of Health.
* Detection of imported products: the importer declares to the competent agricultural authority whether the consignment contains GMOs or products thereof. A certificate giving the result of the analysis issued by one of the three monitoring centers will be attached to the declaration.

- The competent agricultural authority decides on the necessity and the size of genetic protection zones which are placed to separate those fields where traditional and genetically modified plants are cultivated.

4.2.5 Clarification of the rules regarding the permit application procedure

The relationship between the registration of plant and animal species (procedures executed by the National Institute for Agricultural Quality Control) and the commercialization of GMOs is determined. A permit for the latter activity is given by the competent agricultural biotechnology authority. The registration of the species can be applied for after receiving the permit for the release from the competent agricultural biotechnolgy authority. After the species is registered the permit for the commercialization can be applied for. This permit for commercialization contains the competent agricultural authority definitions of those products of the plant or of the animal for which this permit is also valid. The competent agricultural biotechnologyauthority decides on:

- the scale of production of the permitted genetically modified plant species;
- the total number of the permitted genetically modified animals can be bred.

In the course of decision-making, both the scale of production and the number of plants or animals are compared with the proportion of the same traditional plant or animal species bred in the country. However, if the production for agricultural purposes is carried out in a contained system the genetically modified animal species need not be registered. Permits for the breeding and marketing of genetically modified animals and the marketing of their semen have not yet been issued.

4.2.6 Annexes to the decree

- Necessary conditions in case of genetic surgery/engineering;
- Rules regarding the organization, function and operation of the Biotechnology Committee;
- Questionnaires regarding biotechnology activity (genetic modification, contained use, deliberate release and commercialization of microorganisms, plants, animals);
- The list of the detection centers (name and address);
- Labeling (text and pictograph, see below);
- Necessary data for defining the conditions under which to transport GMOs;
- Fees for the printed information which is registered in the database.

4.2.7 Gene technology marking

1. Text and figure to be indicated in case of the Article 15 paragraph (1) of the regulation: "Contains gene technologically produced component". The first fourth of the DNA chain is in national colors (red-white-green).

2. Text and figure to be indicated in case of the Article 15 paragraph (2) of the regulation: "Produced using gene technology method, but does not contain gene technologically produced component". The last fourth of the DNA chain and the triangle is of green color.

5. Present Status

Over the last few years, both 90/219/EEC and 90/220/EEC were revised and the publication of 258/97/EC on novel foods has been considered in the work of the Biotechnology Committee and the competent authorities. The Biotechnology Committee meets every month to evaluate the applications and give advice in written form to the competent authorities to aid in their decision. In the first year of its activity more then ten applications were authorized. All permits and registrations can be found on the Databank under http://biosafety.abc.hu/.

Acknowledgement

The labels of for GMOs and products thereof are designed by József Antal (ABC, Gödöllö).

Further reading

E. Balázs, State of art in biosafety in Central and Eastern Europe. Central and Eastern European Conference for Regional and International Co-operation of Safety in Biotechnology. Keszthely, September 4-6. (abstract), 1995.

E. Balázs, Towards on establishing safety regulation in Hungary. 2nd International Conference in Safety in Biotechnology. Smolenice, October 16-18. (abstract), 1996.

E. Balázs, Slow but definitive progress in the field of regulation of contained used and deliberate releases of GMOs in Hungary. Budapest, August, 22-23, 1997.

E. Balázs, Regulatory framework on biotechnology in Europe: Similarities and differences. II. Encontro Brasileiro de Biotechnologia Vegetal, REDBIO Sub-regiao, Gramado, Brasil, November 24-28. (abstract), 1997.

E. Balázs, The Launch of a Nationwide Project: Hungarian Biosafety Framework. *Hungarian Agricultural Research* **2** (1998) 27.

E. Balázs, Janus face of biotechnology and biosafety in Central and Eastern European Countries. 5th International Symposium: The Biosafety Results of Field Tests of Genetically Modified Plants and Microorganisms Braunschweig, 6-10 September, 1998.

Use of Agriculturally Important
Genes in Biotechnology
G. Hrazdina (Ed.)
IOS Press, 2000

Public Perception and Legislation of Biotechnology in Poland

T. Twardowski

*Institute of Bioorganic Chemistry, Polish Academy of Sciences, 61-704 Poznan,
Noskowskiego 12, Poland*

Abstract. Modern biotechnology is characterised by the value added chain [VAC] of three main factors: VAC = [science & technology] + [law & IPR] + [society & perception]. In this paper the legislation and public perception of biotechnology will be discussed. We found in the Polish society the willingness to accept the novel food. The most significant lesson from the survey is the following: up to two third of Polish society is willing to accept the novel food, however, almost 4/5 expect some kind of governmental supervision and labelling. We did not find a correlation between knowledge and social position. Over the last three years, there has been a tendency to implement a new Polish law, which would be a 'national copy' of the EU directives implemented with the assistance of the international organisations (OECD, UNEP, UNIDO). In November 1997 a team of experts submitted a project of the Polish „gene law" to the government. A revised proposal will be submitted in December 1999. The project is based on Directives 90/219 and 90/220. The new law on environment protection [article 37a] regulates GMO research, releases and trade. In order to support sustainable development, we need to know the molecular background of nature, and the the legislation to be formulated has to be based on solid science. However, only through the understanding how ordinary people accept genetic engineering can we commercialise the success of modern science.

1. Introduction

In order to facilitate the development of modern biotechnology we need a combined efforts from many different fields. Some of these efforts are more significant than others. Modern biotechnology is characterised by a value added chain [VAC] of three main factors: VAC = science & technology + law & IPR + society & perception.

In this particular paper I will focus on the public perception and legislation of biotechnology in Poland. The relationship between research and legal and social aspects is evident. The basic idea of sustainable development, is a top priority for industry, administration and ordinary people. In order to find the best roads to sustainable development, we need to understand the molecular background of nature and the formulation of legislation based on solid science. However, only by understanding how ordinary people accept genetic engineering can we commercialise the success of modern science. The analysis that I will present is related mainly to agrobiotechnology.

We had the "green revolution" and today we are observing "gene evolution". The "green revolution" allowed us to produce more food than ever before. The advantages of this discovery were acknowledged by the Nobel Peace Award granted to Norman Borlaug in 1970. Today, we face a similar, but more universal "evolution" [not "revolution"], with the background in solid basic science. This gene evolution is closely followed by consumers, industry, governments and international organisations. We must not forget about the special nature of discoveries in the field of biotechnology: on the one hand, they stem from basic research and on the other hand, they involve industry and commercialisation. We can talk simultaneously about extremely advanced molecular technologies and laundry detergents. The development of biotechnology entails two entirely different streams of activities, presented in Table 1:

Table 1. Concomitant activities in the development of biotechnology.

A. Research and development	B. Commercialisation
Idea [desired properties]	Intellectual property rights
Finding the gene	Legal aspects
Description of the gene	Biosafety
Multiplication of the gene	Public perception
Expression of the gene	Commercialisation
Genetic modification of the organism	Economic evaluation
Reproduction of the GM organisms	Marketing
Scaling up	Monitoring

It is easy to present several examples of high and low public acceptance. For example, there have been no objections to the production of human insulin in bacteria. On the other hand the introduction of human protein or entire composition of human milk into bovine milk raised strong protests. In any such case public perception is a key factor [1].

The most common (and most controversial) technology used by modern biotechnology is the modification of the existing gene(s). This can be achieved using the following techniques:

1. modification of the existing gene activity, e.g. through antisense strategy,
2. reduction or multiplication of the number of existing genes,
3. modification of the regulatory fragment(s) of RNA/DNA.

In the case of higher organisms, particularly animals, the modification of the DNA or RNA apparatus can be done:

1. for one generation only,
2. for ever (in many generations); the very special case is the "terminator effect," heavily discussed recently.

These gene technologies may be perceived by the public in different ways. Removal of only one gene from the genome or addition of a brand new gene to a genome significantly differs from the lowering of the activity of an existing gene. We have to take into consideration the possibility of chemical synthesis of a yet non-existent gene. Such gene may also be introduced to the existing organism in order to produce a non-existent (today) protein of unknown properties.

I would like to mention two very prominent cases in order to illustrate the ideas presented with concrete examples: fish farming and milk production. Aquatic agriculture is not included in the program of this meeting. In the opinion of many experts the sea offers numerous opportunities for future development of food production. Fish farming is of special importance for at least two reasons: 1. fish is the staple food for one third of the world's population; 2. the genetic modification of fish is relatively easy. It is also very difficult to keep under control after release to the environment. According to recent reports, we should expect a large scale commercial production of transgenic carp in China.

Milk is of special importance in our diet and practically almost everybody used to consume, consumes or will consume it. The composition of human milk was perfected by nature. However, the introduction of a new protein(s) (recombinant proteins, hormones) to animal milk is possible today. Modified milk has already been produced on an industrial scale.

A recombinant protein is a protein obtained through the gene transfer from organism A to organism B and produced in organism B. The best known examples are insulin and growth hormones. The production of the recombinant protein as a milk component is expected to be the cheapest and most efficient way to obtain these expensive and very much wanted hormones. Milk will be considered not only as a nutrient but also as pharmaceutical.

2. Public perception of biotechnology in Poland

The survey was conducted by the Osrodek Badania Opinii Publicznej [OBOP, Centre for Public Perception Research] located in Warsaw, during September 1998 and March 1999. The survey was patterned on the questions contained in the 1997 European Union survey conducted under the "Eurobarometer 46.1" project. Our survey was conducted using a sounding method on people over 15 years old, who were randomly selected. Nine hundred eighty two and 1080 interviews were carried out respectively. Data were evaluated statistically (using the SPSS DOS method) with the error margin of $\pm$ 3% and reliability 0,95 [1-4]. In a separate survey, we asked the same questions of the experts – about 150 biotechnologists [5].

People who were interested in biotechnology had predominantly secondary and higher education. Young people, especially in the 20 year old category, had often heard of, but were not interested in biotechnology. Fourteen percent of the respondents indicated that they were interested in the use of biotechnology methods in food and beverage production; 48% had heard about, but were not interested in it, and 38% have not even heard about it.

The question: "Permit or ban production and sale of transgenic food?" is a very critical one. More than half of the respondents, 57%, think that the production and sale of transgenic food should be permitted. Twenty-three percent are opposed and 20% do not have any opinion. The higher the educational level of the respondents, the better the financial situation and the weaker religious affiliation, the more often they voice the opinion that the production and sale of transgenic foods should be permitted.

The survey's opinions concerning the use of traditional methods for agro-food production in comparison with the methods of genetic engineering are different. Thirty-six percent of respondents favored genetic engineering. A similar percentage (thirty-four) of respondents stated that these methods did not matter; 9% of those surveyed prefers the traditional production methods, and 21% have no opinion about it. The opinions concerning the traditional methods of food production in comparison with genetic engineering methods do not differ among socio-economic groups.

The next question refers to the problem of special labelling requirements for transgenic foods. Five different kinds of these foods were presented. In each case about four fifths of the respondents favored labelling. It is understandable that in the experts' opinion the necessity of labelling products which do not contain GMO's (for example sugar) is judged superfluous more often.

Based on public opinion polls, it appears that about half of the respondents forecast that genetic engineering will affect the different aspects of human life. The largest group (57%) of the respondents think that these methods will improve life for mankind, however this group contains 80% of the experts. Forty-one percent of those polled are afraid that genetic engineering will introduce new diseases, and one fifth of the experts believe in the possibility of such a risk.

In analysing this data we should also pay attention to the high percentage of answers "difficult to say". Generally respondents did not have any opinion in the case of the possibilities of introducing new therapeutical methods (40%) and accelerating the level of natural stock exploitation of the Third World (38%). The question, whether biotechnology will render traditional food production obsolete drew the lowest percentage of responses

(22%). As anticipated, many biotechnologists (from 1/5 to 1/3, according to the problem) abstained with the opinion and forecasting the future. We found in the Polish society a willingness to accept the novel, genetically engineered foods.

The most significant lesson from the survey was that up to two third of Polish society is willing to accept the novel food. However, almost four fifths expect governmental supervision and labelling. The labelling requirement is not thought of exclusively as a precautionary measure, but mostly as the privilege and right of the consumer to fair information. A similar opinion was presented by the experts.

The young and educated people are more open and willing to accept GMO foods. We can speculate that reservation about accepting genetically modified foods is connected with a more orthodox interpretation and reflection about the food production and treatment.

In commenting on the public perception of biotechnology in Poland, we have to take into account the fact that the quantity of currently available data is very limited. The general perception seems to be that Polish people have mixed feelings about biotechnology and their knowledge is a reflection of conventional wisdom. We did not find a correlation between knowledge and social position. The education of society and communication of scientific achievements are key issues for the future of a positive public perception of biotechnology and its products.

3. Legal aspects of biotechnology in Poland

In the 1990s the first legal acts related to biotechnology were passed, with the "Biodiversity Convention" (1995) and "Patent Law" (1993) being the most spectacular [6-8]. The scientific community recognised the new era relatively early. The Biotechnology Committee presented the state authorities with several reports on the future and economic significance of biotechnology [6]. In the 1990s we have observed a significant influence of international conventions on the formation of national legislation, Convention on Biological Diversion (1995), membership in the OECD (1996), the Budapest Treaty and the protection of intellectual property rights (1993). Poland's membership in the EU became the priority of national policy. This also meant acceptance of the European legislation, e.g. Directives of EC (90/219 and 90/220) [6, 9-11].

Over the last three years there has been a tendency to implement a new Polish law, which would be a 'national copy' of the EU directives implemented with the assistance of the international organisations (OECD, UNEP, UNIDO). In November 1997 a team of experts submitted an outline of the Polish "gene law" to the government. However, there was no interest in this initiative. A new and revised proposal will be submitted in December 1999. The outline is based on Directives 90/219 and 90/220 [6].

The Minister of Agriculture in co-operation with the Minister of Environment Protection, State Committee for Scientific Research and the Minister of Health appointed the Experts Committee for Genetically Modified Organisms (July 1996) and a procedure for applications concerning the introduction of GMO into the environment (February 1997). A new law on "protection of the environment" complying with the "Biodiversity Convention" was passed by the Parliament, was signed by the President (September, 1997) and went into effect on Jan.1, 1998. Research, releases and trade are temporally regulated under Article 37a of this law [no technical directives for this law were released until Oct. 10, 1999].

The first field experiments with 3 transgenic plants: potato, corn and beet, were performed in 1997. In 1998 approximately 20 experiments were performed. However, in 1999 the number of the experiments was reduced due to the restrictive policies of the state administration. All field experiments are performed with the permission of the Minister of Agriculture and under the strict supervision of experts.

4. Discussion

Currently, the industry is trying to provide customers with products whose properties meet the expectations of the market. Safe technologies using biological methods are being introduced. The quality of the end product is becoming a top priority for a majority of manufacturers. They are now commonly using organisms and enzyme preparations obtained with the tools of genetic engineering. Modern biotechnology, in particular genetic engineering, plays a very important role in industrial development. On the other hand, we observe that societies are very reluctant to approve transgenic products, especially foods. Very recently the European Parliament discussed a three year moratorium on all GMOs.

A new technology, genetic engineering, will enter the Polish market of almost 40 million people very soon [9,10]. Based on the survey discussed above, we can conclude that ambivalent feelings in the public arise from information from abroad and the commercialisation of biotechnology's achievements. These technologies are meant to make everyday life easier and more comfortable, however, they create concerns about what the unknown will bring to us. The radically new technologies are based on complex scientific achievements. The current position of biotechnology is comparable to the position of space exploration at its commencement in the fifties.

The following specific issues will require our attention: 1. socio-economic consequences (including potentially adverse effect on food security); 2. producers' rights and responsibilities; 3. consumers' right of choice and access to fair information; 4. health consequences; 5. environment aspects.

When discussing the potential risk of transgenic plants and modern food production using biotechnology we should distinguished between the danger stemming from dogmas and potential dangers "based on science". The following risks should be taken into consideration:

- Creation of a new species (in reality it applies only to viruses),
- Effects on biodiversity,
- Increased problems with weeds,
- Probability of unexpected synergetic effects,
- Allergenicity of novel proteins,
- Unknown.

We must remember, however, that dogmas pose the most serious threat to the development of modern biotechnology. The value of the products of modern biotechnology is limited by trade and perception by the potential consumers. In the near future the major impact of biotechnology on agriculture will be related to photosynthesis and the reproductive capacity of biological material.

5. Conclusions

A science based knowledge is required from all participants on the biotechnology scene. The most important of these participants is the consumer. We should avoid a situation where dogmas dictate the development of modern biotechnology. In other words, we should highlight the importance of education of society. Only an educated consumer will be able to assess the value of a biotechnology product.

Recapitulating and formulating the conclusions from the public opinion surveys we could expect that:

- Polish society will accept, but demand the evaluation and labelling of "GMO food";
- The producer will have to demonstrate the superior values of the novel foods.

The future prospects of biotechnology in Poland are connected to agrobiotechnology, health industry and services. These major directions of development are directly related to the sustainable development of biotechnology as covered by the "Convention on Biological Diversity". In the context of that legal regulation, the biological safety of mankind and environmental protection, as well as the protection of intellectual property rights should be of basic significance.

Acknowledgement

The surveys were financially supported by KBN within SPUB project and UNEP project, respectively. The full sets of data were published in the Polish journal "Biotechnologia" [3] and in the proceedings of the "Biosafety workshop" [4] [in Polish].

References

[1]　W. Wagner *et al.*, Europe ambivalent on biotechnology, *Nature* **387**, 26 June 1997, 845-847.

[2]　T. Twardowski, Legislation and public perception of biotechnology in Poland, IX International Congress on Plant Tissue and Cell Culture, Plant Biotechnology and In Vitro Biology in the 21st Century, Jerusalem, Israel, June 14-19. 1998.

[3]　A. Twardowska-Pozorska and T. Twardowski, Odbiór spoleczny nowejzywnosci (Zywnosci GMO w Polsce), *Biotechnologia* **4** (1998) 20-47.

[4]　A. Pozorska and T. Twardowski, Opinia Polaków o zywnosci GMO, w: Przygotowanie Krajowego programu Bezpieczenstwa Biologicznego, warsztaty IV, 1999, 7-19.

[5]　T. Twardowski and K. Ludwiczak, Who-is-Who in Polish Biotechnology, *Biotechnologia* 4, (1995), 59-143.

[6]　Analiza porównawcza z przepisami miedzynarodowymi stanu uregulowan prawnych w obszarze zastosowan genetycznie modyfikowanych organizmów (GMO), ocena zagrozen wynikajacych z rozwoju biotechnologii oraz dostosowanie polskiego prawa do nalozonych na RP zobowiazan, [Biotechnology; proposal for the Polish gene law, in Polish, with extended English summary], ed. T. Twardowski, 1997, Poznan

[7]　T. Twardowski *et al.*, in: Proceedings of the Central and Eastern European Conference for Regional and International Cooperation on Safety in Biotechnology, Keszthely, pp.133-142, 1995.

[8]　T. Twardowski, Polish Biotechnology, *GIT* l, (1995) 29-31 Juni.

[9]　Atanasov *et al.*, Commercial needs and creating new markets in Eastern European countries, IX International Congress on Plant Tissue and Cell Culture, Plant Biotechnology and In Vitro Biology in the 21st Century, Jerusalem, Israel, June 14-19. 1998.

[10]　T. Twardowski, Conditions for commercialisation of GMOs in Central and Eastern Europe, 5th International Symposium on the Biosafety Results of Field Tests of Genetically Modified Plants and Microorganisms, Braunschweig, Germany, 1998.

[11]　6th meeting of the Open Ended Ad hoc Working Group on Biosafety and the Extraordinary Meeting of the Conferences of Parties to the Convention on Biological Diversity, Cartagena, Colombia, 1999.

Appendix I.

Public Perception of GMOs in Poland:

I.   Should research of GMOs be continued and supported?
II.   Are GMOs useful or dangerous?
III.   Should research on GMOs be supported or banned?
IV.   What are the morally acceptable uses of GMOs?
V.   Legislation of GMOs

Summary

Survey of public perception of biotechnology (in Poland)

The majority of Poles support genetic engineering in food technology, pharmacy and environmental protection.

About half of the respondents are afraid that GMOs present a danger to human health and the environment.

The view that the government should supervise and legislate work on GMOs is a common one.

The highest level of support is given for the use of micro-organisms for environmental protection.

The use of biotechnology to produce new drugs is acceptable.

Three quarters of the surveyed Poles (72%) support biotechnology and genetic technology for food production. One fifth (19%) are against its use and 10% have no opinion.

Half of the respondants think that GMOs present a potential danger to human health and/or the environment; one third (32%) don't see a risk and one seventh don't know if there is a risk.

It is a common view (91%) that the government should supervise the research and legislation of GMOs.

I. Should research on GMO's be continued and supported

Question 1: New biotechnology methods and genetic engineering technologies are used for food production. Use of these methods can improve the quality of food and beverages, e.g. increasing protein content, reducing fat level, changing the taste, or extending shelf life. To what extent do you agree with the following statements concerning research on food using biotechnology and genetic engineering?

This research should be continued and supported.

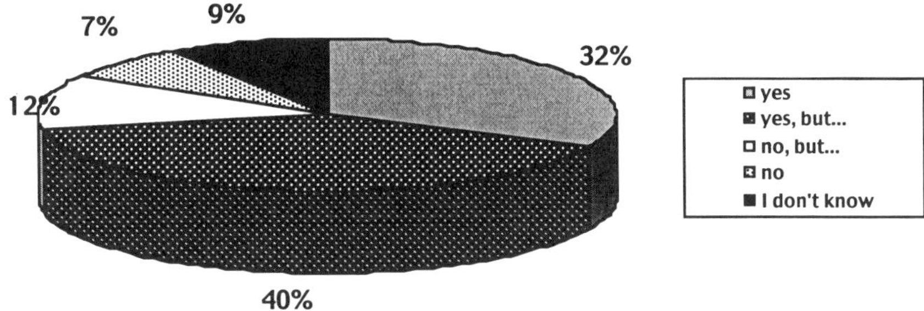

This research poses a threat to human health and the environment.

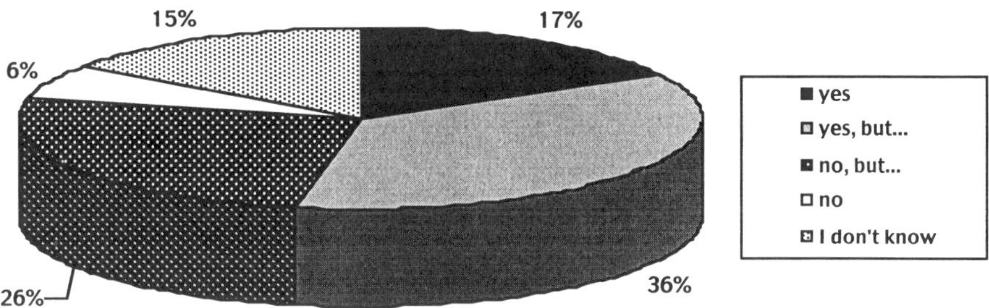

All research should be under governmental control and legislation.

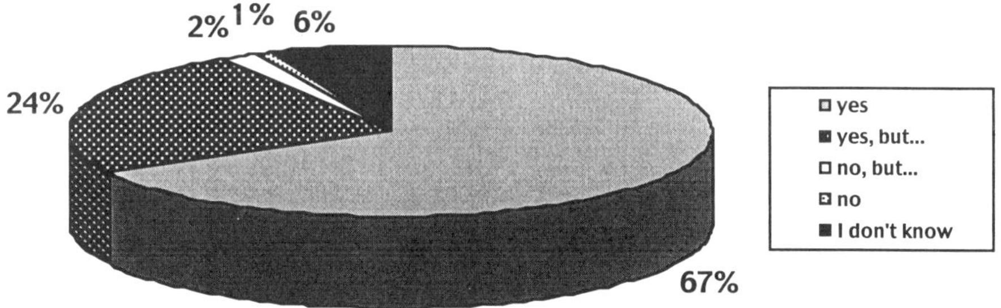

Question 2: Some micro-organisms (bacteria) are used in the production of beer, bread, yoghurt, etc. To what extent do you agree with the following statements concerning the use of micro-organisms in food production?

This research should be continued and supported.

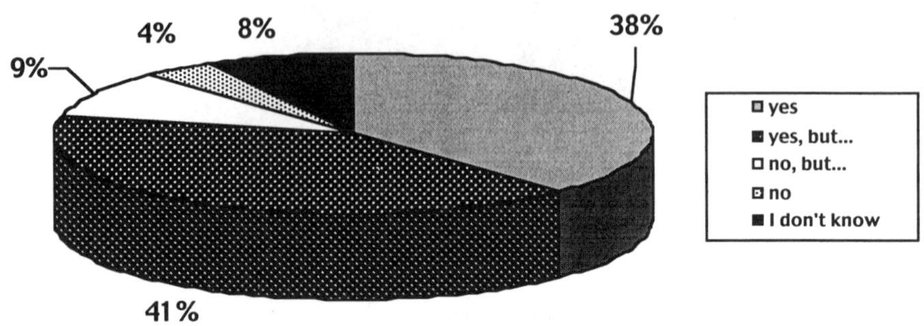

This research poses a threat to human health and the environment.

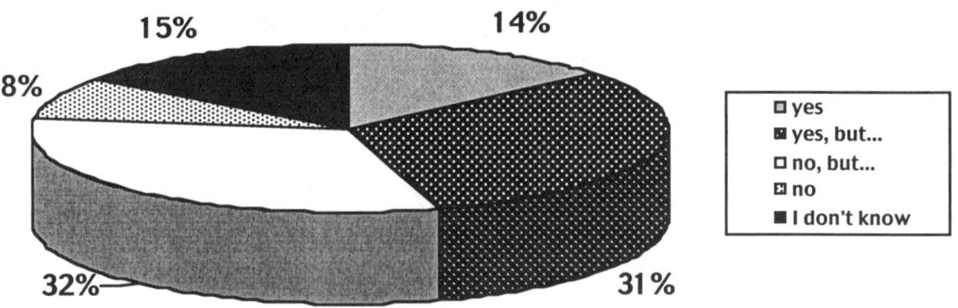

All research should be under governmental control and legislation.

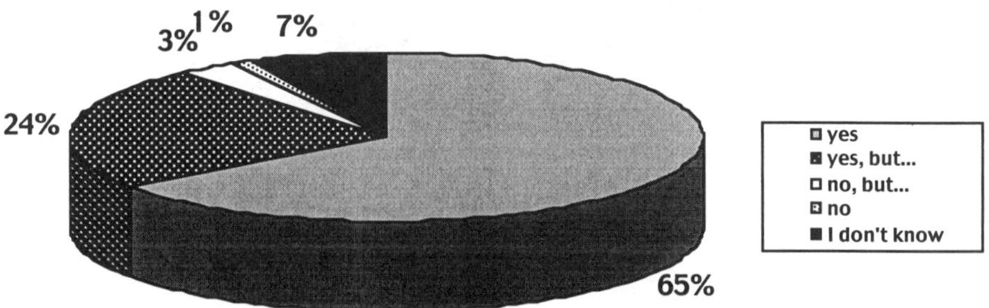

Question 3: Genetic engineering can be used in animal breeding to facilitate faster growth and better meat or milk production. To what extent do you agree with the following statements concerning the application of modern biotechnology to animal breeding?

This research should be continued and supported.

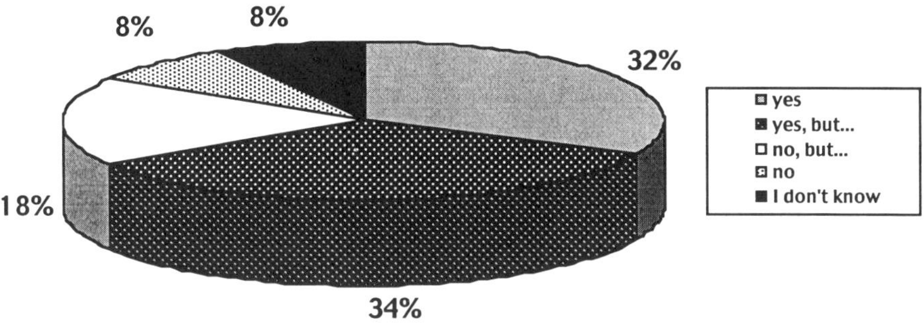

This research poses a threat to human health and the environment.

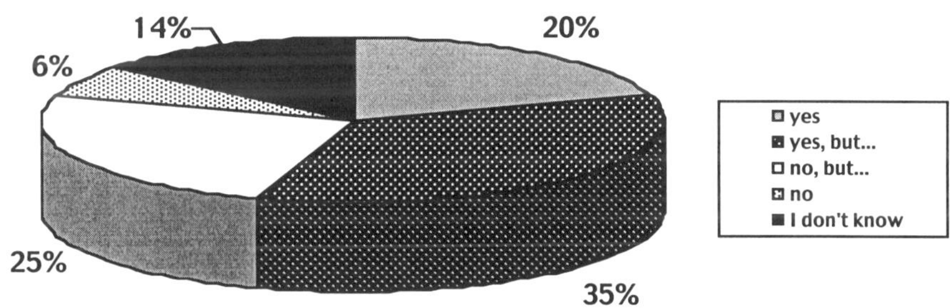

All research should be under governmental control and legislation.

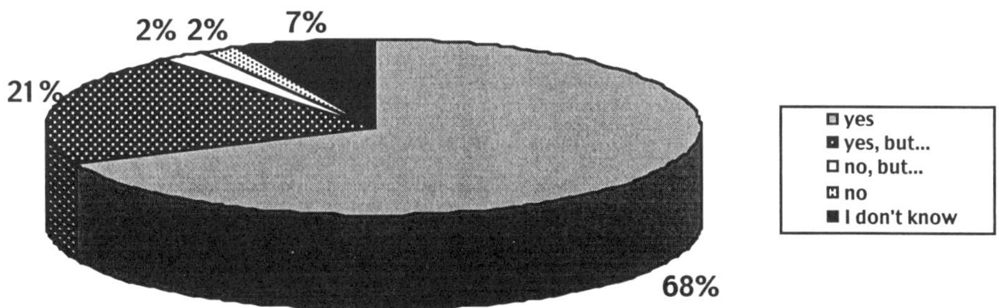

Question 4: Biotechnology and genetic engineering are used in the production of new drugs and vaccines. To what extent do you agree with the following statements concerning research and production of drugs and vaccines using biotechnology and genetic engineering?

This research should be continued and supported.

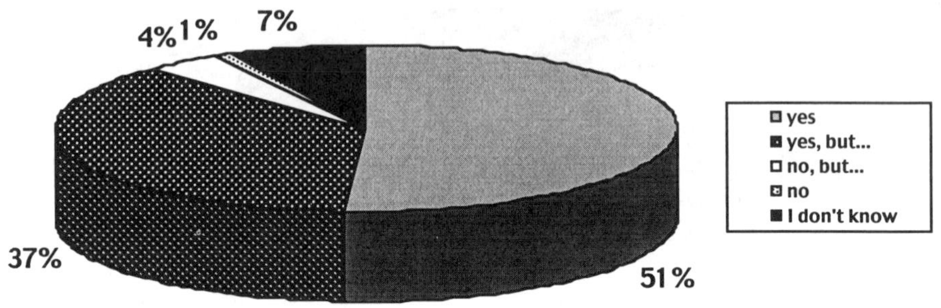

This research poses a threat to human health and the environment.

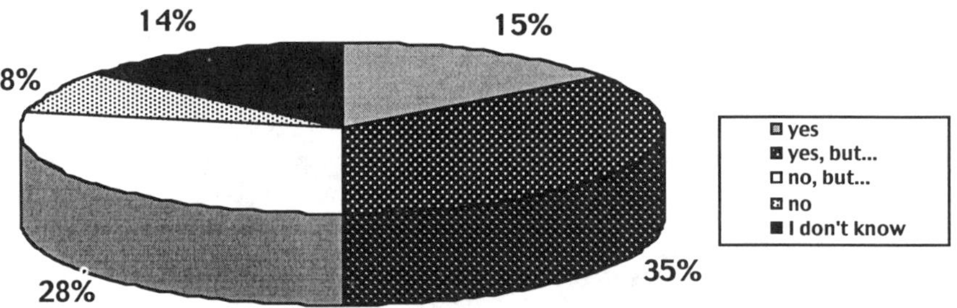

All research should be under governmental control and legislation.

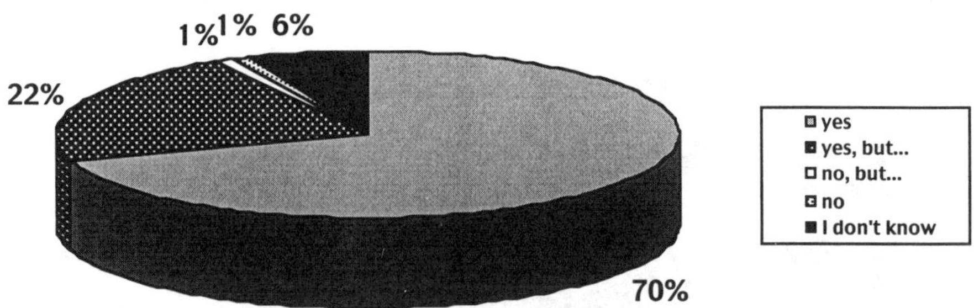

Question 5: Some micro-organisms (bacteria) are used for removing oil and impurities in order to protect the environment. To what extent do you agree with the following statements concerning the use of micro-organisms to clean the environment?

This research should be continued and supported.

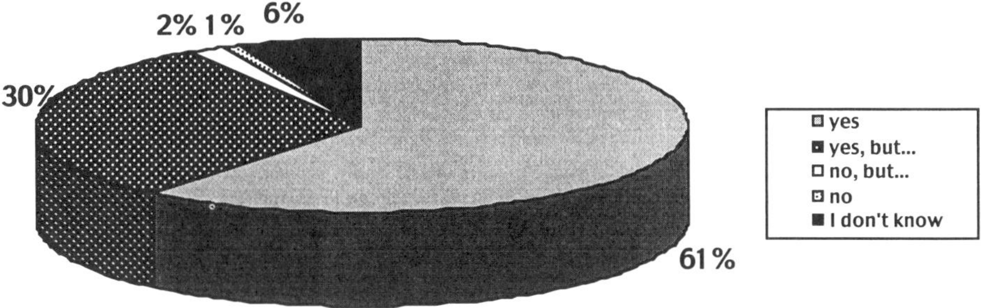

This application is a danger to human health and the environment.

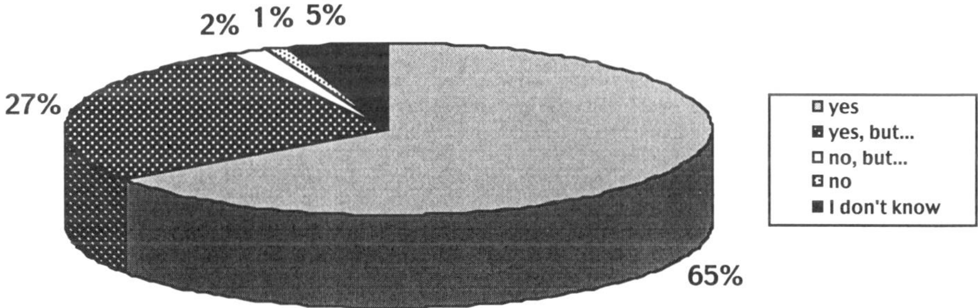

All research should be under governmental control and legislation.

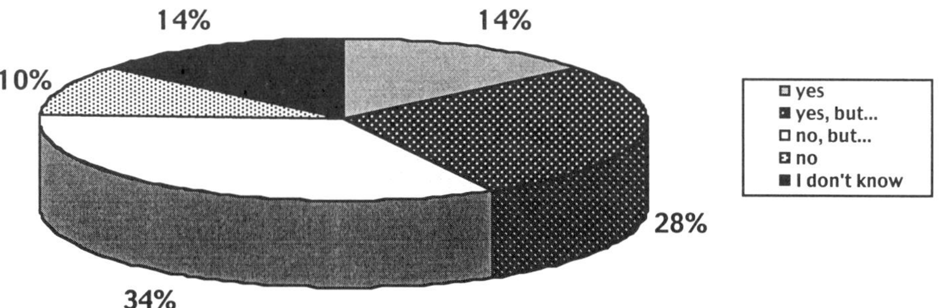

II. Are GMO's useful or dangerous

The majority of Poles think genetically modified micro-organisms are useful for environmental protection (81%).

Almost two thirds of the respondents support the introduction of human genes into bacteria for the production of drugs and vaccines for human therapy (64%).

Similarly, nearly two thirds of respondents support the modification of plants through gene technology to improve resistance to insects and viruses (63%).

Sixty percent of the respondents accept the breeding of genetically modified animals for use in research.

Opinions concerning xenotransplantation are divided: 41% think it is useful and 42% think it is dangerous.

Forty-nine percent of the respondents think biotechnology could be dangerous in food production and 39% expect it to be useful.

Respondent opinions regarding:

1. The application of modern biotechnology for food production, for example for higher protein contents, longer shelf life, improved taste.

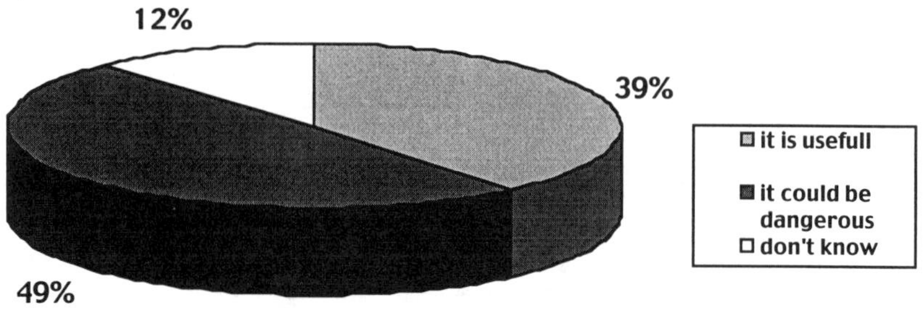

2. The isolation of the genes responsible for insects resistance in plants and production of new plant varieties.

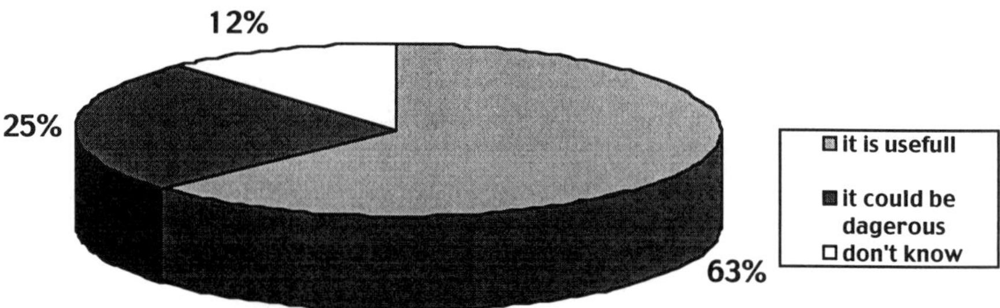

3. The introduction of human genes into bacteria for the production of valuable drugs and medicines for human therapy.

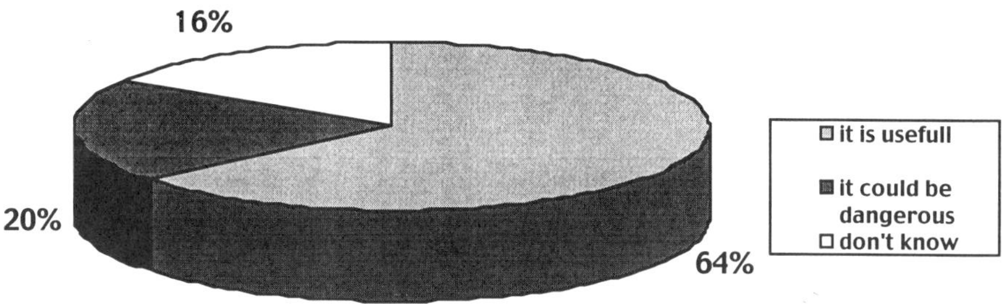

4. The breeding of genetically modified animals for research, e.g. mice.

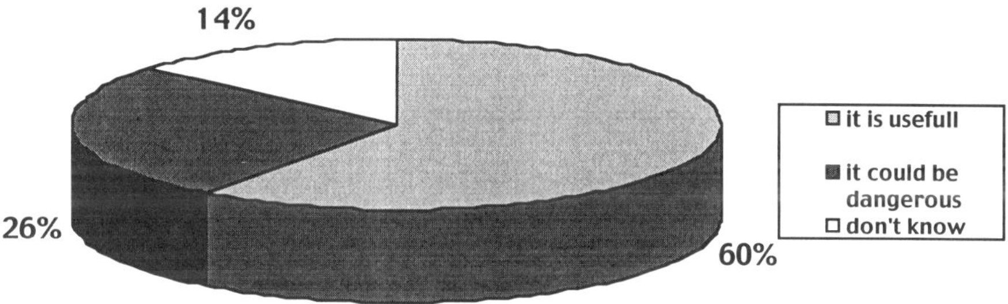

5. The modification of micro-organisms (bacteria) for environmental protection, e.g. to consume lead.

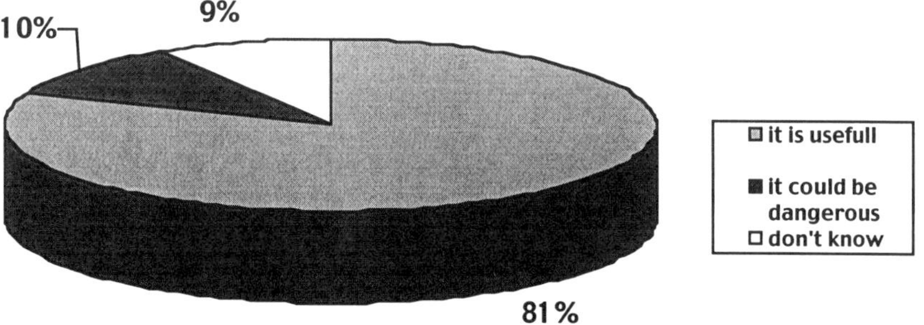

6. The introduction of human genes to animals for xenotransplantation, e.g. pigs for hearts transplants.

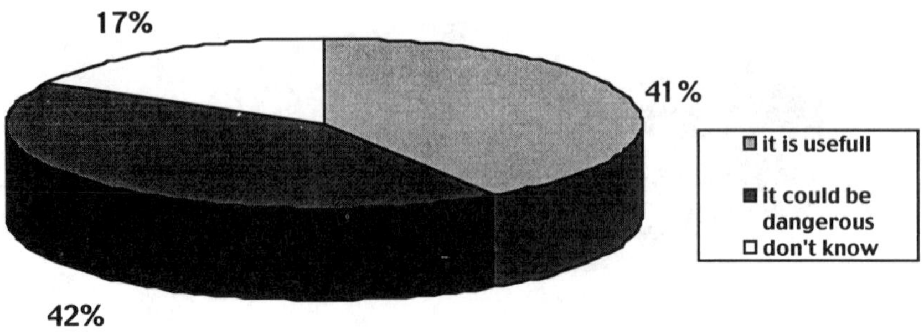

III. Should GMO research be supported or banned

The most controversial area of GMO research is xenotransplantation; 39% of the respondents support such research and 37% are against it.

In all other cases at least half of the respondents support biotechnological applications. The most consistently supported application is ecology (81%), next are plant protection (66%) and the production of new drugs and vaccines (65%), followed by the development of transgenic animals (56%) and improved food production (47%).

Respondents opinions regarding:

1. The application of modern biotechnology for food production, for example, for higher protein content, longer shelf life, improved taste.

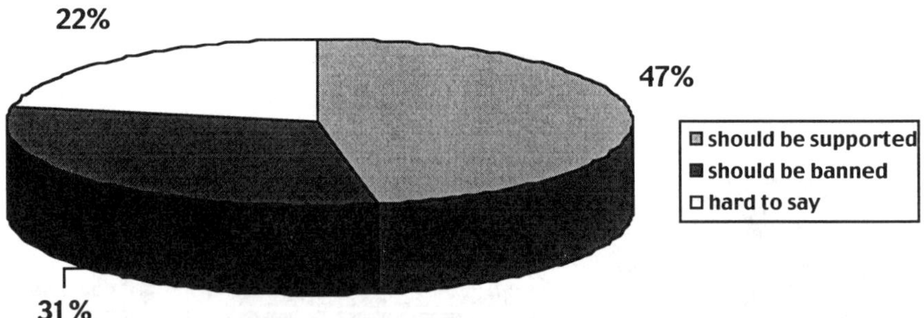

2. The isolation of the genes responsible for insects resistance in plants and the production of new plant varieties.

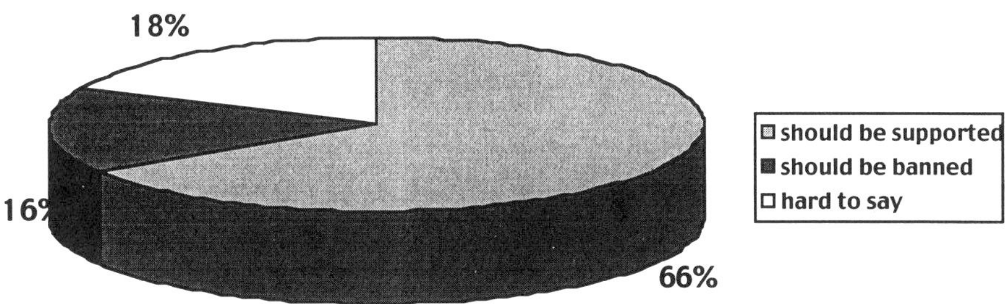

3. The introduction of human genes to bacteria for production valuable drugs and medicines for human therapy.

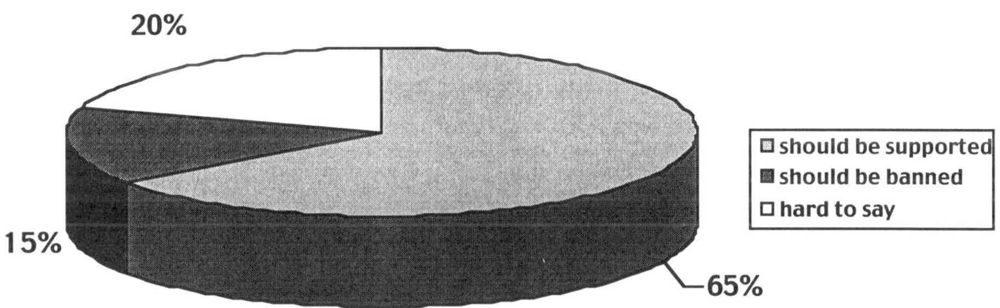

4. The breeding of genetically modified animals for research, e.g. mice.

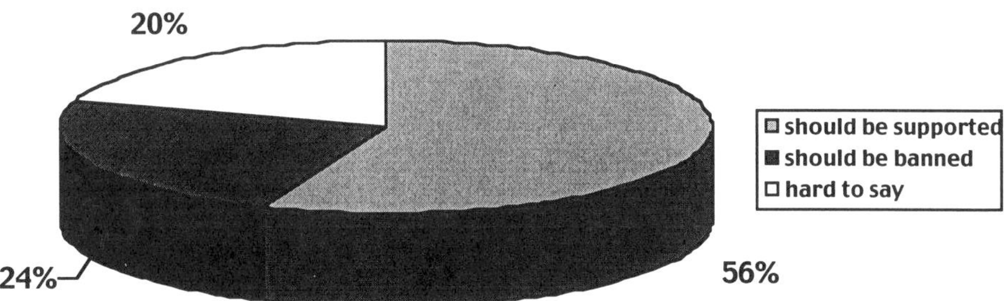

5. The modification of micro-organisms (bacteria) for environmental protection, e.g. to consume lead.

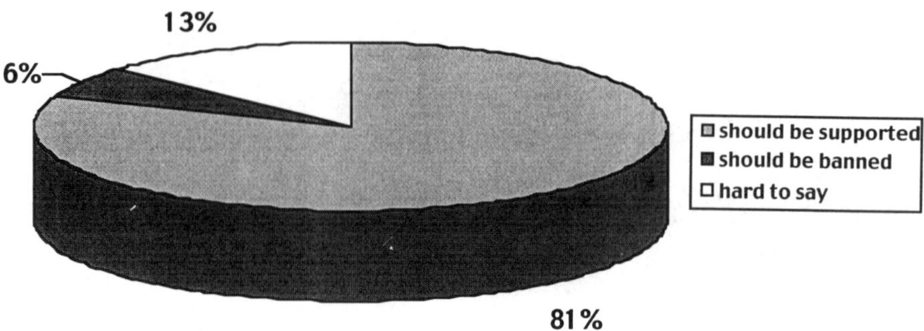

6. The introduction of human genes into animals for xenotransplantation, e.g. pigs for hearts transplants.

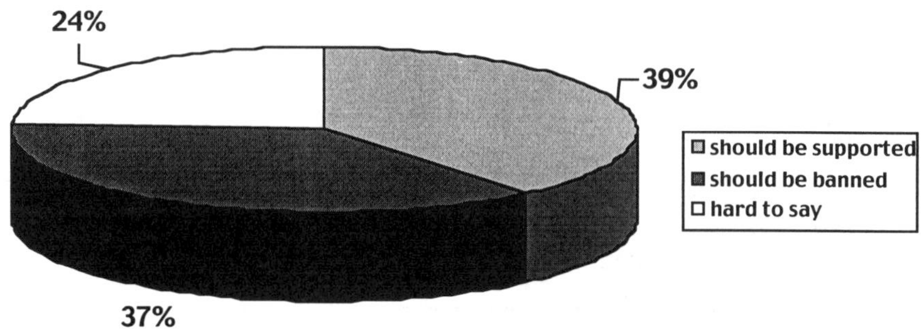

IV. Morally acceptable uses of GMOs

Moral arguments make xenotransplants unacceptable to most respondents: 42% are against and 32% are in favor of xenotransplants.

In all other cases at least half of the respondents find biotechnology morally acceptable: environmental protection (77%), insect resistant plants (67%), the production of new drugs and vaccines (62%), food production (54%), and genetically modified animals for research (50%).

Respondent opinions regarding:

1. The application of modern biotechnology to food production, for example for higher protein content, longer shelf life, improved taste.

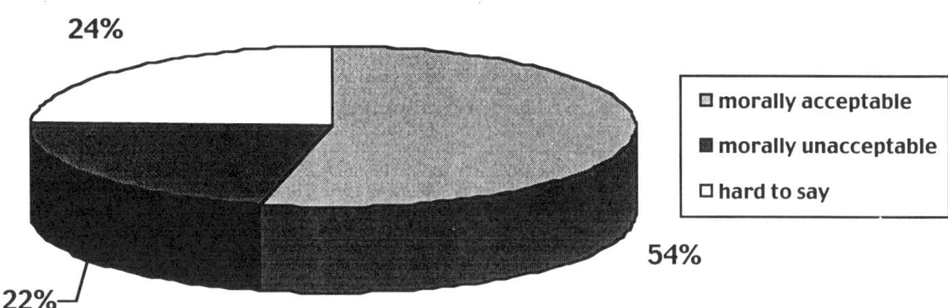

2. The isolation of the genes responsible for insect resistance in plants and the production of new plant varieties.

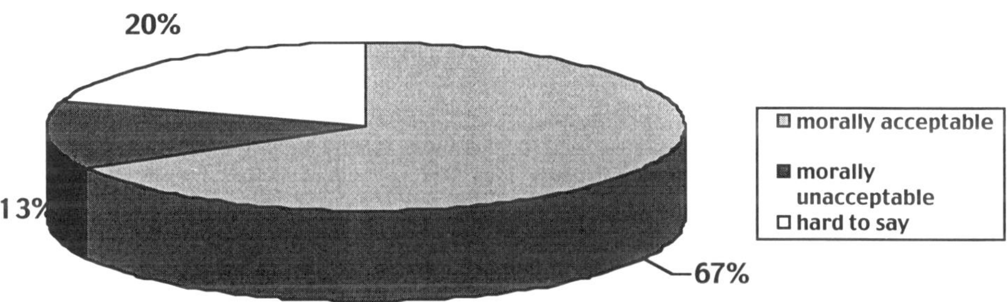

3. The introduction of human genes to bacteria for the production of valuable drugs and medicines for human therapy.

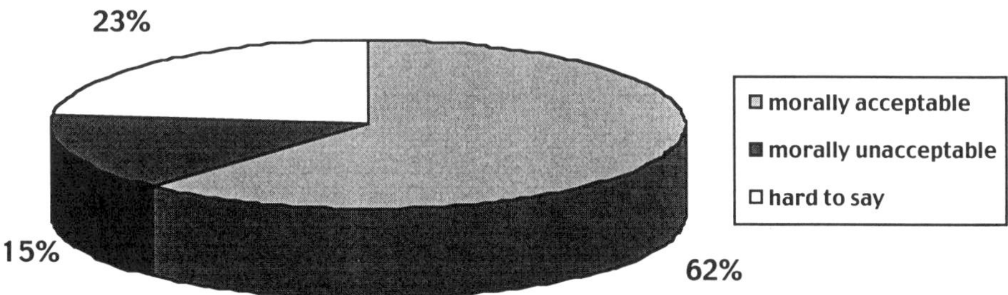

4. The breeding of genetically modified animals for research, e.g. mice.

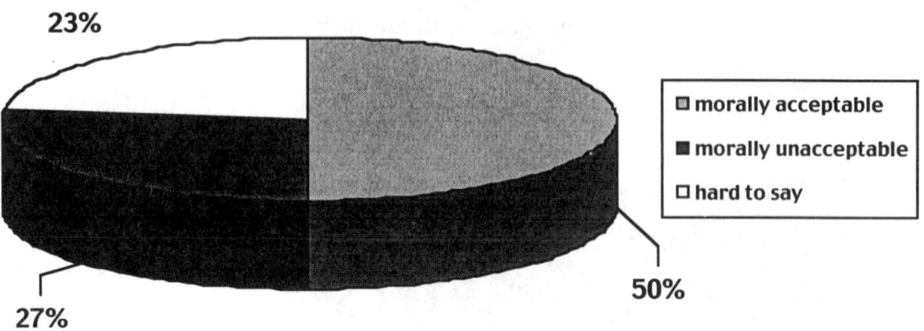

5. The modification of micro-organisms (bacteria) for environment al protection, e.g. to consume lead.

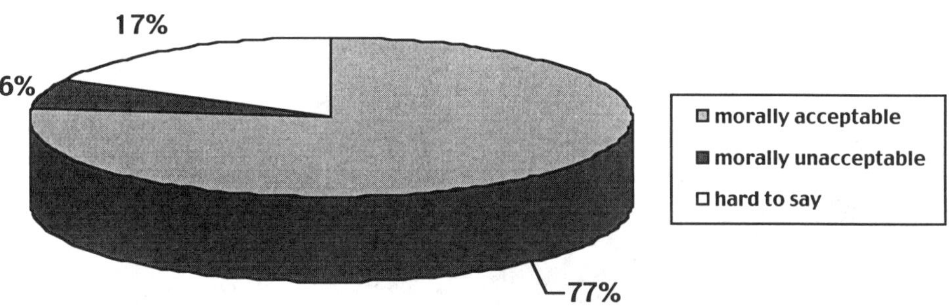

6. The introduction of human genes into animals for xenotransplantation, e.g. pigs for hearts transplants.

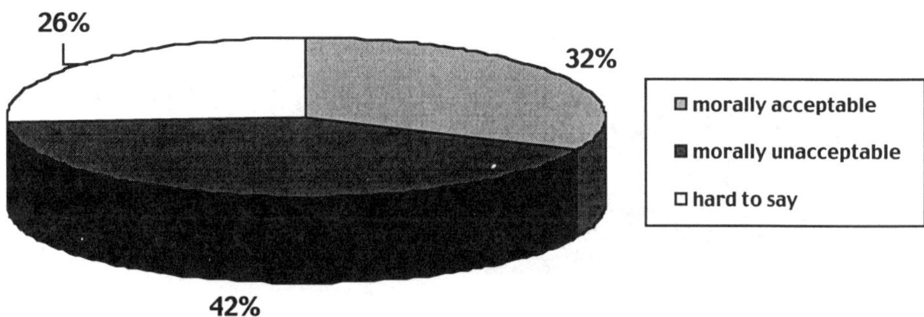

V. Legislation

Many Poles believe that legislation is ineffective for regulating scientist's activities with regard to biotechnology (57%). Many also believe that recent legislation is insufficient for protecting people against the risks related to modern biotechnology (45%).

However, 53% of those surveyed feel that industry should have some influence on the formation of a new law concerning biotechnology. Half of the population (50%) doesn't see the necessity of consulting religious organizations when creating a new biotechnology law. Most of those surveyed (47%) also feel that biotecnological development should proceed only after the public has been consulted.

The majority of Poles surveyed (89%) believe that genetically modified food should be labeled.

Fifty-four percent of the repondents support traditional breeding.

Respondent reactions to the following statements:

1. GMO food should be labeled.

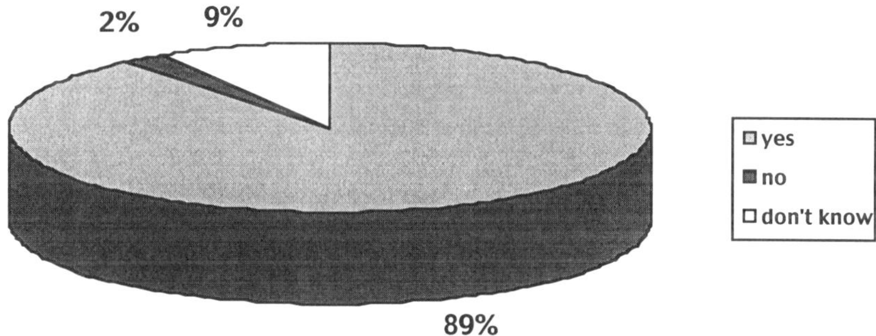

2. Scientists (biotechnologists, physician, physicist's et.) don't care about the law.

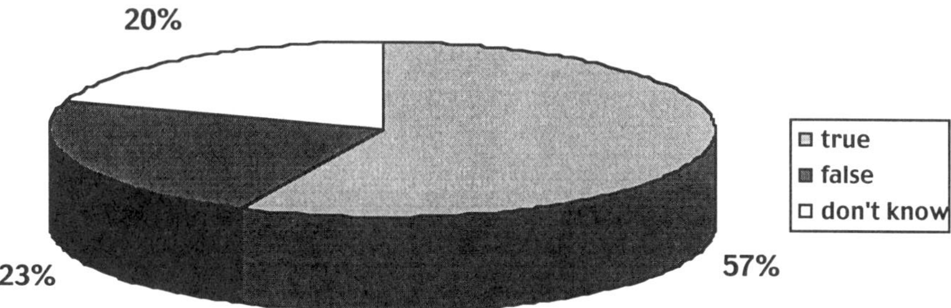

    T. Twardowski / Public Perception and Legislation

3. Traditional breeding should be used first, instead of genetic engineering.

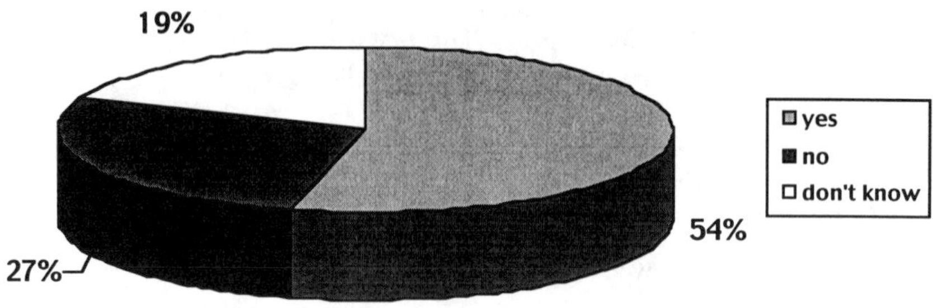

4. When formulating a law for biotechnology, we should first follow opinion of industry.

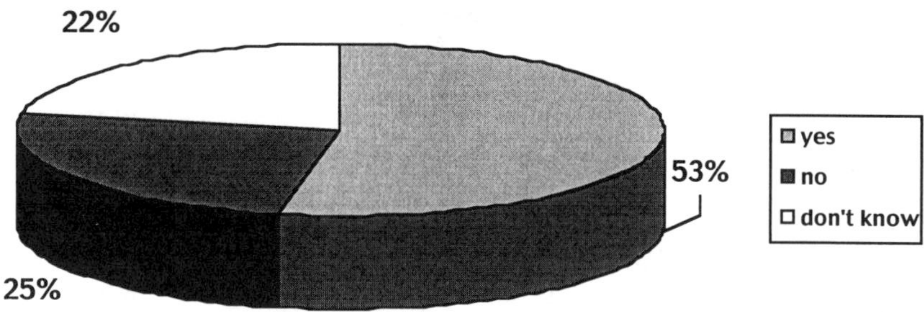

5. Modern biotechnology is so complex that consulting the public is a waste of time.

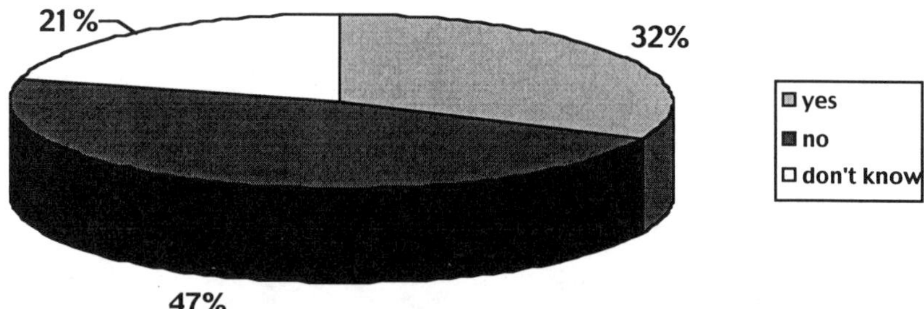

6. When formulating a law for biotechnology, we should follow the opinions of religious organizations.

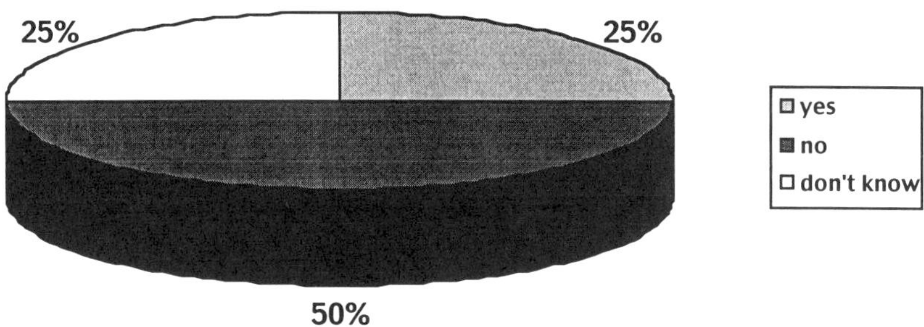

7. The present legislation is sufficient to protect people from risks related to modern biotechnology.

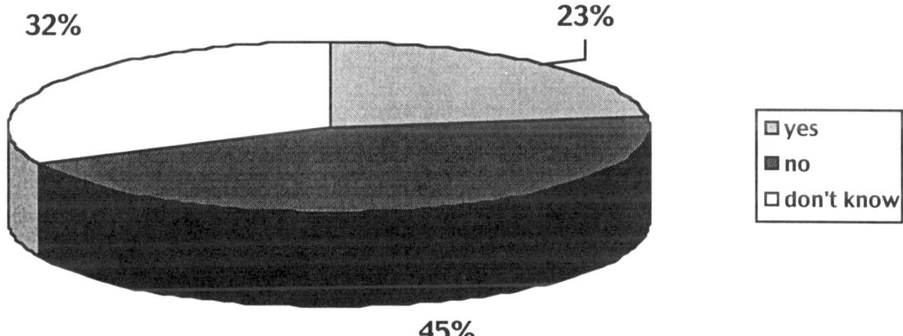

For preparation of new biotechnology law we should cooperate with which of the organization?

- International organizations (UN, WHO)   66%
- Scientific organizations   60%
- Government and parliament   37%
- Industry (e.g. food industry)   32%
- European Union   24%
- Religious organizations   11%
- Others   1%
- Difficult to say   9%

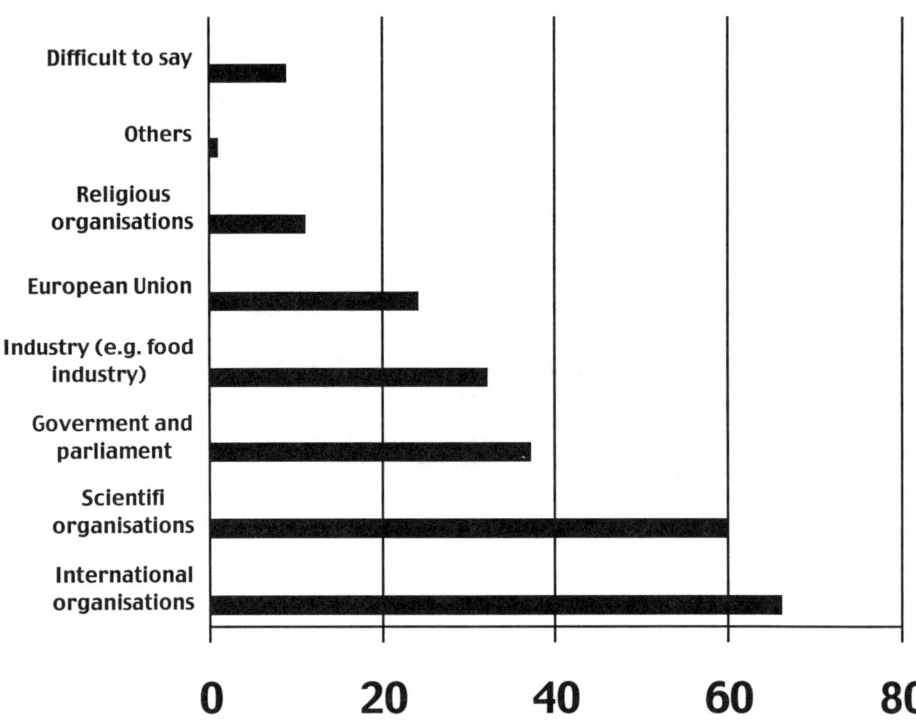

Use of Agriculturally Important
Genes in Biotechnology
G. Hrazdina (Ed.)
IOS Press, 2000

Property Rights
and
Agricultural Biotechnology

R. S. Cahoon

Cornell Research Foundation, Inc., Dept. of Patents & Technology Marketing,
Cornell University, Ithaca, NY 14850 USA

Abstract. Intellectual Property (IP) and ownership of biological materials has a significant impact on the development, implementation and investment in agricultural biotechnology. This paper will provide an overview of the role that such property rights play in inter- and intra- institutional interaction involving research, development and commercialization of agricultural biotechnology. Specific IP mechanisms and their relevance to agricultural biotechnology are discussed and implications of IP for researchers and their institutions described, including issues involving patents, the Biodiversity Convention, "freedom-to-operate", third party rights, academic freedom, contracts and the "ownership mosaic" of transgenic plants.

1. Introduction

This paper describes the relationship between intellectual property rights to biological matter to agricultural biotechnology. Also described is the implication of this relationship for researchers and their institutions. This relation has evolved against a background of property rights in traditional agricultural practices. However, the rise of biotechnology has created a complex new intertwining of property rights, biological matter, science and technology.

The fundamental attribute of property rights is control of access, possession, use and disposition. Property rights are a basic component of any socio-political system, including science and technology. Property rights comprise several distinct rights including: the right to exclusive possession; the right to sole discretionary use; the right to destroy; the right to sell, give away or exchange. Such rights typically combine in a particular mix to define the "ownership" of a thing. It is rare that all property rights are held by a single owner. For example, although one may have possession of property, others may have some rights to use it. In addition, other rights may act to counter and mitigate property rights. For example, a car owner's rights in the car are mitigated by the government's right to assert public safety constraints on the owner's use (e.g. speed limits). Ownership can then be thought of as a "bundle" of rights in property.

There are different kinds of property, including tangible and intangible. Tangible property includes things such as houses, clothing and land. Intangible property includes use rights, insurance policies, share ownership in firms and intellectual property. Biotechnology has led to the development of tangible and intangible property from biota. Tangible property in biota includes any biological matter, while intangible property includes intellectual property. The two types of property which are most relevant to agricultural biotechnology are intellectual property (patents, plant breeders rights, copyrights, trade secrets) and proprietary biological material ("bioproperty"). Specific biotechnologies typically comprise a mixture of intellectual property and bioproperty.

Thus, intellectual property and bioproperty are critical to control the access, development and use of agricultural biotechnology.

Property control is important to agricultural biotechnology researchers and their institutions for several reasons. Intellectual property and bioproperty are essential to the control of access and use of the tools, components and products of biotechnology. Who holds these property rights and how they use them has major implications for research, development and commercialization in agricultural biotechnology. Thus, property rights have a significant impact of property rights on agricultural biotechnology, its developers and users. Ultimately, this control over the agricultural biotechnology enterprise affects the funding of research programs and institutions, freedom to conduct scientific inquiry, ability to develop technology and realize financial benefits from research, ability to motivate and structure economic development and generally enhance the public good.

Elucidation of these issues provides a basis for practical consideration of the management of property rights in the environment of agricultural biotechnology research institutions. These issues also point out a related matter: the importance of the conservation of biological diversity and its relation to property rights and biotechnology.

2. Background and Overview

The discussion of property rights and biotechnology begins with an acknowledgement of the role that property rights have played in traditional agricultural technologies. Plant and animal breeding have been around for many years and the impact of property rights in these areas has not been inconsequential. In many ways, agricultural biotechnology is simply a continuation of the history of the relationship between agriculture, technology and property. In conjunction with traditional technologies, long-standing property right traditions have been in place in most, if not all, societies. For example, ownership of livestock and crops has evolved in societies as a mixture of custom, practice and law. In the United States, property rights in crops have traditionally belonged to the owner of the land on which the crops grow. However, in lieu of a legal agreement otherwise, the planter of the crops owns the produce yield, regardless of whether or not they own the land.

Roman law provides a basis for determining property rights to domesticated animals. Unless otherwise agreed to, the owner of the mother owns her progeny. Stray animals belong to the finder unless the animal is clearly marked. The Roman doctrine of *ratione soli*, originally developed to define the ownership of foraging-but-domesticated bees, provides that whoever owns the soil where an animal nests, owns the animal. These long-standing traditions and common laws cover bioproperty and provide a background for the property rights in biotechnological subject matter. There has been a significant distinction made between wild and domesticated in the property rights pertaining to biota which are the source of bioproperty. Domesticated biota, such as dogs and livestock, have traditionally been owned as personal property. Transient wild biota have historically been freely accessible to all (e.g. insects) or controlled by national or local governments (e.g., fish and game). Intellectual property rights, utility patents, plant patents (in the United States and Europe), UPOV or Plant Variety Protection ("Plant Breeder Rights") and copyrights have been available through national laws for many years and such rights have traditionally been exploited primarily by commercial entities for the obvious purpose of financial return.

Biotechnology, which is particularly well suited to the creation of patentable inventions, has lead to an explosion in the number of utility patents granted, particularly in the U.S. and Europe on a great variety of materials and methods obtained from biota.

Among the many patentable utilities produced by biotechnology, the DNA sequence of genes is particularly notable.

Biotechnology has also accelerated a rise in the utility and value of "micropieces" of organisms, which are biological material (e.g. cell lines, tissue cultures, plasmids, antibodies) which may be held as personal property. Thus, in addition to the value of traditional bioproperty such as herd animals, seeds and tree cuttings, biotechnology has created a new era in which the bioproperty of a test tube of cells or petri dish of tissue can be owned and controlled under laws that govern personal property and can therefore have significant value.

A full complement of property rights allows the holders of such rights (i.e., the owner) to control the access to, possession and use of the materials or methods covered by their rights in these properties for their own particular interest and at their own discretion. In a commercial sense, such property rights are typically used in exchange for a financial or other valuable consideration. However, financial return is not the sole reason to acquire property rights, particularly in the area of agricultural biotechnology. In the hands of a public and/or non-profit research institution, such property rights may be held for a variety of reasons including: assurance of access to other researcher's technologies; motivating research support by for-profit firms; maintaining control over the direction of commercialization of technology; protecting the interest and reputations of the institution and its researchers and to promote the public good. However, at a public, non-profit research institute with a tradition of unfettered intellectual freedom, openness and scientific exchange the control of property rights runs contrary to the institution's philosophy to some degree. Biotechnology has helped to create a new era in which the power and utility of property rights in biological methods and materials must be adapted to the non-profit research environment in which free and easy exchange of information and biological materials has been the traditional *modus operandi.* Conversely, the non-profit research institution is forced by global forces and biotechnology to accommodate the new reality of research enterprises awash with property right matters.

Biotechnology has had a profound impact on the *status quo* of property rights in biota. Before modern biotechnology, value was primarily obtained from the whole organism or its "macro" pieces (e.g., skin, hair, and meat). Since biotechnology can create value in reproducible micro components and allows the conversion of genes into a technological unit capable of being transferred from one organism to another, a complex property right scenario has been created. The rise of the value of bioproperty is attributable to these biotechnical capabilities.

Despite the rising importance and value of bioproperty *per se*, it is intellectual property, which has captured attention in the debate over the issue of ownership and control of biota and its "pieces". Biotechnology allows researchers to create methods and materials from biota which are novel, not obvious, and useful and therefore are patentable. Such biotechnological inventions are readily patentable in the United States and, while the scope of such patentability is more limited in Europe and other countries, combinations of intellectual property and the bioproperty of such inventions can provide significant property right control to the inventor and his/her institution.

In the midst of these technological changes, the global effort to create a structure for the equitable exchange and use of biota has resulted in the signing of the Biodiversity Convention by many countries (not including the U.S.). This global pact, while not binding, represents an initial step in creating a global institutional framework in which the inherent value of biota is acknowledged (wild and domesticated) and bioproperty derived therefrom. It also provides guidelines for the management of international exchanges of bioproperty and intellectual property in biota and its exploitation in an environment of national sovereignty, research exchanges and commercial trade.

3. Biotechnology and Patent Intellectual Property

Patents are a grant of limited rights to an inventor by a national government. Through the patent, the government supports the inventor's right to stop others from making, using and selling the patented invention in exchange for the inventor's complete and accurate communication regarding the best way in which to use the invention. Thus, it grants a limited monopoly-like right in exchange for full public disclosure and is, then, the antithesis of another type of intellectual property – the trade secret. It is essential to understand the fundamental nature of patents, what they are and what they are not. Patents are territorial; that is, a patent in the United States has no force in Europe and *vice versa*. Patents are a negative right in that they only empower the inventor to use legal means to stop others from making, using and selling the patented invention but they do not permit the inventor to practice their own invention. Patents are valid (i.e., enforceable) for a limited term of 20 years from the date of filing the patent application. Patents are costly; one can expect to spend many thousand of dollars to file and prosecute a patent in a single country. Obtaining patent coverage in numerous countries can be very expensive running into the tens or even hundreds of thousands of dollars. Patents may be overlapping in their scope and patents may be dominant or subservient to other patents. It can be frustrating to the owners of a patent to discover that their patent is dominated by other, broader patents and that despite their investment, they cannot practice their own invention.

Biotechnological patents are those utility patents that cover the use of biological material, (for example cDNA, gene sequences, amino acid sequences) as well as methods such as the "gene gun" method of genetic transformation of plants (U.S. Patent No. 4,945,050) and the "replicase" method of plant virus resistance (U.S. Patent No. 5,633,449).

Despite the widespread impression that intellectual property in biotechnological materials and methods is a recent phenomenon, a review of the history of patents shows otherwise [1, 2, 3]. The United States has a long history of patent-based intellectual property rights being granted to materials and methods of a biological nature. For example, the United States Patent Office issued patents on microbial cultures in the late 1800's and by the early 1900's, several patents on vaccines containing biological material were issued. Other biologically related patents were issued during this early period in Germany, the U.K. and France. In the first fifty years of the 1900s, patents on biological subject matter in the U.S., U.K., Germany and France included purified vitamins and the processes for their production, colored animal furs, purified antibiotics and animo acids, biological waste water treatment cultures and systems, microbial cheese cultures and, in the U.S., asexually-propagated and traditionally bred plant varieties. In the 1960s and early 1970s, although modern biotechnology had not yet emerged, biological sciences became more sophisticated, producing yet more patents on biological subject matter. Although some of the most important biotechnologically-based patents did not use genetic engineering (for example, the "chakrabarty bacterium"), it could be argued that the explosion in patents on biological subject matter was initiated by the invention of applied nucleic acid chemistry.

It is important to keep in mind that in spite of this long history of the widespread patentability of biological subject matter, there has always been a widespread prohibition in all countries on the patentability of "products of nature". Except for some exceptions for axenic cultures of characterized bacteria, it is not possible to patent a naturally occurring organism. Despite this prohibition, a large number of patents have been granted by numerous national patent offices on what appear, on the surface, to be "products of nature". Patents on vitamins, antibiotics from natural resources and DNA sequences would appear to violate the "product of nature" prohibition. The justification for granting such patents is based largely upon the doctrine of purification and utility [3]. This doctrine holds that the patentable subject matter did not exist in nature as a purified and characterized substance and that through the inventor's intellectual work, the substance was converted into a non-natural form and made into something technologically useful.

In the United States, the watershed event connecting utility patents and biological materials is the 1980 U.S. Supreme Court case of Diamond *vs*. Chakrabarty [2]. In this case, the Chakrabarty side argued that a naturally occurring bacterium was made patentable by the introduction of xenobiotic-degrading plasmids through technological means. The U.S. Patent Office determined that this bacterium was an unpatentable "product of nature" which initiated a long process of appeals and rejections in the courts. Eventually, the United States Supreme Court, in a 5 to 4 decision, declared that "everything under the sun made by man" is legally patentable in the United States, regardless of whether or not it is of a biological nature, living biological material or even a whole organism.

Despite such defining cases, and the desire for certainty in determining patentability in a particular biological subject, it is important to keep in mind that the legal scope of patentability in any particular country is very dynamic. In addition, since patentablility is a function of national law, it is variable from country to country. What is patentable in Canada may not be patentable in Japan and *vice versa*. Of course, with the European Patent Convention (EPC) there is patentability harmony across the EPC member countries and with GATT and TRIPS, there is momentum toward the "harmonization" of all national patent laws. Nonetheless, there is presently a significant difference between the scope of potential patentability in Europe and the United States for biological subject matter. The United States is considerably more liberal in granting patentablility to biological inventions than is Europe.

The future appears to hold significant uncertainty with regard to scope of patentability of biotechnological subject matter in Europe as well as in the United States. Of particular interest is the role intellectual and bioproperty rights will play in the growing field of genomics. For example, what property right will creators have in databases? Genomics is one of the most interesting opportunities and dynamic areas of property and agricultural biotechnology, since it combines the potential integration of utility and plant patents, bioproperty, "Plant Breeder's Rights," copyrights, trade-secrets and even, perhaps trademarks.

4. Bioproperty and Biotechnology

Humans have owned domesticated biota for millennia, and there is a well-developed body of custom, practice, and common and statutory law that govern these property rights. Regarding wild biota, there has traditionally been a definite distinction between biota considered a resource, such as fish and game, and "trivial species" (trivial in the sense that they have no apparent direct utility and economic value to humans other than ecological niche). Property rights in an organism of domesticated biota or its parts are defined by personal property law. Ownership of a herd of animals which is asserted by physical possession and marking allows the owner to trade, sell, barter or otherwise use and dispose of the animals like any other personal property. Such rights in wild biota are also defined and governed by personal property law but may be mitigated by the possessor's right to initial capture and possession of the wild organism. A possessor's rights in a wild organism are contingent on the prevailing property right regulations for that organism in its geographical context. That property regime is a product of a long history of the evolution of the balance of rights in wild biota between the government, private entities, and communal groups. This balance also included the eventuality of no property rights. For example, in the U.S. property rights in organisms of species listed in the Endangered Species Act and the Migratory Bird Treaty Act are held exclusively by the U.S. Government. Fish and game are "owned" by individual state governments, trees and other rooted plants are owned by the owner of the land on which they grow and no one owns spiders. In the case of no ownership, property rights are obtained by simple capture and possession.

A further level of complexity emerges when examining the property rights in a carcass of a legally-captured organism which is owned by the government. For example, when a deer is free-roaming, it is the property of the state. The hunter obtains a license to capture the deer and obtains physical possession. In this case, the state and the individual share a mixture of property rights in the carcass particularly the right to use, sell or transfer. Determining this mix of rights for a particular biota requires a careful analysis of the national and/or state and local laws and regulations that comprise the government's rights in the biota in a given geography and jurisdiction. Occasionally, taking and possessing a wild organism may be absolutely prohibited by law. In the U.S., the Endangered Species Act prohibits any such acts without permission from the federal government. This is also true for specimens listed in the Convention on International Trade in Endangered Species (CITES) in countries that that are CITES signatories.

In the area of domesticated biota, *ex situ* cultures of biological matter (e.g., crop germplasm) held by individuals and institutions (e.g. CGIAR) are subject to property rights questions such as: who, if anyone, should control access and use of these materials and for what purpose? The issue of importance here is the right to control access, possession, use and transfer of bioproperty. These control rights have profound significance for agricultural biotechnology researchers. Access to and possession of a useful crop germplasm, cell line, plasmid, or cDNA is likely to be critical to a research project. Constraints on possession and use present a significant hurdle, potential problem and in some cases, an opportunity for researchers. Limitations on research use placed by a bioproperty owner can have significant implications for later commercialization efforts. Issues surrounding the control over transfers of bioproperty should be considered when researchers and/or their institutions are attempting to obtain the bioproperty of others or transfer their own bioproperty.

5. Property Rights and Agricultural Biotechnology

Access to bioproperty is essential in order for possession and use to occur. If a research institution wishes to have access to the bioproperty of others and to eventually exploit its own bioproperty for commercial purposes, it must actively manage bioproperty. There are often ownership "strings" attached to bioproperty, including use restrictions and governmental regulations. It is essential to understand the constraints that owners of bioproperty place on the bioproperty that researchers are using in order to determine if and to what extent such constraints may inhibit research activity, its funding or the later commercialization of its results.

There is a complex relationship between bioproperty and patents. Some technologies may be owned simply as bioproperty and not as patents. For example, antibodies and genetic markers are normally not patented but commercialized through contracts. For other biotechnologies, the use may be owned through patents but bioproperty rights may be non-existent or weak. The "35S" promoter is covered by patents but the DNA sequences of "35S" *per se* are widely and freely available. In some cases, both bioproperty and patents may be used to provide a property package over a technology.

For example, the Cornell Research Foundation asserts its rights in its "Rice Actin Promoter" through patents and biological material transfer agreements. For some bioproperty there are no personal property rights beyond the simplest level of mere possession and no patents covering their use. In both cases, the bioproperty would be said to be in the public domain.

Complexity arises when some portions of biotechnology are bioproperty and other portions are covered by patents. There is frequently no one-to-one correspondence between the boundaries of bioproperty and patents on their use. For example, the Cornell Research Foundation owns the technology for an avian cell line useful for improved production of a poultry vaccine. The Foundation asserts property rights over the cell line *per se* by controlling the access and use of the cell cultures in any culture form (e.g., test tubes, petri dishes, bottles, fermenters, etc.) through contracts. That is, no one outside of Cornell may have possession of the cells unless they have signed a Biological Material Transfer Agreement (MTA) that limits the recipients rights to use the cell lines. Such limits include the prohibition of commercial use without a license, the prohibition of distribution to third parties, maintenance of labeling with Cornell's name and accurate record keeping. In addition, the Foundation has filed for patent protection on the use of the cell line in vaccine production. Due to limits of patentability, the scope of the patent is likely to be limited to the use of this cell line for a specific disease. Although this disease is the most economically significant, the cell line can be used in vaccine production for other diseases. Since the patent only covers one disease, the Foundation must use this mix of patent and bioproperty rights to control commercialization activity to its benefit. It does so by using its patent rights and bioproperty rights to control the use of the cell line for the disease within the scope of the patent. It must use only its bioproperty rights to control the use of the cell line for vaccine production in other diseases. In the case where, through some channels outside Cornell, an entity possesses the cell line but has not signed the Foundation's MTA, the Foundation is limited to enforcing its patent rights.

This latter scenario brings up the issue of the severance of rights in bioproperty. Since bioproperty rights are limited by possession and by contract, an owner's rights in bioproperty can be severed by the actions of either one or both of these factors. If the owner is careful that contracts are in place before physical transfer of the bioproperty, he or she need only worry about a breach of contract by the signers of the contract. If one of the signers breaches the MTA by allowing the bioproperty to be possessed by a third party that has not signed the MTA, the owner has legal recourse against the signer of the MTA but not against the third party. The owner's rights have been severed. This demonstrates the weakness in bioproperty rights. It is clear that in order to optimize control of a technology, it is preferable to seek maximum property rights, both bioproperty and patents. However, the details of a technology, the industrial application and cost/benefit analysis will tend to mitigate the simple seeking of maximum property rights. Developing an optimal property right strategy is dependent on the unique factors related to that particular technology.

In order to optimize the conduct of research and to fully realize the fruits of that research, several questions should be asked by the agricultural biotechnology researcher or his/her institution before embarking on the work: Who owns the gene? Who owns the vector? The construct? The germplasm? Are there property rights in the methods to be used in the research and to whom do they belong? Each material or method may have a separate owner. In some cases, a patent may only cover a method used to make an invention (e.g., "genegun"). In others, the method may cover the use of a technology (e.g., "replicase"). Each method and material is likely to have owners who must be dealt with in order to begin to understand the web of ownership "strings" attached to a particular biotechnology and to proactively manage constraints on expected or current use, possession and transfer. Such strings are typically embodied in research contracts or biological material transfer agreements but may be less than explicit. In some cases, informal acquisition of bioproperty can lead to future problems. It is important to determine if transfers of bioproperty have been made by authorized individuals. Either one or both of the researchers involved in a biological material transfer may be unaware they must have the right to make such a transfer. Later, the rightful owners are most certain to assert their rights if they have not been involved in the transfer. In all such transfers, ignorance of someone's rights in bioproperty is likely to present a problem later. Typically the problems arise during commercialization not during research. However, this is not always the case. The Cornell Research Foundation has been called upon to remedy a serious breach of contract created by a post-doc's informal and innocent transfer of a vector and genetic construct to a post-doc colleague. The transfer was made to solve a technical problem. However, that apparently innocent use of the biological materials created a serious legal predicament for Cornell because the vector and its use belonged to one company, the invention made by the vector belonged to another company, and several other companies had rights to commercialize these products due to their sponsorship. In effect, this one simple transfer and use obliged Cornell to give exclusive and conflicting rights to several companies simultaneously. Only time, expertise and diplomacy by the Cornell Research Foundation avoided litigation.

It should be clear that a research institution should establish policies and practices that avoid the problems that arise when property rights are ignored in the process of conducting biotechnological research. Informal transfers between graduate students or post-docs are usually problematic.

5.1 Ownership Mosaics

The relation of property rights and agricultural biotechnology reaches a complex crescendo in transgenic plants. Transgenic plants exhibit an "ownership mosaic" in which each transformed plant has a different combination of intellectual property components and owners. Table 1 shows transgenic apples in which three separate transgenic cultivars, each with a different "mosaic" of genes, promoters, genetic regulators, selectable markers, germplasm and transformation methods are present. In the example of Table 1, the three separate transgenic apple cultivars comprise 84 separate ownership mosaics, each of which must be analyzed and defined in order to manage appropriate research use and to consider eventual commercialization. Table 1 shows that the several intellectual property components that make up each of the transgenic plants includes trait genes, promoters, two types of genetic regulators, a marker gene, a selector gene and a method of transformation. Within each of these components there are one or more types. In trait genes, there are five different types; there are four different types of promoter and four types of one of the genetic regulators. The number of owners is noted in parentheses so that, in the five trait genes used, there are three separate owners (i.e. some of the owners own more than one trait gene). The ownership rights to all of these components is based on patents, bioproperty or some combination of both. Table 1 also indicates that the ownership of some of the components is unknown. For example, it is not knows if the Regulator II component has any ownership strings attached or is fully public domain. This raises an important point: not knowing ownership of a component does not mean one can assume that it is public domain. Thorough evaluation should be done on the ownership situation before it is safe to assume that the component is available without obligation to an owner. A thorough search of the patent literature is necessary as well as a search for any bioproperty contracts or unauthorized transfers.

While Table 1 describes a transgenic fruit example of only seven components, more complex scenarios often arise. An informal discussion with colleagues has revealed that a single transgenic corn plant comprised over sixty separate property components.

Table 1. The Genetic Ownership Mosaic: Intellectual Property Components & Ownership in Transgenic Fruit Crop

Cultivars	Genes	Promoter	Regulator - I	Regulator - II	Marker	Selector	Transform Method
"1"	5 types (3 owners)	4 types (3 owners)	1 type (2 owners) joint	4 types (? owners)	1 type (1 owner)	1 type (1 owner)	1 type (1 owner)
"2"	5 types (3 owners)	4 types (3 owners)	1 type (2 owners) joint	4 types (? owners)	1 type (1 owner)	1 type (1 owner)	1 type (1 owner)
"3"	5 types (3 owners)	4 types (3 owners)	1 type (2 owners)	4 types (? owners)	1 type (1 owner)	1 type (1 owner)	1 type (1 owner)

$\approx$ 84 different "mosaics"

6. Implications of Property Rights for Researchers and Their Institutions

The discussion of property rights in "components" owned by numerous owners of patents and/or bioproperty relates to the issue of "freedom-to-operate". "Freedom-to-operate" is that condition in which an inventor or creator of a technology has complete freedom to make, use and sell an invention. Remember from the discussion above that a patent does not grant the holder the right to use the patented invention. In the example of the transgenic plants, "freedom-to-operate" requires that each of the owners of the separate intellectual properties and/or bioproperties of the transformed plants have been satisfied. Since any one of the owners can decide to prohibit "freedom-to-operate", absolute unanimity of owners acquiescing to obtaining "freedom-to-operate" status. Obtaining "freedom-to-operate" status is a complex and tricky task. In the example depicted by Table 1, obtaining the "freedom-to-operate" for this transgenic fruit crop requires negotiation with all of the owners. While the negotiations may be somewhat staggered chronologically, the agreements reached must ultimately be coherent and coordinated. There can be no conflicting terms or conditions between any of the agreements. In addition, the financial terms of each agreement must not combine to make the commercialization economically non-viable. For example, giving each owner a five-percent royalty would amount to a (5% x 7) thirty-five-percent combined royalty, which is likely to render commercialization unfeasible due to price constraints and compression of the profit margin below the economically viable threshold.

The negotiations are conducted only between each owner and the transgenic plant owner but a mechanism must be established to link the separate negotiations in order to develop an overall comprehensive package. Such complex, multiparty negotiations have been likened to a chess game with as many dimensions as there are parties to the negotiation. In the example of Table 1, a complex chess game must be mastered. A well-crafted strategy is necessary in order to arrive at satisfactory agreements with each of the owners. Two additional factors act to further complicate these negotiations: 1. Each of the owners is a different organization with different missions, goals, and obligations. Some are for-profit companies and others are non-profit research institutions. Thus, each owner may want different outcomes and/or be motivated by different aspects of the project. The strategist must account for all of these factors. 2. The geographical scope of patents and whether the patent covers a one-time use or ongoing method or if the patent covers matter, will greatly affect the strategy. Any researchers making transgenic plants and their institution should be aware of the "freedom-to-operate" situation of their projects and must obtain "freedom-to-operate" status in order to proceed with commercialization. However, the institution may pass this phase to a commercial partner, which is the normal practice for non-profit research institutions in the U.S.

It cannot be overemphasized that researchers and their institutions should be aware of any strings or constraints attached to the possession and use of any third party's bioproperty. These strings may be written in biological material transfer agreements, research agreements, invoices, or any other written contract that the owner required to be signed as a prerequisite of access to the material. These constraints must be defined, analyzed and understood by the researchers and their institution in order to determine the ability of obtaining "freedom-to-operate" status.

Some believe there is a special exemption to patent infringement for researchers. It is important to dispel this notion. A patent gives the holder the right to stop others from making, using or selling. The law does not exempt research use. There is no legal difference between research and commercialization uses with utility patents - both are infringements and can be stopped by the holder of the patent rights. Practically, however, there is a large economic distinction between research and commercialization. Research is not normally stopped by patent owners because unlike commercialization, there are no sales monies involved. While most research institutions hide behind this *de facto* research exemption, at the point of commercialization, research institutions or their commercialization partners must obtain "freedom-to-operate" status.

Because of the rise in importance of bioproperty, biological material agreements have become a critical element of the biotechnology research enterprise. These agreements describe the terms and conditions for receipt, access, use and transfer of materials to and from other institutions, both non-profit and for-profit. These agreements act to assert rights in bioproperty and should be used by institutions for the control of their own bioproperty regardless of whether the material has apparent or immediate value.

License agreements are another factor in access to property of interest to the agricultural biotechnologist. These agreements may control a researcher's ability to use bioproperty and intellectual property rights, which cover key components of a relevant agricultural biotechnology. Such license agreements typically define the right to access and use the properties as well as any financial terms between the parties.

It is important that agricultural biotechnologists and their institutions understand that appreciating intellectual property and bioproperty rights is essential to entering into effective research and commercialization partnerships and collaborations. Such partnerships between non-profit institutions and between non-profit and for-profit organizations are critical to successful research and development in the field of agriculture biotechnology. The saying, "good fences make good neighbors" is true in this context.

In considering this new environment of property rights, freedom-to-operate and commercialization in a research environment, non-profit research institutions must keep the overall goal of scientific advancement, technology exchange and development for the public good firmly in mind. Without these goals, it will be tempting to be distracted by commercialization questions, contract details, legalities, and economic issues. These new issues could diminish the non-profit research enterprise. It is imperative for the well being of the non-profit research endeavor and the global research network that long-term goals guide the accommodation of property rights in the research and development process.

7. Practical Management Considerations

It is suggested here that management of property rights in the context of an agricultural biotechnology research institution is similar to the management of money. Ignore its management and it becomes a tyrant that can hinder or stop ones efforts. Managed wisely, these rights can become a useful and powerful asset to a research institution - an asset that will enhance the financial and technological well being of the institution.

Researchers and their institutions should keep the following in mind as they proactively manage the property issues surrounding their agricultural biotechnology research efforts. The first is the intellectual property "hygiene" process - a process of continual vigilance over any intellectual property, held by others or the institution, which could have an impact on the research. What are the strings attached to this intellectual property? Who has rights? What rights, if any, has the owner conveyed to the institution and the researcher? These questions must be analyzed and not ignored.

There must be active management of bioproperty and intellectual property both of others and of the institution. Passive reaction to events in the arena of property rights and agricultural biotechnology is not an effective strategy to manage these rights and obligations if one desires optimizal control for the purpose of enhancing the non-profit research enterprise.

Despite this strategic emphasis on adherence to property right regimes, it is critical to balance the sometimes-exacting requirements of property rights against the needs of science, technology development and the societal good of a public research institution. To be perfectly property correct is typically a too-legalistic approach that often hinders the development of science and technology. Conversely, technology development without regard to property rights and obligations runs the risk of liability and loss of control and possibly significant loss of opportunity for the researcher and his/her institution.

It is essential to respect property rights, particularly the property rights of others, even when those property rights appear to be blocking or problematic in the desires of the research institution and the researcher. Respect for other's property rights is essential to being a good partner. By being a good property partner, a researcher and his/her institution can participate in the global enterprise of agricultural biotechnology. Regarding property rights and technology, it is essential to do what is right regardless of the cost of time and money, not what is easy, cheap or expedient. In order to make sustainable, long-term collaborations, partnerships, and commercialization of technology a long-term view of the use of property rights is absolutely necessary.

Finally, institutions and their researchers must follow the rules of property rights. However, these rules should be followed thoughtfully, not dogmatically. The rationale for such rules should be understood so that they may be applied flexibly. Without flexibility in the use of property rights, a legal shroud will envelop the research enterprise and inhibit the valuable exchange of technology, information and materials, which are so essential to the development of technology. An overly legalistic approach that reduces risk to infinitesimal levels chills partnering collaboration and interaction between parties – the lifeblood of science or technology.

8. A Note on Biodiversity Conservation, Property Rights and Agricultural Biotechnology

The biota of biodiversity is a basic source of raw materials for biotechnology. Without biodiversity there are no future genes or cell lines, and no biological phenomena with which biotechnologists pursue their science and technology. Given the decline of biodiversity and its critical role in biotechnology, it seems obvious that biotechnologists should be committed to its conservation.

Property rights in biota can be used for conservation purposes. It has been made clear that property rights in biota have utility for researchers and their institutions. Such rights can also be used to enhance biodiversity conservation. For example, the integration of bioprospecting, property rights in bioproperty and patents on biological subject matter can be used to control access and use of wild biota and to generate financial return for conservation purposes [5]. In bioprospecting, property rights in biota are a fundamental link between discoveries made and financial returns directed to conservation.

All biotechnological research institutions should be actively committed to the conservation of biodiversity, through the use of their own property rights and the respect of the property rights of managers of biodiversity conservation. It is important to keep in mind that the biota a researcher possesses may be "owned" by a manager of biodiversity conservation (e.g., in a nature reserve or a national park). Such owners should be respected in terms of the fair exchange that the Biodiversity Convention describes.

There are many examples in which biotechnologists have exploited wild biota diversity without regard to the property rights in the wild biota. It should be said that such property rights in wild biota have either not existed (e.g., in insects) or have not been asserted by the "owner" of the biota. The classic example of this disconnection between the biotechnological value of biota and the cost of conservation is the discovery of polymerase chain reaction ("PCR"). The technology was invented by researchers at Cetus Corporation using biological materials (i.e., cultures of the thermophilic bacterium, *Thermophylus aquaticus*) originally taken from Yellowstone National Park. The patents covering the PCR technology have generated hundreds of millions of dollars in economic activity and value. However, since Yellowstone National Park (i.e., the U.S. Government) did not assert ownership in the biological materials from its property, none of these economic benefits have gone to support the natural resource that produced the biotechnological discovery.

An investigation of intellectual property rights (i.e., patents) reveals many discoveries which originate from wild biota including enzymes from snake venom, secondary metabolites from tree bark, anti-microbials from insects and anti-cancer compounds from marine organisms. Biotechnologists and their institutions should act ethically in their handling of wild biota. At the least, respect of property rights in such biota should be exercised. More proactively, researchers and their institutions may want to partner with the "owner" of wild biota in order to integrate potential biotechnological value and conservation. Cornell University and the Cornell Research Foundation have entered into several such arrangements. These arrangements provide access to biota for biotechnologists and provide for a sharing of the economic benefits by the managers/owners of the biodiversity.

9. Summary and Conclusion

Property rights in biological materials and methods are integral to biotechnology. Anyone involved in biotechnological research, development or commercialization must consider the implications of property rights to their work and its results. Such rights are essential in order to facilitate exchanges of information, bioproperty and intellectual property for the purpose of scientific advancement, technology development and to enhance the research enterprise. Such rights are also essential for commercialization, economic development and the return of a portion of the resulting value to the researcher and his/her institution.

All researchers and their institutions should respect the property rights of others as well as understand and appreciate such rights in their own institution's laboratories. These "bioassets" can be used to enhance the research enterprise and technology development, as well to stimulate economic development. Depending on an institution's willingness and capability to manage biotechnological property rights, its management may facilitate or hinder the objections of the researchers and/or the institution. If managed proactively and correctly these rights can be used by researchers and their research institution to obtain research funds and to control the direction of commercialization. If mismanaged, these rights can hinder or stop research and technology development.

Those who participate in the arena of bioproperty and intellectual property will have a role to play and will wield the power to influence the direction of research, technology and economic development in agricultural biotechnology. Those who do not participate may be ignored and left behind.

Finally, the conservation of biodiversity is a priority which should be kept clearly in mind by biotechnologists and their institutions and the property rights of biological conservation managers should be respected.

References

[1]   S.A. Bent *et al*. Intellectual Property Rights in Biotechnology Worldwide. Publ. By Stocton Press, 1987, NY.
[2]   I. Cooper, Biotechnology and the Law. Clark Boardman Co. Ltd., 1998.
[3]   R.S. Crespi, Patenting in the Biological Sciences J. Wiley & Sons, 1982.
[4]   W. Reid *et al*., Biodiversity Prospecting. World Resources Institute, 1993.
[5]   W. Reid, The economic realties of biodiversity. Issues in Science and Technology. Winter 1993/94.

Use of Agriculturally Important
Genes in Biotechnology
G. Hrazdina (Ed.)
IOS Press, 2000

Opportunities for Successful Collaboration

J. E. Haldeman

Associate Director, International Agriculture Program, College of Agriculture and
Life Sciences, Cornell University, Box 14, Kennedy Hall
Ithaca, New York, USA 14853

Abstract: Cornell University's international involvement began in 1865 when it opened its doors and welcomed it's first international student. Cornell has had strong global ties since those early years. From the days of the Nanjing Cooperative Crop Improvement Program in China (1920s) to the institution building program with the University of the Philippines at Los Banos (1952 - 1972) to the major scientific exchange program with Poland (1970s - 1980s), Cornell has consistently emphasized a close relationship among agricultural instruction, research, extension and outreach. Recently, five focus areas have been identified for future collaborative programs. The areas include Agricultural Biotechnology; Agricultural Market Economics; Environmental Issues Related to Land and Water Management; Quality and Safety Assurance of Agricultural Products and Foods Produced in the Region and Rural Development. The key to this new and important initiative is collaboration, at a level that surpasses the traditional form of the individual scientist to scientist cooperation. Here new avenues of possible arrangements and approaches will be discussed that may provide a unique opportunity to develop substantial collaborative programs, attract significant funding, and to achieve the stated goals and objectives of this important initiative.

1. Background

To set the stage for this paper, some history about how Cornell University became involved in Central Europe is appropriate. Cornell's international involvement began in 1865 when it opened its doors and welcomed at least five international students, including one from Russia, and since that time, Cornell has continued to have strong global ties. Western Europe was the center of advanced agricultural science. Cornell's founders, Ezra Cornell and Andrew D. White, recruited European scholars, adapted methods of agricultural education from the best European colleges, and imported scientific equipment and materials to make Cornell a center for teaching and research in agriculture. By the turn of the century, students from numerous foreign countries were studying agriculture at Cornell, and Cornell faculty and alumni had begun to influence world agriculture. From the very beginning there was a close link between Europe and Cornell. Since that time Cornell has consistently emphasized a close relationship between agricultural instruction, research and extension, and outreach.

In 1925 the College of Agriculture was involved in a pioneering effort in China to improve that nation's food supplies. The Nanjing Cooperative Crop Improvement Program was developed jointly by Cornell's Department of Plant Breeding, the University of Nanjing, and the International Education Board, with the objective of breeding and distributing improved crop varieties. The result was increased yields per acre of more than 50 percent for wheat and 20 percent for barley. This Nanjing project has been regarded as a model for institution-building agricultural development, because of its emphasis on training Chinese scientists to carry on the crop improvement program. Noteworthy was its multidisciplinary nature. The technical assistance in plant breeding was supplemented by Cornell faculty working in entomology, plant physiology, agricultural economics, and other disciplines.

The most comprehensive agricultural development effort undertaken in the middle years by Cornell's College of Agriculture was in cooperation with the College of Agriculture and the University of the Philippines at Los Banos from 1952 to 1962. This project, with financing from the International Cooperation Administration, now the U.S. Agency for International Development, provided for the assignment of 51 American professors, including 35 from Cornell, at the University of the Philippines, during the ten-year life of the project. Essential to the success of the program was close cooperation between the Cornell visiting professors and the Los Banos staff. Research was problem oriented, and demonstrations were carried out on the soils and with the crops common in the Philippines. During this period the focus was on the undergraduate program. As a result of additional funding by the Ford Foundation, this program continued for another ten-year period with a focus on graduate programs [1].

Closer to home, Cornell became more involved in Central Europe starting in 1974. With funding from Cornell's College of Agriculture and the Alfred E. Jurzykowski Foundation, a scientific exchange program with Poland was established. Close to fifty Polish scientists over a nineteen-year period came to Cornell for up to one year to work side by side with a Cornell scientist on important agricultural problems. This was followed by several delegations of Cornell faculty visiting selected institutions in Poland, Hungary and what was then called Czechoslovakia. Again, it was clear to Cornell faculty that a close working-relationship with Central Europe would be useful. During the 1990's Cornell was awarded two Mellon Foundation grants that supported initiatives involving the Slovak Agricultural University at Nitra, and Godollo Agricultural University in Hungary. The focus of the two programs was on training for the transition to free market economies.

Following these experiences of successful cooperation, discussions were initiated to look at our experience in the region and to think how we might make more significant contributions. A more comprehensive program built around collaboration and cooperation seemed to be in order. Successes experienced by an international institute at Cornell further reinforced those conclusions. At the same time, we were approached by an individual who had worked in Poland for several years and felt that Cornell was in a unique position to be a catalyst in Central Europe, during the period of transformation to a market economy and integration in the European Union.

It was concluded that Cornell has the research expertise in both applied and basic research, in biological, physical, as well as the social sciences, and could therefore play a useful role in agricultural and rural communities in Central Europe. It also has well developed extension and outreach capacity that can help bring the knowledge generated at the university to bear on public policy decisions and to make a difference in the lives of citizens.

Recognizing the fact that Central Europe is fortunate to have many outstanding scientists in excellent universities and institutions, there is a good basis for partnership. Cornell's role in the development process can be to act as a catalyst to enhance cooperation among universities and other institutions. Building on our past relationships in the region, we can help to forge productive links among institutions from various countries to solve the difficult problems that face rural Central European economies in the wake of the post-communist revolution in the late 1980's and early 1990's.

2. Mission Statement

The first step in the process was to develop a mission statement for our program [2]. The initial Central and Eastern European Program Initiative stated the following: "We are committed to the development of collaborative programs with educational and research institutions in Central and Eastern Europe. Our overall goal is to develop substantial collaborative programs and secure adequate funding to address issues that will enhance the social, economic, and political development of countries in this important region within the framework of democratic institutions. Our specific objective is to secure adequate funding to address these issues and challenges by the year 2000."

Our guiding principles for program development are the following: (1) work on programs which are institutional strengthening (for Cornell University, as well as for institutions in Central and Eastern Europe); (2) enhance the sustainable development of rural communities with emphasis on agricultural and environmental resources; (3) engage faculty and collaborators in multidisciplinary approaches to solving problems; (4) enhance opportunities for scientist -to- scientist exchanges within the framework of selected issues; (5) and to promote research, education and extension/outreach services that are immediately and visibly useful to both the public and the private sector.

The Dean of our College demonstrated his support for the Central Europe Initiative by providing seed money which was then matched by the Alfred Jurzykowski Foundation. These funds have been instrumental in getting this program off the ground by funding two exchange visits and two major workshops.

A team of Cornell faculty and staff visited Poland in December 1997 to explore the possibility of a planning workshop and to begin the process of identifying key areas for collaboration. As a result of this initial visit and information provided by our friends and colleagues in Central Europe, it was agreed that a planning workshop should be organized that would focus on five priority areas. The specific areas of focus include:

1. Rural Development with a focus on applied research questions that would generate information supportive of rural development policy, education, and outreach.

2. Environmental Issues Related to Land and Water Management with emphasis on the development of a comprehensive set of written educational resources focused on "Environmental Management for Sustainable Agriculture and Rural Development in Central and Eastern European Countries."

3. Quality and Safety Assurance of Agricultural Products and Foods Produced in the Region with a focus on a research project "A Comparative Study on Quality and Safety Management Procedures in the Food System in Hungary, Poland, the Slovak Republic, and the United States - Present Status and Future Considerations."

4. Agricultural Market Economics with a research focus on a systems approach to market economics in agriculture, understanding the consumer market, international trade and cooperation, education in market economics and professional improvement, agriculture and the environment and economics of family farms.

5. Biotechnology with a focus on identification of research areas for collaboration and training graduate and post-graduate students in the new areas of biotechnology.

The first workshop was held in Sielinko, Poland. The site for the workshop, the agenda and topics for discussion, the list of participants and key institutional representatives were identified following close consultations with Cornell faculty and key scientists and administrators in Poland, Hungary and the Slovak Republic. The objectives of the initial workshop included:

1. Provide an opportunity for workshop participants and the institutions they represent to get to know each other.

2. Identify priority issues and concerns in the Central European region that need to be addressed.

3. Identify potential funding sources for future collaboration.

4. Develop an action plan and a network for future collaboration among participants that will lead to successful grant proposals for funding by the year 2000.

Forty-nine participants from 4 countries and 17 institutions, universities, governmental agencies, non-governmental organizations and the private sector attended. All of the objectives were achieved, and the five focus areas listed above were discussed and expanded upon. The workshop discussions were so convergent that the group recommended a follow-up workshop to develop specific proposals for funding. Two of the five groups moved quickly, and before this second workshop took place, they were able to secure funding to support the first phase of their program.

The second workshop was held on the Cornell University campus in April 1999. Participants included the members of an expanded coordinating committee; co-coordinators for each of the five focus groups, plus the three coordinators for the Central Europe/Cornell Initiative. The purpose of the workshop was to develop specific proposals for funding and develop plans for the overall management of the project.

Following the two workshops, the mission statement was expanded to reflect a clearer delineation of each of the five priority areas of focus (refer to revised mission statement in the appendices). The five work groups outlined the following plans: [3]

1. Environmental Issues Related to Land and Water Management: This work group proposes to strengthen the research-outreach continuum for environmental science and technology. A five-year program includes an international conference in 2000 to assess the existing status of research relating to best management practices for environmental protection, and the associated framework for extension/outreach in each country. A curriculum in Environmental Management will be developed, justifying and specifying best management practices associated with key environmental issues (water quality management for streams, lakes and ground water; ameliorating effects of soil metal contamination, organic matter losses, acidification, and erosion; and watershed-based land use planning). The proposal would also establish demonstrations and field-based training workshops for landowners, government decision-makers, and extension workers in each of the three countries to assist in local community development.

2. Quality and Safety Assurance of Agricultural Products and Foods: This work group will engage in collaborative research to identify, compare, and evaluate existing quality standards and safety assurance activities as well as management procedures in partner countries. An action plan for improving the quality and

strengthening the safety of food will be designed to reduce the risk of foodborne illness. The action plan includes six train-the-trainer workshops (two each in Poland, Slovakia, and Hungary) and development of a resource manual of training aids and references. Small-to-medium-sized food companies will be the target audience for the proposed program.

3. Rural Development: This work group will develop an applied research network that collaborates to produce information supportive of rural development policy, education, and outreach. A "state-of-the-knowledge:" conference will be held in Slovakia in December 1999 to identify the high-priority issues that will shape the collaborative research agenda for this focus area. A strategic plan to link institutions and scholars with responsibilities for rural development research and education in Central and Eastern Europe will be established. Small teams of researchers will prepare proposals for collaborative research on high-priority topics. Annual research conferences will be held, and proceedings of conference papers will be published. At the end of the five-year plan, a scholarly book on rural development in Central and Eastern Europe, co-edited by the program planning committee, will be published.

4. Agricultural Market Economics: This work group will conduct a workshop to determine the current state of the dairy industry in Central Europe. Researchable issues facing the industry are being determined. Participants in the workshop will identify common interests and synergies that will result in joint research proposals after the workshop.

5. Biotechnology: This group is pursuing collaborative initiatives on the use of agriculturally important genes in agricultural biotechnology. Their proposed work includes the areas of plant transformations, plant genomics, biopreservation and artificial seed biotechnology, and legal aspects of biotechnology, including risk assessment. The workshop conducted in Szeged in October 1999, will provide the opportunity to develop a common platform for establishing scientific cooperation among interested scientists to help solving common agricultural problems using the tools of biotechnology. Topics covered at the workshop include genetic engineering of crops, genome analysis, plant transformations, breeding for resistance, and legal aspects of biotechnology.

3. Funding

Our proposed collaborative effort is generating considerable interest among potential donors. The seed money provided in 1998 has helped to attract money from NATO, USDA and the Farm Foundation. Additional funding from these organizations is anticipated and will be used to support the specific activities of each of the five work groups.

This conference, organized by the Biotechnology Working Group, was the first to be funded, mainly by NATO, and with important support coming from Hungary, Cornell and others. The Rural Development Working Group was able to secure funding from the United States Department of Agriculture, the Farm Foundation, Slovakia and Cornell for an important conference that will take place in December of this year. The Agricultural Market Economics Group and the Environment Group have also submitted proposals to

NATO to fund two important workshops scheduled for June 2000. Each of the five work groups is responsible for securing funding to support their respective agendas.

Core funding that will serve as umbrella funding for the CEE Initiative is now needed in order to keep the Central Europe Initiative on track, to support the framework for the five focus areas, and to secure program funding from other donors. This funding will be used to support the following activities:

- Travel coordination, reporting, supplies, and communications;
- Travel and per diem to support travel by Cornell faculty and CEE colleagues in order to maintain joint activities and opportunities for dialogue among participants from all work groups with emphasis on inter-group linkages;

- Funds to bring the work groups together each year;

- Support faculty/scientists to take sabbatic leaves to further the CEE Initiative;

- Support joint teaching/distance learning initiatives (non-degree and selected graduate programs;

- Short term grants to serve as bridging funds to carry groups from one phase of their work to the next; and

- A competitive grants program for new and innovative initiatives.

    A proposal has already been submitted for this important funding.

4. Why Collaborate?

It is our strong belief that the synergy that is developing and the success to date are due to the fact that this is a collaborative effort that is building on momentum where significant results are anticipated. This collaboration is somewhat different from the traditional scientist-to-scientist cooperation; it is a more expanded cooperation that can lead to significant contributions toward the development of new knowledge that will be of tremendous benefit to our colleagues in Central Europe and the U.S. Collaboration goes far beyond the traditional international development assistance, and international development cooperation. In the case of international development assistance it was assumed that someone already had the answers to problems which needed attention. Projects were designed to deliver economic assistance and technical assistance. Technical cooperation did at least recognize that the outsiders did not always know the answers, or even the right questions.

The scientist-to-scientist cooperation is an essential and important form of cooperation since it can serve as a base for a more wide ranging collaboration. Even at this level, a biotechnologist collaborating with another biotechnologist, mutual benefits must be realized or the relationship will not last. However, there can be even more significant social-based benefits to collaboration that result from, for example, a biotechnologist working with a food quality specialist or a market economist. Later in the paper we will consider where scientists need to become involved at another level, linking what they have learned and applying the new knowledge to the larger development issues.

What does the word collaborate mean? According to Websters Dictionary, there are several definitions. The first one that hopefully will not reflect our cooperation states:

Collaboration is "to cooperate treasonably, as with an enemy occupying one's country." The definition that is preferred and more clearly represents our efforts states: "to work together, especially in a joint intellectual effort." However, one must recognize that there are costs and benefits to both groups, and only when each group considers the benefits of collaboration to be greater than its costs does the interaction continue. [4] In the case of the CEE Initiative, there are significant costs and benefits to Central Europe as well as the U.S. Being equal partners, there is sharing information, sharing in the generation of new knowledge, and sharing in the identification and securing of funding.

An important component of the collaborative process includes community participation. In addition to involving scientists, universities, research institutions, government agencies, and non-government agencies, we must recognize the importance of the end user, in our case, the rural households of Central Europe. After all, only today is the United States recognizing the importance of bottom-up vs. top-down management. Even in the work place, management is seeing that factory workers do in fact have some good ideas and can contribute to the overall enterprise.

The Cornell International Institute for Food, Agriculture and Development (CIIFAD), founded in 1990, is a program that operationalizes collaboration. CIIFAD and its partners have three immediate goals:

1. knowledge generation--producing as well as adapting and disseminating ideas, concepts, theories, technologies, and techniques that will contribute to sustainable development.

2. Human resource development--enhancing the capabilities of people to create and apply new knowledge and to establish and lead institutions that can foster sustainable development.

3. Institutional strengthening-building collective capacities to plan and implement appropriate innovations for sustainable development, from national to local levels [5].

All three goals are relevant to any of your institutions concerned with the advancement of knowledge and the promotion of economic and social development.

Although CIIFAD's approach is not the only way to undertake international collaboration, it has been tested and offers promise for lasting beneficial results, if persevered. The criteria that were established for CIIFAD support of a faculty initiative to start a collaborative effort overseas included:

a. "a critical mass of faculty from different disciplines who are prepared to take responsibility from our side for the planning and management of joint activities;

b. sufficient like-minded and similarly dedicated colleagues in one or (preferably) more institutions in the other country who will be responsible for carrying forward what is mutually agreed upon; and

c. a significant problem that (a) and (b) can agree on as important for that country and also of general importance for sustainable agricultural and rural development elsewhere, so that the learning and experience gained in that country can have beneficial spread effects" [6].

"A collaboration starts with an initial partner institution overseas that offers good colleagues to work with. What brings us together, in addition to mutual respect, is a concern for the chosen problem. A natural consequence of this is that the collaboration expands to include other like-minded persons in other institutions who can bring additional, particularly complementary, knowledge and skills to the problem-solving effort. Thus what starts as a partnership becomes a network" [7].

One example of a CIIFAD success story is in the Dominican Republic. Working with colleagues at a national university, in government agencies and NGOs, as well as in communities, an interdisciplinary team of Cornell faculty and students has built up a base of knowledge about one of the Dominican Republic's largest national parks and the surrounding area. There was a need to reconcile both conservation and sustainable development goals through a system of adaptive co-management by major stakeholders. The park was studied from many different perspectives including agronomic, biological, economic, and social.

Key accomplishments of this effort include institutionalizing a geographic information system with extensive baseline data on the park; a soil foundation map; comparative social and biological surveys inside and outside the park; conflict management workshops dealing with natural resource utilization; enhancement of agricultural and economic activities for households around the park; and reorienting land reform resettlements to have a higher degree of community involvement in land use management [8].

This effort was accomplished as the result of full involvement and cooperation of many parties including local communities. Each person brings his/her special strengths and contributes to the big picture.

Before making a commitment to the collaborative process, one must be aware of the fact that this is not simple or easy. There are challenges and impediments involved such as the time investment that is required; continuity of personnel; interdisciplinary work can be more difficult overseas than at home; and different decision-making styles.

Now, what is the plan for the Central Europe Initiative, a plan that has evolved as a result of close cooperation with scientists and key administrators from a variety of institutions in both public and private sectors. Serious collaboration is already underway as a result of the existing scientist-to-scientist relationship; the growing list of partner institutions; and the existing interactions between and among the five work groups that developed from the initial planning workshop.

Within the CEE Initiative, our goal is to develop a series of problem-focused, interdisciplinary collaborations which bring together a number of partners in selected countries. Collaborations are two-way and mutual, unlike the technical assistance which has been a prominent feature of the international development scene. We are developing a program to work with a network of knowledge-generating institutions (universities and some national and international research institutions), and action agencies, (both governmental and non-governmental). We aim to develop a closer working relationship between the producers and users of knowledge.

As stated earlier, we are committed to the development of collaborative programs with educational and research institutions in Central and Eastern Europe. Our overall goal is to develop substantial collaborative programs and secure adequate funding to address issues that will enhance the social, economic, and political development of countries in this important region within the framework of democratic institutions and with special emphasis on rural communities and agricultural and environmental resources. Our specific objectives include the following:

1. Work on programs that promote institutional strengthening (for institutions in Central and Eastern Europe as well as for Cornell)

2. Engage faculty and collaborators in multidisciplinary approaches to solving problems;

3. Promote research, education and extension/outreach services that are immediately and visibly useful to both the public and the private sector.
   a. Develop human capital to enhance the social and economic welfare of rural communities and rural residents.

   b. Develop agricultural technologies (i.e., biotechnology, and food quality and safety) that will lead to more efficient producers and processors in Central Europe and assure an adequate supply of safe and nutritious food products for consumers.

   c. Develop analyses and processes that bring the countries of Central Europe into conformance with European Union regulations and place Central European agricultural producers and agribusinesses in positions of competitive advantage.

   d. Ensure sustainable use of agricultural and environmental resources and rural communities.

5. General Agreement

To this end, a general agreement was developed and distributed among a number of potential partners. From the previous two workshops there was a strong interest in the development of a consortium of institutions in Central and Eastern Europe with Cornell and other selected institutions in a collaborative project. The General Agreement Between Selected Institutions in Central Europe and Cornell University for Collaboration in Agriculture and Rural Development in Central Europe and the U.S. states the following: With the transformation of social and economic systems in Central and Eastern Europe to democracy and free market economies and with the upcoming entrance into the European Union, the restructuring, modernization, and competitiveness in agriculture and the rural areas of Central Europe is a high priority. Environmental degradation, inefficient economics, the changing rural landscape, vast changes in social and economic opportunities facing rural people and communities and issues related to quality and safety of food, offer unique opportunities for collaboration involving institutions in Central Europe and the U.S.

Building on "A Planning Workshop on Agriculture and Rural Development: Restructuring - Modernization - Competitiveness held in Sielinko, Poland May 28 - 30, 1998 the institutions affix their signature to this agreement. In the interest of establishing a collaborative relationship of mutual benefit and for the purpose of jointly undertaking research, education and scholarly projects, we agree to work together to build linkages and opportunities among all the institutions to strengthen teaching, research and outreach in agriculture and life sciences, rural development, and related fields and to broaden the international experiences of all the signing institutions and their staff.

Science and technology is increasingly seen as a major driver of economic advancement. This requires modern agriculture, industrial systems and education to adapt and develop technologies appropriate to local circumstances and to help strengthen education. We agree that five scientific areas should be the focus of collaborations:

Environmental Issues Related to Land and Water Management to strengthen the research-outreach continuum related to environmental science and technology. Specific objectives include: evaluate and strengthen institutional capacity to address environmental problems; develop an outreach curriculum for environmental management; pilot an outreach infrastructure for improved environmental management in each of the three Central European countries.

Quality and Safety Assurance of Agricultural Products and Foods Produced in the Region with a focus on a research project "A Comparative Study on Quality and Safety Management Procedures in the Food System in Hungary, Poland, the Slovak Republic, and the United States - Present Status and Future Considerations."

Rural Development will develop an applied research network that would collaborate to produce information supportive of rural development policy, education, and outreach.

Agricultural Market Economics Agricultural Market Economics with a research focus on the further changes and adjustments necessary to assist the integration of Central Europe agriculture into world economies. Biotechnology with a focus on identification of research areas for collaboration and training graduate and post-graduate students in the new areas of biotechnology.

We agree to work together on these areas and are committed to a collaborative effort consistent with the recommendations of the May 1998 Workshop held in Sielinko, Poland. This includes the development of a comprehensive and integrated proposal to attract funding for a major collaborative initiative. Furthermore, assuming that funding becomes available, we agree to be responsible for those activities that our institution has included in the proposal.

In witness whereof the undersigned, representing their respective institutions, hereby sign and approve the Memorandum of Agreement.

To date, the following institutions have signed this agreement:

<u>Central European Universities</u>: Adam Mickiewicz University, August Cieszkowski Agricultural University at Poznan, Olsztyn University of Agriculture and Technology, Slovak Agricultural University at Nitra, Warsaw Agricultural University, Agricultural University at Krakow and the Godollo Agricultural University, Hungary.

<u>Central European Institutions</u>: MLOCHOW Research Center/IHAR, Polish Academy of Sciences, Cereal Research Non-Profit Company in Szeged, Hungary.

<u>Other Institutions</u>: Triple F Incorporated/INSTA-PRO International, Poznan.

<u>U.S. University</u>: Cornell University.

Other institutions in Central and Eastern Europe, Western Europe and the United States are welcome to join.

6. Linkage

The programs being developed by each of the five focus groups are impressive and are quickly getting the attention of donors. Based on an April 1999 visit to NATO headquarters in Brussels, it became clear that the Central Europe Initiative was of considerable interest to them. Since that visit, NATO and other donor agencies have demonstrated their interest in these initiatives by providing seed money. It is anticipated that this seed money will attract significant funding from other donors. However, if we are going to secure major funding, this initiative has to have some characteristics that set it apart from all other initiatives. One of the unique characteristics of the Central and Eastern Europe Initiative is the linkages among the five focus areas. If the goals of enhancing sustainable development of rural communities in the region are to be achieved and if funding is going to be realized, it is the numerous linkages between and among the five areas that need to be maintained and strengthened. However, this linking of activities will depend on a true collaborative mode of operation.

Members of the CEE Initiative Coordinating Committee have provided several examples of possible links. They include the following:

<u>Environmental Issues Related to Land and Water Management (EMW)</u>
(Prof. Rebecca Schneider, Department of Natural Resources, Cornell University and Co-chair, Environmental Issues Related to Land and Water Management)
There are several opportunities for integrating the efforts of the EMW with those of the other Workgroups, but within the goals of the Environmental Management Work Group:
a.　to improve the overall quality of the land and water resources of Central and Eastern Europe while building the economic viability of its human communities, and

b.　to expand the capacity for environmental stewardship at the local level by involving landowners and local governments. Good environmental stewardship is recognized as critical to the long-term sustainable development of the Central and Eastern European society.

Current efforts within the EMW are directed towards building stronger connections between researchers who identify best environmental management practices and extension educators who train farmers and other direct users of these practices. This environmental outreach approach can easily be broadened to encompass the subject content of the other Workgroups. For example the Food Safety Workgroup is working to identify and correct dairy practices in order to improve the safety of the milk, cheese and other dairy products. Many dairy practices can potentially impact soil and water quality, e.g. manure management, milk house cleaning, high-density cattle operations. By prioritizing environmental protection as part of the dairy operations, best management practices can be chosen which maximize both a safe environment and healthy products. Further, these multiple benefits could be highlighted by the Marketing Workgroup in a "green" dairy promotion. Similar programs could be developed for vegetable and fruit crop industries as well.

Promoting environmental stewardship is also key to building a sense of self-pride, independence, and psychological security in the lower income rural communities. A healthy environment can provide alternative, local sources of food through fishing and hunting opportunities. Careful logging practices can provide a sustainable source of fuel wood, timber, and income. With its aesthetic appeal, a healthy environment may also provide a basis for building an eco-tourism industry. However a powerful outreach program is critical to build the local capacity for such environmental protection. The EMW can work with the Rural Development Workgroup to develop and implement an outreach program, which integrates environmental stewardship with local capacity and sustainability.

Biotechnology

(Prof. Geza Hrazdina, Department of Food Science, New York State Agricultural Experiment Station, Cornell University, and Co-chair, Biotechnology).

Biotechnology plays an important role in and serves as a linkage to not only Agricultural Production, but also to such important areas as Food Quality and Safety, and Environmental Quality. The tools of Biotechnology permit us to make significant improvements to yield of crops, storage life of fresh fruits and vegetables, changes in the composition of carbohydrates, lipids, amino acids, pigments and other secondary metabolites, and as a whole improves the quality of our foods. Rapid and precise methods in the identification of spoilage and disease causing microorganisms enables us to prevent food spoilage and the outbreak of epidemics that have plagued humanity from prehistoric times on.

The tools of Biotechnology also provide us with alternatives to chemical sprays in combating economically important plant pathogens that presently require costly chemical treatments that also contaminate our environment. Production of transgenic plants with heavy metal accumulating capacity will enable us to clean up and improve our environment. In general, there is barely any area that is related to agriculture, food and fiber production and environmental issues that will not positively benefit from such an important linkage.

Quality and Safety Assurance of Agricultural Products and Foods

(Prof. Robert Gravani, Department of Food Science, Cornell University and Co-chair, Food Quality and Safety Group)

The activities of the food quality and safety group link with each of the other working groups since food quality and safety are inherently of interest to all sectors of the food system, from production agriculture to the ultimate consumer. The availability of high quality and safe foods is of utmost importance to everyone. Examples of linkages to other groups are provided below.

Biotechnology: Today, consumers are very interested in the safety of biotechnologically derived foods. This interest has been sparked by activist groups in Europe and the US. There has been a great deal of misinformation about these foods that has been disseminated through television and radio reports, magazine and newspaper stories and other media outlets. Several research studies that have been conducted have been taken out of context and consumers are concerned about biotechnologically derived foods that they have called, "franken foods". Their worries are broad based and include environmental concerns, allergens that may be created or enhanced, the violation of religious dietary laws, and the ethical concerns about "tampering with mother nature", as well as others. The quality and safety assurance of all products derived from biotechnology, including foods must be carefully and thoroughly evaluated to be sure that they are safe.

Environmental Issues: In all sectors of the food system, from farm to table, there are concerns about how the production, processing, distribution and preparation of foods, influences the environment. From the way that crops are rotated, the microbiological quality of the water used for irrigation, to the handling and application of manure to crops all impact the environment and have an influence on the safety of foods. There is a clear linkage between the production of good quality and safe foods and the environment.

Agricultural Market Economics: The transportation, distribution and merchandising of foods certainly influences their quality and safety. Time and temperature control, packaging, the way that foods are handled throughout the food chain have a profound effect on their shelf life and therefore on their quality and safety. In the US, we've had a major foodborne outbreak caused by cross contamination between a raw and processed food in a transport vehicle. This is only one example of how food quality and safety links to the marketing of food products.

Rural Development: In order to improve market economies and add value to foods through processing and preparation, a knowledgeable and skilled work force is needed. With the small and medium sized food businesses that make up the industry in central and eastern Europe, trained workers are needed to provide this value added technology. The linkage of food quality and safety to rural development may be somewhat remote, but by providing higher quality and safe foods that are marketed to areas outside towns and villages, more people are needed and rural development can be enhanced.

Rural Development
(Prof. David Brown, Department of Rural Sociology, Cornell University and Co-chair, Rural Development Group)

The strongest link between rural development and another group is with environmental management. Human activities (social and economic) are effected by and effect the natural environment. Hence understanding the interaction between rural community and economic development: utilization of new technology, development of new infrastructure, establishment of new businesses, etc. and natural processes such as water quality and quantity, land use, air pollution, etc. is important for both development and the environment. If biotechnology helps to remediate the negative impacts of human activities on the environment, then you have three groups involved.

Agriculture is still an important part of the rural economy in CEE, especially in Poland where there are still almost two million small holders active in farming. Hence, structural changes in agriculture have direct and indirect impacts on employment, incomes, and other economic and quality of life parameters in rural communities.

Food Science has a connection to rural development in that food manufacturing is an important source of rural jobs. Enhancing the opportunities for entrepreneurial activity related to food manufacturing in rural economies could have important impacts on rural community viability and population retention.

Agricultural Market Economics
(Prof. Loren Tauer, Department of Agriculture, Resource, and Managerial Economics, Cornell University and Co-chair, Agricultural Market Economics Group)

Biotechnology is an applied science in biology. Some of it is pure or basic science, but what they ultimately would like to end up is products that have utility or value to mankind. The job of economics is to determine the economic value of products. Thus economics can be used to determine the value of a biotechnology product based upon the lost reduction that occurs with the product or the value added. Too often products are developed that have little commercial value. These are doomed from the start. This economic research entails more than simply multiplying the increased yield by a price. When there are significant yield impacts there are large price and market effects. Thus econometric models that estimate the full impact of a change is necessary. What economics can do is help determine a value for many traits that are considered important in biotechnology. This includes pest resistance, input substitution, and yield increases. Strategic planning is important in biotechnology. Certain products are more acceptable to farmers and consumers. Thus BST for milk production is less acceptable to the public than

BT corn that resists pests. Acceptability is a role for psychologists, but economics plays a role also. The key is to find a product that has value, requires few undesirable changes on the part of farmers and consumers, and that is socially and ethically acceptable.

Adoption rates is very important in biotechnology and economists have done a large amount of estimating adoption rates for various products. Some of this has been *ex ante* (before the product has been marketed) to *ex post* (after the product is marketed, i.e. actual rather than hypothetical adoption). This research can help determine the extent and speed of adoption and thus determine economic payback. If you have to wait a long time to recover your money the investment is less attractive.

Economics is important in the environment. There is a large number of environmental economists. Most pollution occurs because of undefined property rights. Economists have developed pollution rights that can be sold to the highest bidder. Environmentalist often feel uncomfortable with this, often on ethical grounds. They feel that all pollution should be eliminated, but that is not possible. Pollution can be reduced but not eliminated. By selling the right to produce and defining a market for pollution, the firms that have production processes that produce less pollution can afford to buy the right to pollute, driving out the higher polluting firms. Even if markets for pollution do not exist, pollution might be controlled by taxing, using restrictions, or by the legal system (suing for damages). Economists can determine the cost and benefits of these various approaches.

Again, there is no such thing as a completely safe food. It is a matter of degree. At some point the value of reducing food induced illnesses will be lower than the cost of accomplishing that. It is the job of economists to determine this cost/benefit ratio.

Much of rural sociology requires economics. Plant location is an economic decision in a market economy. Many plants in Eastern Europe were placed because of political decisions rather than economic decisions. That is why many of those plants are in difficulty. They are often only in business because of low labor cost or subsidized inputs, such as energy.

7. Distance Learning as an Opportunity to Support Collaboration

Where do we go from here? What steps can be taken now to support research and educational collaboration? As stated earlier, one of the essential factors in good collaboration is the willingness of all sides to invest time and money in sufficient communication. Distance learning or electronic communication networks offers new and innovative opportunities for collaboration [9]. This can serve as a new and exciting dimension to the Central Europe Initiative. The world is in the process of redefining the traditional means of gaining and providing knowledge. Distance learning is growing rapidly in popularity and offers unique opportunities to exchange information and ideas, co-teach courses that involve several institutions, facilitate international conferences, etc. A very important concept that must be sustained for maximizing learning and research collaborations is person to person interactions, something that can be accomplished through distance learning technologies.

A framework is already in place for major international cooperation. Systems already in use in the United States include cable television, the Internet, local area networks, satellite television, teleconferencing, wide area networks, World Wide Web, and electronic mail. With limitations of expense, personnel, travel distances and time it is unlikely that true collaborations can be established between scientists in Central Europe and the U.S. without telecommunication networking. Some places to start include:

1.	Establish a list-serve involving all of the collaborators. Someone must be willing to channel and maintain communications, an important ingredient for momentum, to direct the network's activities.

2 .	Through electronic networking it is possible to develop joint research collaborations and joint courses. In the case of courses, joint teaching can be accomplished through one of several mechanisms. Such joint efforts could involve one or more institutions in Central Europe and Cornell or it may involve two institutions in one or more countries in the region.

a.	Web-based course: through the use of teleconferencing and the WorldWideWeb, it is possible to develop and deliver a synchronous exchange allowing students and faculty to interact in real time. Through advance preparation, material is presented in the form of a slide presentation. This is a less expensive and a useful way to engage faculty and students from different countries. Much of the technical content of a course can be learned by students from such a web-based course. This tool allows a course taught at one site to be taken by students anywhere.

> An example of this is the Global Seminar on The Environment and Sustainable Food Systems: Changing the Shape of the Earth. The Global Seminar, organized by the College of Agriculture and Life Sciences at Cornell University, is an international collaborative effort to offer a fully interactive distance education course to undergraduate and graduate students around the world on The Environment and Sustainable Food Systems using interactive compressed video and satellite technology.
>
> The institutions responsible for developing and hosting the seminars are:
>
> Agricultural College at Wageninen, The Netherlands
> Cornell University, Ithaca, New York
> E.A.R.T.H. College, Costa Rica
> Open University of the Netherlands
> Swedish University of Agricultural Sciences, Sweden
> Melbourne University, Australia
> Uppsala University, Sweden
> Zamorano, Honduras
> Royal Agricultural College, Chichester, United Kingdom

b.	The ultimate connectivity is a synchronous exchange allowing students and faculty to interact in real time and to see each other, face to face. A pilot course is now being developed that will provide students at Cornell University and students at the Agricultural University, Poznan and Olsztyn University of Agriculture and Technology to enroll in a seed physiology course taught by Prof. Ralph Obendorf. The course is expected to be presented starting in January 2000. While optimum, live synchronous transmission is costly.

> c.	A collaborative learning environment can be achieved where a course is developed jointly by a colleague in a Central European institution and a U.S. faculty member. It is when a colleague at a Central European institution and U.S. faculty cooperate fully on a course that he/she can represent U.S. faculty at a distance and by the same token U.S. faculty can represent a Central European colleague when his/her material is delivered to U.S. students. This suggests that U.S. faculty members will meet once a

week in person with the students in Ithaca and the Central European colleague would do the same with students in Central Europe. The emphasis is on course content of mutual agreement and understanding, not the form of delivery.

    d.   Through distance learning technology, degree programs can be offered by Cornell and made available to, for example, mid-career professionals in Central Europe. Such an initiative is being considered involving agribusiness professionals in India.

3. The development of a homepage on the WorldWideWeb is a first step towards linking the partner institutions in Central and Eastern Europe and Cornell. Items to include on the web site would be the reports from the two original workshops (Sielinko and Ithaca), relevant papers and a description of each of the partner institutions. The homepage will promote and encourage the addition of materials from the participants and provide links to the specific entities collaborating in the Central Europe initiative for sharing with a larger community of scholars.

At Cornell, we are able to draw on the resources of the Office of Distance Learning. This office was created to coordinate and support the development of distance learning programs and serve as the central resource of knowledge and expertise in distance learning techniques and technology for the university community. Resources available to us include staff support, the Cornell "Virtual Classroom" Amphitheaters, the distance learning classrooms available for use free of charge to all Cornell faculty for distance learning classes, and Public Videoconference Rooms.

8. Other mechanisms to promote collaboration include:

1. <u>Faculty/Staff</u>: Linkages will be strengthened by exchanges/visits to develop specific collaborative projects that involve a mix of scientists from Central Europe and other institutions. Postdoctoral fellows/visiting staff will be a major component to advance technical and conceptual skills and to prepare returning persons for leadership positions.

2. <u>Graduate Students</u>: A significant key to the success of research projects will be the inclusion of graduate students in the projects. Students provide the "glue" to this research initiative by serving as "junior" scientists who will participate in active research in both countries. Sometimes referred to as "sandwich programs", a graduate student from Central Europe will carry out a majority of graduate level course work in a U.S. university, return to Central Europe to conduct thesis research on a topic, jointly guided by faculty of collaborating U.S. and Central European universities and defend the thesis before the joint committee in either the U.S. or Central Europe. Likewise a student from the U.S., interested in international studies, after completing graduate level course work in the U.S., would go to Central Europe to conduct thesis research, with guidance from the joint committee, and likely return to the U.S. for a thesis defense before the joint committee. This concept has a proven track record and can be a principal element in advancing agricultural science of both nations.

3. <u>Undergraduate Students</u>: Undergraduates from both countries are increasingly looking at opportunities to learn more about the culture and science of the other. Student exchanges can be a significant component of the collaborative program. The very popular NSF program for undergraduates, "Research Experience for Undergraduates" (REU) serves as a model for extending a research opportunity to an international setting by including undergraduates in the research.

4. <u>Other</u>: short-term visits, individual or group: CU to CEE (one - three weeks); Short-term visits, individual or group: CEE to CU (one - three weeks); Longer term visits to CU, typically for younger faculty: (four months minimum); Six-month sabbaticals from CU to CEE (cover expenses except for salary); Nine to twelve months joint research CEE to CU.

9. Conclusion

Our guiding principles for program development are the following: (1) work on programs which are institutional strengthening (for Cornell University, as well as for institutions in Central and Eastern Europe); (2) enhance the sustainable development of rural communities with emphasis on agricultural and environmental resources; (3) engage faculty and collaborators in multidisciplinary approaches to solving problems; and (4) enhance opportunities for scientist -to- scientist exchanges within the framework of selected issues; (5) to promote research, education and extension/outreach services that are immediately and visibly useful to both the public and the private sector. If we are to attain these goals and have a significant impact on the region and learn from our experiences, serious collaboration is necessary. A well thought out integrated program will lead to positive results.

Cornell's role will continue to be that of a catalyst, i.e., identifying faculty who have an interest in and commitment to working on topics of mutual interest in the CEE region; contribute to your teaching programs through guest lectures, sabbaticals, etc. and pursue distance learning opportunities; assist in writing joint proposals for external funding; and pursue opportunities for faculty and students from our partner institutions to come to Cornell as visiting professors, visiting fellows, degree students, or non-degree students.

Appendix 1
Mission Statement
Central and Eastern European Program Initiative

We are committed to the development of collaborative programs with education and research institutions in Central and Eastern Europe. Our overall goal is to develop substantial collaborative programs and secure adequate funding to address issues that will enhance the social, economic, and political development of countries in this important region within the framework of democratic institutions. Our specific objective is to secure adequate funding too address these issues and challenges by the year 2000.

Our guiding principles for program development are the following: (1) work on programs which are institutional strengthening (for Cornell University, as well as for institutions in Central and Eastern Europe); (2) enhance the sustainable development of rural communities with emphasis on agricultural and environmental resources; (3) engage faculty and collaborators in multidisciplinary approaches to solving problems; and (4) enhance opportunities for scientist-to-scientist exchanges within the framework of selected issues; (5) to promote research, education and extension/outreach services that are immediately and visibly useful to both the public and the private sector.

The specific areas of focus are:

1. **Environmental Issues Related to Land and Water Management** to strengthen the research-outreach continuum related to environmental science and technology. Specific objectives include: evaluate and strengthen institutional capacity to address environmental problems; develop an outreach curriculum for environmental management; pilot an outreach infrastructure for improved environmental management in each of the three Central European countries.

2. **Quality and Safety Assurance of Agricultural Products and Foods Produced in the Region** with a focus on a research project "A Comparative Study on Quality and Safety Management Procedures in the Food System in Hungary, Poland, the Slovak Republic, and the United States - Present Status and Future Considerations."

3. **Rural Development** will develop an applied research network that would collaborate to produce information supportive of rural development policy, education, and outreach.

4. **Agricultural Market Economics** with a research focus on the further changes and adjustments necessary to assist the integration of Central Europe agriculture into world economies.

5. **Biotechnology** with a focus on identification of research areas for collaboration and training graduate and post-graduate students in the new areas of biotechnology."

Appendix 2
General Agreement

GENERAL AGREEMENT
Between
SELECTED INSTITUTIONS IN CENTRAL
EUROPE AND CORNELL UNIVERSITY
For
COLLABORATION IN AGRICULTURE AND RURAL
DEVELOPMENT IN CENTRAL EUROPE AND THE U.S.

With the transformation of social and economic systems in Central and Eastern Europe (CEE) to democracy and free market economies and with the upcoming entrance into the European Union, the restructuring, modernization, and competitiveness in agriculture and the rural areas of Central Europe is a high priority. Environmental degradation, inefficient economics, the changing rural landscape, vast changes in social and economic opportunities facing rural people and communities and issues related to quality and safety of food, offer unique opportunities for collaboration involving institutions in Central Europe and the U.S.

Building on "A Planning Workshop on Agriculture and Rural Development: Restructuring - Modernization - Competitiveness held in Sielinko, Poland May 28 - 30, 1998 the institutions affix their signature to this agreement. In the interest of establishing a collaborative relationship of mutual benefit and for the purpose of jointly undertaking research, education and scholarly projects, we agree to work together to build linkages and opportunities among all the institutions to strengthen teaching, research and outreach in agriculture and life sciences, rural development, and related fields and to broaden the international experiences of all the signing institutions and their staff.

Science and technology is increasingly seen as a major driver of economic advancement. This requires modern agriculture, industrial systems and education to adapt and develop technologies appropriate to local circumstances and to help strengthen education. We agree that five scientific areas should be the focus of collaborations:

1. Environmental Issues Related to Land and Water Management: to strengthen the research-outreach continuum related to environmental science and technology. Specific objectives include: evaluate and strengthen institutional capacity to address environmental problems; develop an outreach curriculum for environmental management; pilot an outreach infrastructure for improved environmental management in each of the three Central European countries.

2. Quality and Safety Assurance of Agricultural Products and Foods Produced in the

3. Region with a focus on a research project "A Comparative Study on Quality and Safety Management Procedures in the Food System in Hungary, Poland, the Slovak Republic, and the United States - Present Status and Future Considerations."

4. Rural Development will develop an applied research network that would collaborate to produce information supportive of rural development policy, education, and outreach.

5. Agricultural Market Economics with a research focus on the further changes and adjustments necessary to assist the integration of Central Europe agriculture into world economies.

6. Biotechnology with a focus on identification of research areas for collaboration and training graduate and post-graduate students in the new areas of biotechnology.

We agree to work together on these areas and are committed to a collaborative effort consistent with the recommendations of the May 1998 Workshop held in Sielinko, Poland. This includes the development of a comprehensive and integrated proposal to attract funding for a major collaborative initiative. Furthermore, assuming that funding becomes available, we agree to be responsible for those activities that our institution has included in the proposal.

In witness whereof the undersigned, representing their respective institutions, hereby sign and approve this General Agreement.

Appendix 3
Coordinating Committee for CEE/Cornell Initiative

Prof. Mieczyslaw Adamowicz, Director, College of Agricultural Market and Cooperatives, Warsaw Agricultural University, Poland

Prof. David Brown, Department of Rural Sociology, College of Agriculture and Life Sciences, Cornell

*Prof. Piotr Golinski, Vice Rector, August Cieszkowski Agricultural University, Poznan, Poland

Prof. Robert Gravani, Department of Food Science, College of Agriculture and Life Sciences, Cornell

*Mr. James Haldeman, Associate Director, International Agriculture program, College of Agriculture and Life Sciences, Cornell

Prof. Geza Hrazdina, Department of Food Science and Technology, College of Agriculture and Life Sciences, Cornell

Prof. Dusan Huska, Dean, College of Gardening and Landscape Engineering, Slovak Agricultural University, Nitra, Slovak Republic

Dr. Janos Pauk, Head, Wheat and Tissue Culture Lab, Cereal Research Non-Profit Company, Hungary

Dr. Tomasz Radomyski, INSTA-PRO International, Poland

Prof. Rebecca Schneider, Department of Natural Resources, College of Agriculture and Life Sciences, Cornell

Prof. Michael Sznajder, Head, Department of Agricultural Economics, August Cieszkowski Agricultural University, Poznan, Poland

Prof. Loren Tauer, Department of Agricultural, Resource and Managerial Economics, College of Agriculture and Life Sciences, Cornell

*Prof. Gerald White, Department of Agricultural, Resource and Managerial Economics, College of Agriculture and Life Sciences, Cornell

*Committee co-chairs

References

[1]　Cornell University International Agriculture Program - Meeting the Challenge of World Agriculture. International Agriculture Program, 1986.

[2]　A Planning Workshop on Agriculture and Rural Development: Restructuring - Modernization - Competitiveness. International Agriculture Program, 1998.

[3]　Sustainable Development of Rural Communities - An Initiative for the Development of Human, Agricultural and Environmental Resources in Central Europe. International Agriculture Program, 1999.

[4]　Collaboration in International Rural Development. George H. and Nancy W. Axinn. 1997.

[5]　CIIFAD Annual Report 1993-94. Cornell International Institute for Food, Agriculture and Development, 1994.

[6]　N. Uphoff, "Collaborations as An Alternative to Projects: Cornell Experience with University-NGO-Government Networking." Journal of the Agriculture, Food, and Human Values Society. (1996) p43.

[7]　Ibid. p43-44.

[8]　CIIFAD Annual Report 1997-98. Cornell International Institute for Food, Agriculture and Development, 1998.

[9]　Telecommunication Linkages Support Research and Educational Collaborations between China and the U.S. in Sustainable Agriculture and Development. Dr. Norman R. Scott, Department of Agricultural and Biological Engineering, Cornell University, 1999.

Key speakers and participants

• *Aldwinckle, H.*
Cornell University
Department of Plant Pathology
Geneva, NY 14456, USA
Phone: 315-787-2369
Fax: 315-787-2389
e-mail: hsa1@cornell.edu

• *Aniol, A.*
Plant Breeding and Acclimatization Institute
(IHAR) Radzikow,
 05-870 Blonie, Poland
Phone: 48-22-7253611
Fax: 48-22-7254714
e-mail: a_aniol@ihar.edu.pl

• *Balazs, E.*
Agricultural Biotechnology Center, Gödöllö, P.
O. Box 170
H-92101, Hungary
Phone: 36-28-420-521
Fax: 36-28-420-096
e-mail: balazs@abc.hu

• *Bánfalvi, Z.*
Institute for Plant Science
Agricultural Biotechnology Center, Gödöllö
P. O. Box 170
H-2101, Hungary
Phone: 36-62-430-600
Fax: 36-62-430-482
e-mail: banfalvi@abc.hu

• *Barnábas, B.*
Agricultural Research Institute of
The Hungarian Academy of Sciences
Martonvásár
P. O. Box 19
H-2462, Hungary
Phone: 36-22-569-500
Fax: 36-22-460-213
e-mail: bea@fsnew.mgki.hu

• *Bastar, M.*
Biotechnical Faculty
University of Ljubljani
61111 Ljubljana, Slovenia
Phone: +38 661/123-11-61
Fax: +38 661/123-10-88
e-mail: tina.bastar@bf.uni-lj-sl

• *Blaszczyk, A.*
Institute of Biochemistry and Biophysics
Polish Academy of Sciences
Ul. Pawinskiego 5A, 02-106 Warsaw, Poland
Phone: +48 22 659 70 72 ext. 1213
Fax: +48 39 121 623
e-mail: blaszczyk@ibb.waw.pl

• *Bohanec, B.*
Centre for Plant Biotechnology and Breeding
Biotechnical Faculty
Agronomy Department
Jamnikarjeva 101
1000 Ljubjana, Slovenia
Phone: +386 61 123 1161
Fax: +386 61 123 1088
e-mail: borut.bohanec@uni-lj.si

• *Bona, L.*
Cereal Research Np. Co., Szeged
P. O. Box 391
H-6701, Hungary
Phone: +36-62-435-235
Fax: +36-62-434-163
e-mail: lajos.bona@mail.gk-szeged.hu

• *Cahoon, R.*
Cornell University
Dept. of Patents & Technology Marketing
Ithaca, NY 14853 USA
Phone: 607-257-1081
Fax: 607-257-1015
e-mail: rsc5@cornell.edu

• Dudits, D.
Institute of Plant Biology
Biological Research Center
Hungarian Academy of Sciences, Szeged
P. O. Box 521
H-6701, Hungary
Phone: 36-62-433-388
Fax: 36-62-433-188
e-mail: dudits@nucleus.szbk.u-szeged.hu

• Farago, J.
Research Institute of Plant Production
Piestany, Slovak Republic
Phone: +421-838-7722311
Fax: +421-838-7726306
e-mail: lbmb@vurv.sk

• Flis, B.
Plant Breeding and Acclimatization Institute
(IHAR)
05-870 Rozalin, Poland
Phone: 72 99 248
Fax: 72 99 247
e-mail: b.flis@ihar.edu.pl

• Gálová, Z.
Slovak Agricultural University
Dept. of Biochemistry & Biotechnology
Tr. Andreja Hlinku 2
949 76 Nitra, Slovak Republic
Phone: 42187/601596
Fax: 42187/513097
e-mail: galova@afnet.uniag.sk

• Griga, M.
AGRITEC Ltd., Biotech. Dept.
Zemedelska 16
CZ-78701 Sumperk, Czech Republic
Phone: +420-649-382126
Fax: +420-649-382126
e-mail: griga@agritec.cz

• Haldeman, J.
Cornell University
International Agriculture
Box 14 Kennedy Hall
Ithaca, NY 14852 USA
Phone: 607-255-3037
Fax: 607-255-1005
e-mail: jeh5@cornell.edu

• Hemmat, M.
Cornell University
Department of Horticultural Sciences
Geneva, NY 14456 USA
Phone: 315-787-2245
Fax: 315-787-2397
e-mail: mh20@cornell.edu

• Heszky, L.
Godollo University of Ag Sciences
Godollo
P. O. Box 391
H-6701, Hungary
Phone: +36-62-522-069
Fax: +36-62-430-618
e-mail: heszky@fau.gau.hu

• Horvath, G.
Institute of Plant Biology
Biological Research Center
Hungarian Academy of Sci., Szeged
P. O. Box 521
H-6701, Hungary
Phone: +36-62-432-080 x303
Fax: +36-62-433-188
e-mail: horvath@nucleus.szbk.u-szeged.hu

• Hrazdina, G.
Cornell University
Institute of Food Science
Geneva, NY 14456 USA
Phone: 315-787-2285
Fax: 315-787-2284
e-mail: gh10@cornell.edu

• *Jenes, B.*
Institute for Plant Science
Agricultural Biology Center, Gödöllo
P. O. Box 170, H-92101
Hungary
Phone: 36-62-430-600
Fax: 36-62-430-482
e-mail: jenes@abc.hu

• *Király, Z.*
Plant Protection Institute
Hungarian Academy of Sciences, Budapest
P.O. Box 102
H-1525, Hungary
Phone: 36-1-355-8722
Fax: 36-1-356-3698
e-mail: zkir@nki.hu

• *Kiss, E.*
Gödöllö University of Agricultural Sciences
Gödöllö
P. O. Box 391
H6701, Hungary
Phone: 36-62-522-069
Fax: 36-62-430-618
e-mail: kisse@fau.gau.hu

• *Kiss, G.*
Institute of Genetics
Biological Research Center
Hungarian Academy of Sciences, Szeged
P. O. Box 521
H-6701, Hungary
Phone:36-62-432-232
Fax: 36-62-433-503
e-mail: kgb@nucleus.szbk.u-szeged.hu

• *Kotai, E.*
Cereal Research Np. Co., Szeged
P. O. Box 391
H-6701, Hungary
Phone: +36-62-435-235
Fax: +36-62-434-163
e-mail: janos.pauk@mail.gk-szeged.hu

• *Kraic, J.*
Research Institute of Plant Production
Piestany, Slovak Republic 92168
Phone: +421 838 7722311
Fax: +421 838 7726306
e-mail: jankraic@hotmail.com

• *Kubek, A.*
Slovak Agricultural University
Nitra, Slovak Republic
Phone:
Fax:
e-mail:

• *Lenndvai, A.*
Institute of Plant Biology
Biological Research Center
Hungarian Academy of Sci., Szeged
P. O. Box 521
H-6701, Hungary
Phone: +36-62-432-232
Fax: +36-62-433-188
e-mail: picur@hotmail.com

• *Luthar, Z.*
Biotechnical Faculty
University of Ljubljani
BF, Jamnikarjeva 101
61111 Ljubljana, Slovenia
Phone: +38061/123-11-61
Fax: +38661/123-10-88
e-mail: zlata.luthar@bf.uni.lj.sl

• *Mórocz, S.*
Cereal Research NP Co., Szeged
P. O. Box 391
H-6701, Hungary
Phone: 36-62-435-235
Fax: 36-62-435-235
e-mail: h9863mor@helka.lif.hu

• *Oravecz, S.*
Ministry of Ag & Regional Development
Kossuth Lajos ter 11
Budapest, Hungary
Phone: +36-1-301-4520
Fax: +36-1-301-4668
e-mail:

• *Pauk, J.*
Cereal Research NP Co., Szeged
P. O. Box 391
H-6701, Hungary
Phone: 36-62-435-235
Fax: 36-62-435-235
e-mail: h10151pau@ella. hu

• *Ponya, Z.*
Agricultural Research Institute
Hungarian Academy of Sciences
Martonvasar
P. O. Box 19
H-2462, Hungary
Phone: +36-22-569-500
Fax: +36-22-434-163
e-mail: ponyazs@fsnew.mgki.hu

• *Purnhauser, L.*
Cereal Research NP Co., Szeged
P. O. Box 391
H-6701, Hungary
Phone: 36-62-435-235
Fax: 36-62-434-163
e-mail: h10256pur@ella.hu

• *Rakosy-Tican, L.*
University of Babes Bolyai
Dept. of Ecology-Genetics
3400 Cluy-Napoca, Romania
Phone: 00 40 64 431 878
Fax: 00 40 64 191906
e-mail: lrakosy@biolog.ubbcluj.ro

• *Sadowski, J.*
Institute for Plant Genetics
Polish Academy of Sciences
Ul. Strzeszynska 34
60-479 Poznan, Poland
Phone: 48/61/82-33-511
Fax: 48/61/82-33-671
e-mail: jsad@igr.poznan.pl

• *Szarka, B.*
Cereal Research Np. Co., Szeged
P. O. Box 391
H-6701, Hungary
Phone: +36-62-435-235
Fax: +36-62-434-163
e-mail: bela.szarka@mail.gk-szeged.hu

• *Szwacka, M.*
Warsaw Agricultural University
Dept. of Plant Genetics, Breeding & Biotech
Warsaw, Poland
Phone: 48 22 843 09 82
Fax: 48 22 843 09 82
e-mail: szwacka_alpha.sggw.waw.pl

• *Talpas, K.*
Cereal Research Np. Co., Szeged
P. O. Box 391
H-6701, Hungary
Phone: +36-62-435-235
Fax: +36-62-434-163
e-mail: janos.pauk@mail.gk-szeged.hu

•*Toldi, O.*
Institute for Plant Science
Agricultural Biotechnology Center, Gödöllo
P. O. Box 411
H-2101, Hungary
Phone: +36 28 430 600
Fax: +36 28 430 482
e-mail: toldi@abc.hu

• *Twardowski, T.*
Institute of Biorganic Chemistry
Polish Academy of Sciences
Ul. Noskowskiego 12
61-704 Poznan, Poland
Phone:
Fax:
e-mail: twardowsk@ibch.poznan.pl

• *Wolko, B.*
Institute for Plant Genetics
Polish Academy of Sciences
Ul. Strzeszynska 34
60-479 Poznan, Poland
Phone: 48/61/82-33-511
Fax: 48/61/82-33-671
e-mail: bwol@igr.poznan.pl

• *Zagórski, W.*
Institute of Biochemistry and Biophysics
Polish Academy of Sciences
Ul. Pawinskiego 5A
02-106 Warsaw, Poland
Phone: 48 22 658 4924
Fax: 48 22 658 4636
e-mail: secretariate @ibb.waw.pl

• *Zimnoch-Guzowska, E.*
Plant Breeding & Acclimatization Inst.
IHAR
Mlochow Research Center
05-832 Rozalin, Poland
Phone: 4822 72 99 089
Fax: 4822 72 99 247
e-mail: e.zimnoch-guzowska@ihar.edu.pl

• *Zimny, J.*
Plant Breeding and Acclimatization Institute
(IHAR)
Radzikow
05-870 Blonie, Poland
Phone:
Fax:
e-mail:

Subject Index

C

D

E

F

Author Index